OPTICAL INFORMATION PROCESSING

OPTICAL INFORMATION PROCESSING

Edited by

Yu. E. Nesterikhin
*Institute of Automation
Academy of Sciences of the USSR
Siberian Branch, Novosibirsk*

George W. Stroke
State University of New York at Stony Brook

and

Winston E. Kock
University of Cincinnati

PLENUM PRESS • NEW YORK AND LONDON

Library of Congress Cataloging in Publication Data

US—USSR Science Cooperation Seminar on Optical Information Processing, Washington, D. C., 1975.
 Optical information processing.

Sponsored by the National Science Foundation and held June 16-20, 1975.
Includes bibliographical references and index.
1. Optical data processing—Congresses. I. Nesterikhin, IŬ. E. II. Stroke, George W. III. Kock, Winston E. IV. United States. National Science Foundation. V. Title.
TA1630.U54 1975   621.38'0414   75-42415
ISBN 0-306-30899-1

Proceedings of the US—USSR Science Cooperation Seminar on Optical Information Processing, sponsored by The National Science Foundation, held in Washington, D. C., June 16-20, 1975

© 1976 Plenum Press, New York
A Division of Plenum Publishing Corporation
227 West 17th Street, New York, N.Y. 10011

United Kingdom edition published by Plenum Press, London
A Division of Plenum Publishing Company, Ltd.
Davis House (4th Floor), 8 Scrubs Lane, Harlesden, London, NW10 6SE, England

All rights reserved

No part of this book may be reproduced, stored in a retrieval system, or transmitted in any form or by any means, electronic mechanical, photocopying, microfilming, recording, or otherwise, without written permission from the Publisher

Printed in the United State of America

Foreword

This volume contains the complete set of papers presented at the First U. S.-U. S. S. R. Sciences Cooperation Seminar on "Optical Information Processing" held at the U. S. National Academy of Sciences in Washington, D. C. from 16 - 20 June 1975 under the sponsorship of the National Science Foundation in cooperation with the U. S. S. R. Academy of Sciences.

The papers present the latest theoretical advances and experimental state of the art in the newly developing field of "optical information processing", with particular emphasis on applications to communication, information storage and processing. Digital as well as optical systems are discussed in terms of concepts and implementations. Included are coherent and incoherent optical processing systems (for images and signals), materials and devices for optical computing, acousto-optic signal processing, memories (optical, digital and holographic), optical logic and optically-accessed digital stores, non-linear optical processing, as well as an analysis of the information capacity of optical processing systems and a report on new extensions of information processing in synthetic aperture radar. Detailed configurations and new manufacturing techniques for several components are presented, including such topics as "asymmetric interference fringes in reflected light" and 'kinoform optical elements" of very high quality; these are phase plates having a carefully controlled thickness, somewhat comparable to the famous Schmidt plates and which could have an important role in many optical computer and communications systems.

Some questions have been asked about the new optical approaches in the recent years, comparable to those first asked in the early 1960's about the usefulness and practical applications of the newly developed lasers. In particular, there have been some valid questions regarding the technological contributions which are likely to result from some of the rather striking scientific advances which have been demonstrated in the recent decade or so, from about 1962-1975, in the development of fundamental concepts for the fields of "optical information processing" and "optical computing", including some aspects of "holography". The present volume, in addition to the presentation of two recent symposia (1974 and 1975) on "optical computing" published by the Institute of Electrical and Electronic Engineers should satisfy, we believe, the reader, including responsible science research administrators, industrial executives and managers, among others, that practical and useful implementations in these fields are likely to develop now with an acceleration comparable to that which followed the somewhat earlier work on the development of fundamental concepts in the field of "lasers", and this probably with a comparably great industrial and societal impact. In fact, many of the developments in the two fields, lasers and optical information processing (and optical communications) are closely interdependent in many ways which are further clarified in this book. One of the most important achievements in optical computing, illustrated by the new results recently obtained in image deblurring (for example, in electron microscopy and space photography) has been the appreciation that the still unmatched perfection in optical image processing is a result of some fundamental advantages including those related to the enormous speed and parallel processing capacity for tens of millions of bits which are easily achievable in optical computers, such as those consisting of lenses and lasers in combination with holographic "filters", without the scanning and quantizing degradation losses which tend to plague digital image processing. On the other hand, there exists considerable merit in constructing some parts of the holographic filters with the aid of digital computation and photomechanical realization, in view of the subsequent use of the filters in the powerful optical computers. In the same context, consideration of detailed problems (theoretical and experimental) solved in the optical computers should help in bringing about an earlier implementation of the digital computer solutions which have long been sought for many of the image processing problems, because of some very desirable advantages and flexibility, but which have so far failed to match the success of the optical implementations: several helpful fundamental concepts may be found, it is hoped, under the title "optical foundations of digital communication and information processing". Laser printing of newspapers from a master original, at several widely distributed locations,

FOREWORD

linked by communication satellites, and the use of holographic image deblurring methods in image improvement and in image interpretation of space photographs, in view of improved capability for locating natural oil and gas reservoirs, are just two examples of societal areas of great impact which are coming into being at an increased speed, we believe, thanks to the mutual benefits of a continuing, even though healthily and naturally competitive, scientific cooperation such as that which formed the basis of this First U. S. - U. S. S. R. Science Cooperation Seminar on Optical Information Processing.

Science cooperation between the Soviet Union (then Russia) and the West was actually initiated in the early 1700's by the great Russian westernizer, Peter the Great. During his trips to Western Europe he collected physical and chemical apparatus and placed them in a museum in St. Petersburg (now Leningrad) called the Kunstkamera: this was the first laboratory in Russia. Just before his death, Peter the Great had planned to establish the Imperial Academy of Sciences (today the U. S. S. R. Academy of Sciences) but the Academy was not actually organized until one year later in 1725 by his successor, Catherine I.

Since there were no native scientists in Russia at that time, the Imperial Government invited a number of German, Swiss and French scientists of great renown and others to come and work in Russia with the new generation of Russian scholars. Most famous among these foreign scientists were Leonhard Euler and also several of the Bernoulli family. It was not long thereafter that one of the great Russian scientists, Michael Lomonosov, son of a peasant and educated already in Moscow, founded the University of Moscow in 1755. Lomonosov continued the tradition of science exchanges and cooperation by his travels to the Western Europe centers of learning, helped by traveling fellowships (we would say "grants") which he was awarded in Moscow.

It seems to be particularly fitting, therefore, that this great universal tradition of continuing science cooperation between the Soviet Union and the West was also symbolized further now, in this year of the 250th anniversary of the Academy, by the First U. S. - U. S. S. R. Science Cooperation Seminar on "Optical Information Processing" which took place, as mentioned at the National Academy of Sciences, Joseph Henry Building in Washington, D. C. 16 - 20 June 1975 under the sponsorship of the National Science Foundation. Many other fascinating and probably not commonly known details about the developments of "Soviet Science" may be found in the excellent little monograph published under that title in 1952 by the American Association for the Advancement of Science.

The crucial initiatives, most effective guidance and counsel throughout the entire planning and execution stages given to the U.S.A. co-chairman (Dr. G. W. Stroke) by Dr. John R. Thomas, Director of the National Science Foundation U.S.S.R. Program at the NSF Office of International Science Activities, was vital in every way to the success of this conference. We are pleased to single out Dr. Thomas with a special mention of commendation and immense gratitude to him and to all concerned at the National Science Foundation.

The Seminar brought together what we believe was not only a group of most active leading contributors in the field of "optical information processing" in the U.S.S.R. and U.S.A., but we think that the readers will find also that the papers delivered at the conference, and reprinted herewith, represent in toto probably the most advanced summary of the state of the art which may be currently available in a single book. The three editors are pleased to acknowledge the contributions of all the authors who have worked hard and on schedule in order to make it possible to produce a volume of such completeness and high scientific standard so shortly after the conference took place: essentially all the papers, including notably all those from the Soviet Union, were brought to the conference in Washington in IBM-typed camera-ready form, according to plan, but probably somewhat of an exception for busy scientists. For this reason also, in consultation with the authors and with the Publisher, it was decided to maintain the speed and lowest possible cost of bringing this new material to interested readers at the expense of leaving some perhaps unusual technical expressions in some titles, proper names and text, notably in some of the Soviet manuscripts: informed readers, we found, have no difficulty in correcting the terminology and language, and they may in fact be interested and pleased, as we were, by the knowledge that each paper represents, in fact, the high degree of competence in English of the responsible Russian authors, in contrast, no doubt, to probably a more customary lack of corresponding fluency in Russian by most English-speaking authors. The editors, in any event, acknowledge full responsibility for what may appear to some to be "unedited" language. They will welcome appropriate comments, of course, perhaps to be included as "appendices" in future printing, and in view of helping

FOREWORD

further in the establishment of accepted terms in this newly developing field.

The readers may be interested in a few more details about the seminar which forms the basis of this book.

The conference opened with an address by Dr. Leo S. Packer, Director of Technology Policy and Space Affairs of the U. S. Department of State, who addressed the U. S.-U. S. S. R. delegates and a select group of officials from the National Science Foundation (including Dr. Goetz Oertel, Director of the Astronomy Division) and Dr. J. C. R. Licklider, Director of Information Processing and Technology Office of ARPA, at a luncheon held at Washington's famous Cosmos Club. We continue to be grateful to Dr. Packer for his brilliant and stimulating address on fundamental ingredients in the development of technology policy which are so central to our vital concerns at this time and with which he is particularly familiar due to his long high-level service with the U. S. Government. He was in fact the first Presidential appointee to head up the Post Office Department Bureau of Research and Engineering as Assistant Postmaster General, a position established by Act of Congress in 1966. Another highlight of the meeting was an exclusive U. S. presentation by the Soviet delegation of a newly completed film on advanced Soviet computer technology, featuring production lines for the latest versions of LSI (large scale integration) circuits, disks, and tape devices, data processing lasers, fiber optic systems and optic communications, holographic displays and memories. The projection of the 35 mm sound film was made possible on short notice through the extraordinary efforts of Miss Nancy Carlile of the National Academy of Sciences, who had kindly assumed the responsibility for what turned out to be outstanding conference facilities at the National Academy of Sciences, in keeping with an official request made to this effect to Dr. Philip A. Handler, President of the National Academy of Sciences: the organizing U. S. co-chairman (G. W. S.) wants to express his official acknowledgement to Dr. Handler, Dr. L. C. Mitchell, Miss Nancy Carlile and all concerned at the National Academy of Sciences for this outstanding support of this conference.

The formal sessions of the conference in Washington were followed by a post-conference visit to a select group of leading U. S. East Coast institutions by the six Soviet delegates escorted by Dr. George W. Stroke and Dr. Winston E. Kock in coordination with Mr. Ray Pardon of the U. S. Department of State, whose most effective guidance and kind advice are deeply appreciated. The visits were carried out according to a plan worked out by the U. S. co-chairman (G. W. S.) based on his conference-planning visit with the Soviet organizing co-chairman (Yu. E. N.) in Akademgorodok (Novosibirsk) in August 1974. Included were visits to Bell Laboratories in Murray Hill, New Jersey, the IBM Thomas J. Watson Research Laboratories in Yorktown Heights, New York, and to MIT in Cambridge, Mass. where the group was also additionally honored by the presence at dinner of MIT's President (and former Science Advisor to President John F. Kennedy) Dr. Jerome B. Wiesner. As examples, among the highlights of these visits, the group was shown the most recent status of magnetic bubbles memory technology at IBM (under the direction of Dr. P. Chaudhari) the latest stage of thermo-nuclear fusion energy research at MIT in the form of the "Alcator", a high field tokamak-type machine at the National Magnet Laboratory which is directed by Professor Benjamin Lax (who was responsible for the visit at MIT together with Dr. Donald Stevenson, Associate Director of the National Magnet Lab) and also at the MIT Magnet Laboratory a number of high-power infrared lasers for the required plasma diagnostics (temperature measurements which are being developed for the one hundred million degree range needed in this field). Finally, also at MIT, in the laboratory of Professor William F. Schreiber, the group could see a demonstration of the latest types of laser facsimile devices, in the form used by the Associated Press via ordinary low-frequency (2400 Hz) telephone circuits with scan-line resolutions in excess of one hundred lines per inch.

More recently it has become known that high-frequency facsimile systems (in the 4GHz and 6 GHz range) are becoming operational based on experiments which have been successfully carried out in late 1974 by the COMSAT laboratories with the INTELSAT IV F-7 Atlantic satellite to transmit laser-scanned pages of the Wall Street Journal with 800 line/inch resolution at 360 million bits/journal page (average, at 50 kilobits/second) in a remote platemaking system for regional printing and distribution, using a laser scanning and reproduction system developed by the Dest Data Corporation of Sunnyvale in California. Automated remote printing using lasers and electro-optical communications technology, together with such areas as computerized X-ray diagnostic imaging (e.g. as pioneered by EMI) and image deblurring and image synthesis (e.g. as applied in X-ray crystallography and X-ray astronomy, radar and sonar, as well as in scanning electron microscopy) represent only a few among the several newly developing fields in which optical information processing is likely to continue to contribute in a fundamental way.*

>
> GEORGE W. STROKE
> WINSTON E. KOCK
> YURII E. NESTERIKHIN

---

*General background, notably for additional historical details and the necessary mathematics, may be found in several articles and books such as:

G.W.STROKE, "An Introduction to Coherent Optics and Holography" (Second Edition,1969,New York:Academic Press)

D.GABOR,W.E.KOCK and G.W.STROKE, "Holography", Science, Vol. 173, pp. 11-23 (2 July 1971) (Dennis Gabor was awarded the Nobel Prize in Physics in 1971)

G.W.STROKE,"Optical Computing", IEEE Spectrum,Vol. 9, No. 12, pp. 24-41 (December 1972)

W.E.KOCK, "Engineering Applications of Lasers and Holography" (1975,New York:Plenum Publishing Corp.)

L.M.SOROKO, "Principles of Holography and of Coherent Optics" (1971,Moscow:Nauka)(in Russian:English translation, edited by G.W.Stroke, in preparation:Plenum Publ.Corp.)

Contents

OPTICAL REALIZATION OF A FOUCAULT-HILBERT
    TRANSFORM 1
    V. A. Arbuzov and V. A. Federov

MATERIALS AND DEVICES FOR COHERENT OPTICAL
    COMPUTING 13
    David Casasent

INFORMATION PROCESSING IN OPTICAL SYSTEMS OF
    HOLOGRAPHIC MEMORY DEVICES 47
    I. S. Gibin and P. E. Tverdokhleb

GENERATION OF ASYMMETRIC INTERFERENCE
    FRINGES IN REFLECTED LIGHT 75
    N. D. Goldina and Yu. V. Troitsky

NOISE IN COHERENT OPTICAL INFORMATION
    PROCESSING 85
    J. W. Goodman

INFORMATIONAL CAPACITY OF COHERENT OPTICAL
    PROCESSING SYSTEMS 105
    S. B. Gurevich

EXTENSIONS OF SYNTHETIC APERTURE RADAR
    INFORMATION PROCESSING 117
    Winston E. Kock

CONTROLLED TRANSPARENCIES FOR OPTICAL
   PROCESSING 129
    I. N. Kompanets, A. A. Vasiliev and A. G. Sobolev

KINOFORM OPTICAL ELEMENTS 153
    V. P. Koronkevitch, G. A. Lenkova, I. A. Mikhaltsova,
    V. G. Remesnik, V. A. Fateev, and V. G. Tsukerman

ACOUSTOOPTIC SIGNAL PROCESSING 171
    A. Korpel

SOME PECULIARITIES OF PHYSICAL REALIZATION OF
   OPERATION OPTICAL MEMORY 195
    E. G. Kostsov, V. K. Malinovski, Yu. E. Nesterikhin,
    and A. N. Potapov

NON-COHERENT OPTICAL SYSTEM FOR PROCESSING OF
   IMAGES AND SIGNALS 203
    B. E. Krivenkov, S. V. Mikhlyaev, P. E. Tverdokhleb,
    and Yu. V. Chugui

OPTICAL LOGIC AND OPTICALLY ACCESSED
   DIGITAL STORAGE 219
    Rolf Landauer

NONLINEAR OPTICAL PROCESSING 255
    Sing H. Lee

OPTICAL FOUNDATIONS OF DIGITAL COMMUNICATION
   AND INFORMATION PROCESSING 281
    George W. Stroke and Maurice Halioua

THE INFORMATION CONTENT OF OPTICAL DIFFRACTION
   PATTERNS 313
    Brian J. Thompson

HOLOGRAPHIC MEMORIES 347
    A. Vander Lugt

BIOGRAPHIES OF AUTHORS 369

AUTHOR INDEX 381

SUBJECT INDEX 389

OPTICAL REALIZATION OF A FOUCAULT-HILBERT TRANSFORM

V.A. Arbuzov, V.A. Fedorov

Institute of Automation and Electrometry
Novosibirsk, USSR

ABSTRACT

The schemes of two optical devices realizing the isotropic Foucault-Hilbert transforms have been described. Formulas are given relating the parameters of the schemes with those of the output signal. The influence of light asymmetry of the light source are being discussed. The recommendations on the method of control of the transformer characteristics in the process of operation have been given.

An isotropic Foucault-Hilbert transform is determined by the expression

$$|\dot{\varphi}(x,y)|^2 = \alpha |\dot{g}(x,y)|^2 + \beta |\dot{\chi}(x,y)|^2 \qquad (1)$$

where α and β are numerical coefficients, $\dot{g}(x,y)$ is an input monochromatic "signal-pattern", $|\dot{\chi}(x,y)|^2$ is an isotropic Hilbert image pattern of the function $\dot{g}(x,y)$ related with its one-dimensional Hilbert transforms by the relation

$$|\dot{\chi}(x,y)|^2 = |\dot{\chi}_X(x,y)|^2 + |\dot{\chi}_Y(x,y)|^2 \qquad (2)$$

and the $\dot{\chi}$- transform [1] is

$$\dot{\chi}_x(x,y) = \frac{1}{\pi} P \iint\limits_{-\infty}^{\infty} \frac{-\dot{g}(x',y')\delta(y-y')}{x-x'} dx'dy'$$

Applied significance of the operator $|\dot{\chi}(x,y)|^2$ and systems of its realization were reported in [2,3]. The operator $|\dot{\varphi}(x,y)|^2$ is a generalization of $|\dot{\chi}(x,y)|$. The optical systems realizing this operator widen a range of problems being solved by utilizing data processing optical systems. Taking this into account the authors have developed two optical schemes permitting isotropic and one-dimensional Foucault-Hilbert transforms to be realized for an image $\dot{g}(x,y)$.

Figure 1 represents the first of these schemes. It comprises the following components: a laser - 1; a microobjective - 2; a polarization light - dividing cube - 3; polarization quarter - wave phaseplates - - 4,5; cylindrical mirrors - 6,7; collective lenses - - 8,8'; a rotative clouded glass - 9; a projective lens - 10; a cross-shaped diaphragm - 11; a collimation objective - 12; a direct Fourier-transform objective - 13; a quadrantal amplitude-phase filter - 14; an inverse Fourier-transform objective - 15. Components 1-11 as a whole form a monochromatic spatially non-coherent cross-shaped light source. Element 14 in odd quadrants (with respect to even ones) has transmission equal to $\tilde{\tau}e^{i\delta}$. Such a quadrantal mask is brought

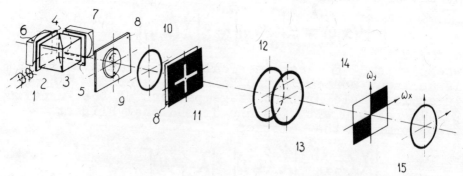

Figure 1. The Scheme of Device Realizing Isotropic Fuco-Hilbert Transform.

into coincidence with an image-pattern of the cross-shaped source. Each its leaf (half-slit) undergoes one-dimensional space filtering. Frequency characteristics of the partial filters are of the form

$$\dot{H}_{V1}(\omega_x) = \begin{cases} 1, & \text{at } \omega_x < 0 \\ \tau e^{j\theta}, & \text{at } \omega_x > 0 \end{cases} = 0.5\left[(1+\tau e^{j\theta}) - (1-\tau e^{j\theta})\text{Sgn}\,\omega_x\right]$$

$$\dot{H}_{V2}(\omega_x) = \begin{cases} \tau e^{j\theta}, & \text{at } \omega_x < 0 \\ 1, & \text{at } \omega_x > 0 \end{cases} = 0.5\left[(1+\tau e^{j\theta}) + (1-\tau e^{j\theta})\text{Sgn}\,\omega_x\right]$$

$$\dot{H}_{H1}(\omega_y) = \begin{cases} \tau e^{j\theta}, & \text{at } \omega_y < 0 \\ 1, & \text{at } \omega_y > 0 \end{cases} = 0.5\left[(1+\tau e^{j\theta}) + (1-\tau e^{j\theta})\text{Sgn}\,\omega_y\right] \quad (3)$$

$$\dot{H}_{H2}(\omega_y) = \begin{cases} 1, & \text{at } \omega_y < 0 \\ \tau e^{j\theta}, & \text{at } \omega_y > 0 \end{cases} = 0.5\left[(1+\tau e^{j\theta}) - (1-\tau e^{j\theta})\text{Sgn}\,\omega_y\right]$$

where $\omega_x = \frac{2\pi}{\lambda f_o}p$, $\omega_y = \frac{2\pi}{\lambda f_o}q$, λ - is a wavelength of the laser beam, p and q are spatial frequencies, f_o is a focal distance of objective 13.

It can be shown that the light intensity $\overline{I}(\tilde{x},\tilde{y})$ in the system output plane is determined by the relation

$$\overline{I}(\tilde{x},\tilde{y}) = M^2 A\left[\alpha|\dot{g}(\tilde{x},\tilde{y})|^2 + \beta|\chi(\tilde{x},\tilde{y})|^2\right] \quad (4)$$

wherein

$$\alpha = \begin{cases} \dfrac{(1-\tau)^2}{2}, & \text{at } \theta = (2k-1)\pi \\ \\ \dfrac{(1+\tau)^2}{2}, & \text{at } \theta = 2k\pi \end{cases}$$

$$\beta = \begin{cases} \dfrac{(1+\tau)^2}{4}, & \text{at } \theta=(2k-1)\pi \\ \dfrac{(1-\tau)^2}{4}, & \text{at } \theta=2k\pi \end{cases} \quad (5)$$

A is intensity of the light beam passing through a slit of the cross-shaped source; M is a magnification coefficient of reconstructing objective 15 $(M=f_o/f_v)$ $\tilde{x}=-Mx, \tilde{y}=-My$. The focal distance of objective 15 is designated by the symbol f_v.

Having compared (4) and (1), it can be concluded that the system described does realize the isotropic Foucault-Hilbert transform. It can process slides with amplitude, phase and complex coefficients of the transmission $\dot{g}(x,y)$. The parameter θ can be changed by varying the wavelength λ. When developing the system it is desirable to fulfil the condition $A_{v1}=A_{v2}=A_{H1}=A_{H2}=\dfrac{A}{2}$, where A_{ij} is light power arriving from each half-slit of the source. When this condition is broken in Expression (3) disturbing summands apper. For example, at $A_{H1}-A_{H2}=\Delta A_H$ (a horizontal slit), $A_{v1}-A_{v2}=\Delta A_v$ (vertical slit) and $A_{v1}+A_{v2}=A=A_{H1}+A_{H2}$ the noise components can be estimated by the following relation

$$\dfrac{\Delta A_v}{A} Re\left[j(1+\tau e^{j\theta})(1-\tau e^{-j\theta})\dot{g}(\tilde{x},\tilde{y})\dot{\chi}_x(\tilde{x},\tilde{y})\right] + \dfrac{\Delta A_H}{A} Re\left[j(1+\tau e^{j\theta})(1-\tau e^{-j\theta})\dot{g}(\tilde{x},\tilde{y})\dot{\chi}_y(\tilde{x},\tilde{y})\right] \quad (6)$$

Provided that one light source channel is cut off, on the output of the system under consideration a one-dimensional Foucault-Hilbert transform is obtained when varying the parameter τ to the extent of $1 \geqslant \tau \geqslant 0$, the coefficients α_o and β_o change in the opposite directions with in the range of (0 ÷ 1) and (1 ÷ 0), respectively. In this case the parameter θ should successively take the values π and 0 (transfer at $\tau=0$).

A one-dimensional Foucault-Hilbert transform can be obtained, e.g. on the Belvaux system base [4] as well. This system utilizes a combination of a slit-shape light source and a complex one-dimensional Foucault knife. Such a scheme can be considered as a particular case of the system proposed when only one half-slit of the cross--shaped light source is illuminated. Under such a mode the following equalities take place

$$\Delta A_v = A_{v1} - A_{v2} = A,$$

$$A_{v1} + A_{v2} = A_{v1} + 0 = A \quad \text{and} \quad \frac{\Delta A_v}{A} = 1$$

Taking them into account on the basis of (6) it may be stated that in the Belvaux system the Fuco-Hilbert transform is imposed with maximum-possible distorting terms. In the system proposed $\Delta A/A \ll 1$, hence on its output only the Foucault-Hilbert transform is obtained.

The second scheme is shown in Figure 2. It also represents the system of space filtering and is a generalization of the first one. It comprises the following components: a circular non-coherent light source - 1, a collimation objective - 2, a slide to be processed - 3, a direct Fourier - transform objective - 4, a coaxial phase plate - 5, an inverse Fourier - transform objective - 6, a recording photomedium - 7.

The central region of plate 5 relative to its outer circular area shows transmission $\tau e^{j\theta}$. The circular source image is brought into coincidence with the contact boundary of disk and circular components of plate 5. For each pair of diametrically-spaced points the above boundary surface is equivalent to two amplitude-phase knives with the functions of transmittance of Type (3). The output signal from each pair of monochromatic light source can be represented as

$$\underline{I}_n(\tilde{x}, \tilde{y}, \varphi) = M^2 E_\varphi^2 \left[\alpha_0 |\dot{g}(\tilde{x}, \tilde{y})|^2 + \beta_0 |\dot{\chi}_\varphi(\tilde{x}, \tilde{y})|^2 \right] \quad (7)$$

where

$$\dot{\chi}_\varphi(\tilde{x}, \tilde{y}) = \frac{1}{\pi} P \int_{-\infty}^{\infty} \frac{\dot{g}(x'\cos\varphi - y_0'\sin\varphi, x'\sin\varphi + y'\cos\varphi)}{x' - x_0'} dx'$$

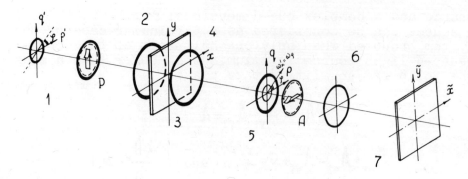

Figure 2. The Scheme of Device Realizing Isotropic Foucault-Hilbert Transform (averaging in respec of   angle).

is a one-dimensional Hilbert transform of the function $\dot{g}(\tilde{x},\tilde{y})$, along the diameter chosen. The coordinate, x' and y' are related with the coordinates $\tilde{x}, \tilde{y}$ by the relation

$$\tilde{x} = x'\cos\varphi - y'\sin\varphi, \quad \tilde{y} = x'\sin\varphi + y'\cos\varphi$$

Integrating (6) according to the angle φ we obtain

$$\bar{I}(\tilde{x},\tilde{y}) = \frac{M^2 B}{2}\left[\alpha|\dot{g}(\tilde{x},\tilde{y})|^2 + \beta|\dot{\chi}(\tilde{x},\tilde{y})|^2\right] \qquad (8)$$

where $\quad B = \int_0^{2\pi} E_\varphi^2 \, d\varphi \quad$ is a summarize

brightness of the input light field.

From Expression (8) it is seen that the second scheme also realizes the Foucault-Hilbert transform. The effect of brightness assymetry of the pairs of diamet-

OPTICAL REALIZATION OF FOUCAULT-HILBERT TRANSFORM

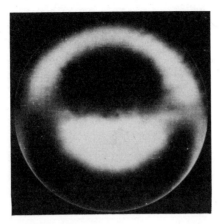

Figure 3a. Spherical Aberration of Photoobjective (Tepler device with slit light source and one-dimensional amplitude Foucault knife).

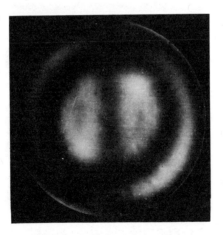

Figure 3b. (slit light source; quadrantal amplitude knife).

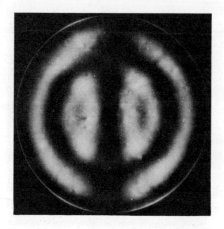

Figure 3c. (slit light source, quadrantal amplitude-phase knife; $\tau \simeq 0.9$; $\theta \simeq 180 \pm 5°$).

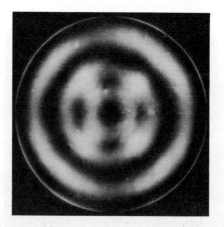

Figure 3d. (cross-shaped light source; amplitude-phase knife; $\tau \simeq 0.9$; $\theta \simeq 180 \pm 5°$).

Figure 4a-b represent the responses to the amplitude function $g(x,y)$.

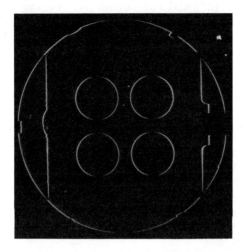

Figure 4a. One-Dimensional Hilbert Transform of Amplitude Function $g(x,y)$ (vored metallic plate).

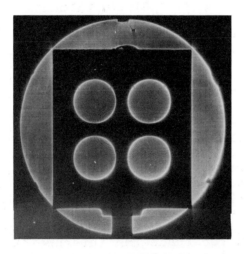

Figure 4b. Isotropic Foucault-Hilbert Transform of Amplitude Function , $\tau = 0$.

rical points of the circuit can be estimated with the help of relation (6). The partial coefficients α and β are related with the parameters τ and θ by expression (5). If necessary to change α and β in the process of work a Kastler coaxial phase plate should be introduced into the scheme instead of dielectrical coaxial plate 5. It consists of coaxially-positioned double-refracting plates of the same thickness. Their main optical axes are set at an angle of 45°. In addition to the Kastler plate a polaroid P and an analyzer A schematically shown in a dotted line are introduced into the scheme. Their planes of transmission are orthogonal. For a smooth variation in the parameter τ and a discrete variation in θ (0 or π) the rotation of the pair $(P \div A)$ by the angle ψ is carried out, their orthogonality being maintained. The count of the angle ψ is performed from a bisectrix of the angle formed by the main optical axes of the components of plate 5. The parameters τ and θ are related with the angle ψ by the relation

$$\tau = |tg^2\psi| \, , \, -\frac{\pi}{8} \leq \psi \leq \frac{\pi}{8}$$

$$\theta = \begin{cases} 0 \, , \, \text{at} \, -\frac{\pi}{8} \leq \psi \leq 0 \\ \pi \, , \, \text{at} \, 0 \leq \psi \leq \frac{\pi}{8} \end{cases}$$

The Kastler plate is broadband. Therefore the system can realize the Foucault-Hilbert transform in white light as well. Figure 3 represents the responses of the systems of four types to the axis-symmetrical function $\dot{g}(x,y)$ describing spherical aberration of the photoobjective.

From figures it follows that the isotropic Hilbert and Foucault-Hilbert transforms are free of "curvatures" (contrast inversion on the half-plane) 5 and are of an isotropic nature with an axis-symmetrical character of the input signal $\dot{g}(x,y)$. This feature can be used, e.g. when developing shade devices. In this case a shadegram is vivid and has no blind zones. The electronic devices for shadegram interpretation are reasonably simplified since they are rid of the work on eliminating the shadegram curvatures.

REFERENCES

1. Titchmarsh E.C. Introduction to the Theory of Fourier Integrals, Oxford University Press, 1937.

2. Eu J.K.T., Lohmann A.W.- "Opt. Communs.", 1973, 9, N 3.

3. V.A. Arbuzov, V.A. Fedorov. -"Avtometria", 1975, N 5.

4. Y. Belvaux, J.C. Vareille. - "Nouv. Rev. Opt. Appl.", 1971, v. 2, N 3, 149.

5. L.M. Soroko. "Osnovy golographii i kogerentnoi optiki". M., "Nauka", 1971, p. 409.

MATERIALS AND DEVICES FOR COHERENT OPTICAL COMPUTING

DAVID CASASENT

Carnegie-Mellon University, Dept. Elec. Engr.

Pittsburgh, Pa. 15213

INTRODUCTION

The tremendous potential offered by optical data processing has been well reviewed in numerous texts [1-3], articles [4-7], conferences [8-10], and special issues [11-15]. The ability to perform complex spatial filtering, correlation, linear and non-linear transformations optically in parallel and in real-time is well recognized. While tremendous advances have been made in recent years in optical processors, film is still generally used to input data into and store reference data in these processors. The long development and chemical processing required with film cause considerable delay between the occurrence of the information to be processed and when it enters the optical processor. The second major problem with the use of film is that the data recorded on it cannot be changed once the film is developed, rather a new piece of film must be used. Other problems of noise, granularity, dynamic range, etc., are also of importance.

This paper summarizes various recent developments in real-time, reusable devices and materials for optical processing. This field is so large (fostered by the large display market [16], research in optical memories [11] for digital computers, etc.) that all potential devices and materials and research efforts cannot be listed, let alone discussed. A liquid crystal bibliography [17] with over 3000 entries and several display bibliographies [18-19] exist. This discussion will concentrate on general considerations, representative devices, recent advances, the present state-of-the-art, and is flavored with critical comments on problems and limitations (fundamental and technological) of various devices and materials. Any omissions are due to incomplete or only preliminary data, the proprietary nature of many recent developments, but more generally to reasons of time and space. Optical storage and memory devices and page composers for digital computers are not considered per se. The discussion is restricted to devices for use in coherent optical systems (not displays [16]), two-dimensional devices (spatial not 1-D light modulators), and recyclable devices useable in real-time or near real-time. Several device and material survey articles exist [20,33-39], Some are addressed to the non technical audience; others are restricted to OALMs [20,35,36] or EALMs [33] only concentrate heavily on non-real-time and non-reuseable devices [37], etc.

CLASSES OF SPATIAL LIGHT MODULATORS

Two general classes of spatial light modulators exist, distinguished by the nature of the input data: optical or electronic. We refer to these as optically addressed light modulators OALMs and electronically addressed light modulators EALMs. The OALMs can be further divided into two classes, those that are point-by-point addressed (as with a scanning laser beam) and those that are parallel addressed by focusing an ambient scene or the output of a CRT display onto them. This characterization only affects the system not the device. Two further classifications relate to the nature of the recorded data. Is it spatial or holographic? Holographic (Fourier not Fresnel) recordings require higher resolutions (Sin $\theta)/\lambda$ = 800 lines/mm for θ = 30 (the angle between reference and image beams) than those from spatial scenes. Single holograms, several spatially separated ones, several superimposed encoded ones, or combinations of the last two can be used. Now recording material and system properties must both be considered. These distinctions are listed in Table 1.

The classes of EALMs are fewer and distinguished by the addressing scheme used: matrix of strip electrodes or electron beam addressed. The electroded devices have less resolution (generally 32 x 32 or 128 x 128) while the electron beam addressed devices offer 10^3 x 10^3 point resolution and 30 frame/sec. operation. The contrast of the electroded devices is usually less because of crosstalk and they consist of discrete rather than continuous modulating elements. Their use is principally as page composers and will not be extensively discussed here. OALM devices are always of higher resolution than EALMs. For this reason one of the most versatile arrangements for the use of both devices employs an EALM in the input plane (fed by electronic input signals from a sensor, the output of a high resolution camera, etc.) and an OALM in the matched filter plane (where holographic matched filters are recorded or stored) of an optical processor as shown in Fig. 1. The holographic matched filter can now be formed on the OALM from data on the EALM and updated as needed.

Table 1. Classes of Spatial Light Modulators.

OPTICALLY ADDRESSED LIGHT MODULATORS
Addressing scheme - point-by-point vs. parallel Recording - spatial vs. holographic

ELECTRONICALLY ADDRESSED LIGHT MODULATORS
Addressing scheme - electroded vs. electron beam

Various other configurations too numerous to list exist. Dual in-line image correlation (Fig. 2) is quite useful with lower resolutions (50-100 $\ell\ell$/mm) OALMs. In this case the two images to be correlated are placed side-by-side in the input plane and their transform formed on an OALM. Reading out this recorded OALM data and transforming it yields the correlation of the two input images. A major reason for the use of such a scheme is that many OALMs operate in reflex, are written from one side and read from the other using two different wave-lengths of light, etc. Dual in-line image correlations removes problems of scaling for two wave-lengths, does not require the same lens system to operate at two different wave-lengths and easily accomodates many actual OALM devices as they would have to be used. The system in Fig. 1 is simpler conceptually to demonstrate the point but must be modified in practice.

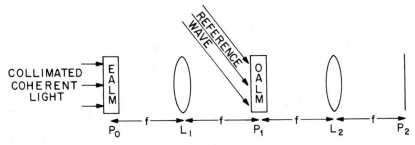

Fig. 1. Typical real-time coherent optical computer for correlation.

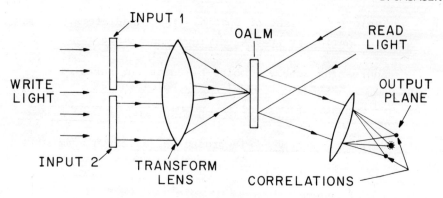

Fig. 2 Dual in-line image correlator.

Other factors that must be remembered are the use of lenslet arrays with spatially multiplexed holograms. A most important factor is that deflectors and scanners as well as lens systems with proper specifications must exist. Often such elements are simply drawn in an optical system and not analyzed. One problem in the dual in-line image correlator is how to accurately position the two input images <u>in real-time</u>. In other words, care must be given to system tradeoffs in selecting EALMs and OALMs.

BASIC PERFORMANCE FACTORS

Many parameters are used to characterize and compare devices and materials for coherent optical computing. Several good descriptions of these for spatial light modulators [20] and holograms [21-22] exist. For a sinusoidal input at a spatial frequency f the exposure is $E(x) = E_o (1 + m_1 \cos fx)$ and the output amplitude is $A = T(a_o + a_1 E + a_2 E^2 + ...)$ if the exposure is outside the device's linear range. For a phase modulator

$$A = T \exp[i\phi(E)] = T[1 + i\phi(E) - \phi^2(E)/2 + ...]$$

Thus even if $\phi(E)$ is linear the response will not be for phase modulators. The importance of this and whether it should be corrected for must be answered in individual cases.

Resolution is a subjective concept. It is the highest spatial frequency that can be passed by a device. For display applications this definition may be adquate but for processing of data it is essential that the MTF (output modulation m_o divided by input modulation m_i as a function of spatial frequency f) be provided. Usually m_i is 1 and m_o/m_i is normalized to 1 at f = 0. Device resolution is often specified at the visible limit

(MTF = 0.05) simultaneously with the contrast at low spatial frequencies. Such specifications are of little use. Care must be taken that the contrast and resolution specified in comparing devices are simultaneously obtainable.

Sensitivity S is usually specified for OALMs. It is the reciprocal of the exposure necessary to change the output intensity by a given amount, usually by a factor of 1/e. Care should be made that S is not given for a 10% change or at the optimum (usually nonlinear) part of the response curve.

Both holograms and modulating devices are characterized by whether the amplitude or phase of the output read light is changed [22]. This depends on the modulation mechanism used by the recording device or material. Changes in absorption index (e.g. film and photochromics) are examples of amplitude modulation. The first is an effect within the volume of the material, the second a surface effect. Refractive index changes (e.g. lithium niobate and SBN) and surface relief patterns (e.g. thermoplastics) are examples of phase modulation. Again the first effect occurs in the material's volume and the second on the surface.

The diffraction efficiency η is usually quoted for phase modulation devices. The MTF of these devices and the deformable surface type devices which have bandpass characteristics cannot easily be defined. η can be related to M = MTF for amplitude modulation devices [20]

$$\eta = \frac{(1-\sqrt{1-M^2})^2}{4(1 + M + \sqrt{1-M^2})^2} \sim \frac{M^2}{16}$$

where the approximation holds for small MTF. The maximum obtainable η as a percent for various types of holograms recorded on OALMs is given [22] in Table 2.

Table 2. maximum diffraction efficiency as a percent for various types of holograms. [22]

		AMPLITUDE	PHASE
THIN HOLOGRAMS	Linear	6.25	33.8
	Binary	10.13	40.5
VOLUME HOLOGRAMS	Transmission	3.7	100
	Reflection	7.2	100

Table 3. Representative Factors in the Comparison of Spatial Light Modulators.

Diffraction efficiency	Type of Modulation
Resolution and contrast	Type of addressing
Sensitivity	Lifetime
Cycle Time	Optical quality
Response Time	Ease of fabrication
Erase mechanism	Special features
Storage capability	Complexity

Diffraction efficiency may not be as crucial in coherent processing applications as it is in display, reconstruction, and optical memory cases unless many superimposed or spatially modulated holographic filters are used for correlation. The importance of grey scale and linear recording in processing applications is of course greater than in optical memory cases. In all cases of interest in this paper, Fourier not Fresnel holograms are assumed since the output is a correlation not a display.

Many other factors (Table 3) must be noted in any spatial light modulator survey. The erase mechanism is of great importance since we are considering real-time devices. Devices which rely on charge leakage (e.g. electron beam on oil) for erasure produce residual images and are of limited use in applications where data enters continuously in real-time. For display applications, many of these problems are not of paramount importance since the eye is rather tolerant; however, signal and information processing requires more precision. Similarly, since applications requiring 2-D transforms are of concern the entire input must be formed before the transform is valid. In a sequentially addressed device (electron beam or scanned laser addressing, etc.) storage much greater than one frame time is required. In other cases where a detailed analysis of the input data is required, this feature is also needed. For OALM materials used in the spatial filter plane, storage and insensitivity to numerous readouts is essential.

The type of device (OALM or EALM), modulation (amplitude or phase), and addressing (see Table 1), and the ability to store volume holograms are of obvious concern. Device lifetime and optical quality are often the determining factors in selecting one device over another. In some applications, a device lifetime of 20 cycles (missile guidance) may be adequate. In some cases fast cycle times may not be needed (a fixed filter bank or one changed rather infrequently) but resistance to many readouts may be a major concern. In other cases the filter is changed every few frames (change detection), fast cycle times are crucial, and storage and readout are of minor concern. Optical quality refers to the fact that greater flatness and homogeneity and lower noise and scatter levels are

COHERENT OPTICAL COMPUTING

necessary in coherent processing than in noncoherent displays. Ease of fabrication and complexity are obvious factors referring to the device's reliability and reproducibility and the need for special care such as cryogenic operation, vacuum enclosures, high voltage supplies, etc. Special features of interest may include contrast reversal or level slicing of images.

DEFORMABLE TARGET DEVICES

There are four types of spatial light modulators that use the actual deformation of the recording surface to modulate light.

THERMOPLASTIC DEVICES

If an input wave of uniform amplitude and phase is incident on a thermoplastic, the output wave can be approximated by

$$A(x,y) = A_o(x,y) T_o \exp[j\theta(x,y)]$$

where a constant transmission T_o is assumed and $A_o(x,v)$ is the aperture. Dropping a constant phase term, $\theta(x,y) = k(n-1)d(x,y)$ where $k = 2\pi/\lambda$ and d describes the thickness deformations in the thermoplastic of index n. For a sinusoidal deformation $d = d_o + d_1 \sin f_x x$

$$A(x) = A_o(x) T_o \sum_{m=-\infty}^{+\infty} j^m J_m [k(n-1)d_1] \exp(jm\omega_x x)$$

Plane waves $\exp(jm\omega_x x)$ with amplitude J_m thus emanate from the modulator. The optical Fourier transform of $A(x)$ causes each plane wave to focus to a spot size dependent on the size of the modulating aperture $A_o(x)$ with a peak value proportional to the Bessel function $J_m[k(n-1)d_1]$ and its position proportional to the spatial frequency ω_x.

Because of the Bessel transfer function, linear operation restricts the maximum deformation to 0.2λ and diffraction efficiencies of 10% result. The amplitude of deformation varies as $1/\omega_x^2$ so that compensating electronic filters may be needed to preserve signal fidelity.

The thermoplastic TP used is a low melting temperature plastic (polystyrene, chlorinated polyphenal, alpha-methylstyrene, Stayblite ester-10, Foral-105, etc.) 0.4-1.5 μm thick. A plane electrode is placed on one large surface and the modulating signal as a charge distribution on the other. The charge distribution can be produced by a modulated scanning electron beam (EALM) or by an input light

(OALM) in which case a photoconductor PC layer is placed between the TP and grounded electrode. Operation of both devices requires heating and cooling of the TP while the charge distribution is present.

In the OALM device, a uniform charge layer is first placed on the PC. Corona charging has resulted in damage and excessive heating [23]. Parallel plate charging [24] and use of an aperture grid [23] over the TP have produced reproducible results. The PC is then exposed to the input light distribution containing the desired information. From 5-100 $\mu J/cm^2$ are required and depends on the PC used. This causes ionization and charge migration to the PC-TP interface. The surface charge density on the free TP surface is still uniform however, and a constant electrostatic force results. Charging is continued during exposure (to produce field variations in the TP). The TP is now heated to its development temperature by joule or rf heating. (Typical heat powers [23] are 30 W/cm^2 for 100 msec and 45 W/cm^2 for 4 msec development). Precise control of this is necessary to lock in the pattern since erasure is also accomplished by heating the TP to its softening temperature (50-100 $^{\circ}$C).

During development, the TP deforms by electrostatic forces. Subsequent cooling freezes in these deformations. A stored image can be read out indefinitely. During erasure heating, the conductivity of the PC and TP increase, the stored charge decays, the TP softens and again becomes flat.

One of the main problems with TP devices has been obtaining reproducible results. Emphasis is thus on this aspect rather than speed, although the required heating and cooling will certainly dominate. A typical cycle [24] yielding rather reproducible results requires charging for 25 msec, 100 msec exposure, 150 msec development at 4.5 W (0.7J), reverse charging for 50 msec (this greatly improves performance) and a 450 msec erasure at 4.5W (2J). Reproducible results can now be somewhat obtained but the complexity of the recording apparatus (and monitoring equipment needed for reproducible results) is a disadvantage.

The device resolution is high (2000 lines/mm) but is a band pass centered at 1/2d where d is the TP thickness. Obtaining predictable uniform thickness is still difficult; this makes fabricating devices with predictable performance a problem. The bandwidth about the center frequency is generally only several hundred lines/mm. For holographic recording, this center frequency can be used to advantage [25] to determine the reference beam angle. To record spatial rather than holographic data, a grating

COHERENT OPTICAL COMPUTING

whose spatial frequency is matched to the bandpass center frequency of the device is needed. Reproducible devices of uniform and predictable thickness are needed however. Selective erasure of spatially multiplexed holograms recorded on one TP has been demonstrated by heating from a CO_2 laser [24].

Fatigue remains the major problem in TP devices. Fatigue manifests itself as a drop in η, resolution, or modulation, and a change in the development of softening temperature. The onset of fatigue is irreversible and occurs after 10-1000 cycles. Many possible factors enter: polymerization and crosslinking between molecules, oxidation at the surface, impurities and contaminants and a tearing of the TP under growing electrostatic forces. Use of improved charging techniques and operation in the presence of argon gas have increased lifetime of some units to 1000 cycles. The optical quality of any deformable surface after many cycles is still a major question. Other problems are the difficulty in completely erasing images stored for long times. In Stayblite TP, when one image is erased and a second one written, the first reappears and grows due to residual deformations not charge. The device does form good thin phase holograms that can be read out indefinitely; but its lifetime limits it use to special applications where the stored data need not be changed rapidly nor a large number of times.

TP can also be electron beam addressed [26]. The TP is written on by the scanning electron beam. At room temperature, the high resistance of the TP results in storage of this deposited charge. When heated to the development temperature (by a heating pulse to the rear electrode of the TP) the TP deforms. The heating current is then removed, the TP returns to room temperature, and the deformations are stored. Other heating cycles are possible. The TP can be erased by heating it above its softening temperature ($100°C$) by a pulse of heating current in the conducting layer. Surface tension forces will exceed electrostatic forces and pull the surface flat. The device's optical quality after successive cycles is again unproven. The time constant for deformations is on the order of tenths of seconds while the decay time constant is over one second due to the resistivity of the TP. Eight shades of grey and 1800 line resolution have been demonstrated. The long 1 sec erase time limits device use. Lifetime of the TP as well as cathode contamination are a problem as before. One of the more dramatic demonstrations of its potential is the real-time hologram reconstruction [26] in Fig. 3. In another version of this device [27], the central annular ring of a rotating 14" disc of TP is written on by an electron beam.

Fig. 3. Real-time remote reconstruction [26] of hologram imaged onto a TV, written onto the TP, and subsequently reconstructed. Courtesy Applied Optics, 11, 1261(1972).

ELECTRON BEAM ON DIELECTRIC

One of the most remarkable engineering achievements is the electron-beam-addressed oil film target spatial light modulator [28] shown in Fig. 4. The device is a modification of a projection TV version of the Eidophor. The target surface consists of a thin layer of dielectric fluid on which a scanning electron beam deposits charge. Strong grooves due to the scan lines result. Brightness is controlled by varying the depth of the scan line grooves. The disc target continuously rotates slowly, refreshing the oil layer from a reservoir. Cathode contamination and device lifetime have been overcome and a 1000 hour lifetime obtained with a sealed tube. The target surface deforms and produces phase modulation with a Bessel transfer function as in the thermoplastic system. Surface deformations must be kept less than 0.6 μm or 0.1λ for linear modulation so that only 1% diffraction efficiency results.

The device can write about 10^3 x 10^3 points at 30 frames/sec and claims a 40-50 dB dynamic range. Specifications such as this are quite misleading. The device does have 10^6 point resolution, but the modulation mechanism needs description. The fluid deforms by electrostatic forces that are opposed by surface tension and viscous forces. Throughout this process, the deposited charge decays. The fluid's mechanical properties determine the deformation while its electrical properties determine the decay time. Erasure is thus by charge decay. In transform applications this decay must be much longer than one frame time since the transform cannot be taken until the entire input is written. The decay time cannot be simultaneously both at the frame rate and much longer. The 30 frame/sec rate must thus be qualified. Electronic preemphasis

COHERENT OPTICAL COMPUTING

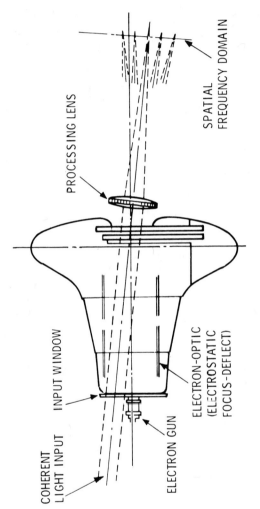

Fig. 4. Schematic [28] of the electron-beam-addressed dielectric fluid or oil film light valve.

and refresh is needed. The 40 dB dynamic range is from peak to
noise level of the Bessel function. This range is reduced for
linear operation. A 100:1 contrast ratio is easily obtained. The
folded spectrum output is most useful in extracting signals
from background clutter. The oil can apparently be kept quite
flat, but this requires a uniform beam current raster and an
elaborate carrier modulation scheme to correct for surface
irregularities to maintain optical quality [29]. The oil film
light valve is operated at 50°C by electric heaters under the shell
holding the tube, an aluminum casting is used for temperature
distribution. Once these clarifications are made, a remarkable
achievement in engineering results with many special uses for the
folded spectrum present. This device is useful as an example of
what technology can do and apply to other devices.

ELASTOMER OR RUTICON

The third deformable target OALM is the Ruticon [30] or
elastomer [31]. The structure for this device is shown in Fig. 5.
It is quite similar to the oil film light valve with a thin 4.5 μm
elastic medium replacing the oil film and a photoconductor used to
produce an OALM. The gamma Ruticon is the most appropriate device
for coherent optical processing. The device is initially charged
by applying a voltage between the electrodes. This divides
capacitively between the elastomer E and photoconductor PC. When
the device is exposed to the input information, charged pairs are
created in the PC, migrate to the PC-E interface, and are trapped
there. The forces due to these charges cause the elastomer and
flexible metal electrode to deform. Its sensitivity is 300 ergs/
cm^2 (determined by the PVK photoconductor and voltage used). Thin
phase holograms with η = 15% result.

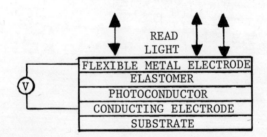

Fig. 5. Schematic structure of elastomer OALM.

The device operates in reflection from the metal electrode layer. Erasure is accomplished in 10 msec by removal of the dc field, a distinct advantage over erasure by heating in thermoplastic devices. The device's frequency response is bandpass centered at about 100 lines/mm, although resolutions of over 500 lines/mm have been demonstrated.

The device does not exhibit permanent storage. The holographic and image recording comments made for bandpass thermoplastic OALMs apply here also. Details on reproducibility and ease of fabrication are not known. The device's optical quality is also not certain, but the ability of any deformable device to retain optical quality to better than a wavelength after many cycles of operation seems questionable [20]. The device's lifetime should also be limited as it is for all deformable OALMs; a 10,000 cycle lifetime is quoted [31]. It is still superior in lifetime, ease of erasure, and cycle time to many other deformable target devices.

MEMBRANE LIGHT MODULATOR

One matrix addressed spatial light modulator of use in some cases is the membrane light modulator [32] or MLM shown schematically in Fig. 6. It offers the highest resolution of any electroded EALM, with 500 element per inch devices fabricated. A thin dielectric area is deposited on a glass substrate on which stripe chromium electrodes have been deposited. A regular array of 40 µm (5 µm possible) perforations on 50 µm centers are made in the dielectric layer. A thin 0.1 µm reflective goldcoated collodium membrane mirror is then stretched across all perforations and grounded. A 20-40 V signal to the electrode causes the addressed membrane to deform, thus modulating light incident on the membrane in reflection. The membrane surface can be fabricated flat to 0.1λ and 0.1-1 nsec response and a lifetime of over 10^{12} cycles obtained.

The MLM has been used extensively in coherent light, but it is only a row-addressed 1-D modulator. It also lacks storage (< 1 sec.), and since only a small part of each membrane is used for reflection has low optical efficiency.

A photoactivated or OALM type of 2-D spatial MLM light modulator [36] has been fabricated from a 50-100 µm thick n-type semiconductor crystal. An array of p-n junction diodes (one per membrane element) is diffused into the lower side of the MLM. A perforated resistive layer and collecting electrode are then deposited and a metallized polymer membrane applied over the entire structure. A transparent electrode is formed by diffusion below the p-n diodes. In operation, a voltage is applied between

the collecting and diffused electrode to back-bias all p-n junction
diodes. When write light enters the device, hole-electron pairs
are produced, a potential develops between the p-region of the
diode and the collecting electrode, and the membrane deforms.

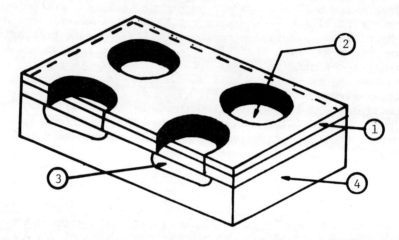

Fig. 6. Schematic diagram [33] of section of electroded MLM:
(1) reflecting membrane, (2) perforation, (3) stripe electrode,
(4) glass substrate.

Several other deformable target devices exist: deformable
mirror light valve, deformagraphic tube, electron beam on metal
film, etc. They have not seen extensive use in coherent optical
processing and are generally more research devices and less further
developed than the above four.

LIQUID CRYSTAL DEVICES

The large volume display market has fostered considerable
research and development into liquid crystal LC devices and effects.
Over 15 different electro-optic effects [39] and over 3000 LC
references [40] exist. Research activity has been prolific since
Heilmeir's classic paper [41]. Many LC devices use the dynamic
scattering mode. This effect is not recommended for use in a
coherent optical processor because of the halo effects, noise, and
scatter that result since the very nature of the effect is to
destroy the coherence of an incident laser beam. The use of LCs
has even been suggested [42] as a replacement for ground glass in
partially coherent applications.

The electro-optic effects in LCs are of far more use in coherent optical processing applications. Both current and field effects, in which an applied field converts one molecular ordering to another, are possible. Current effects involve the transport of charge and produce hydrodynamic instability in the LC. Field effects involve the tilting and turning of molecules and are more promising; conduction currents are negligible in these cases. Current effect LC devices usually have low lifetimes and are slower than field effect LC devices. This discussion will thus concentrate on the field effects and will emphasize the use of coherent light and the need for lifetime. These factors together with the historically slow response and low contrast of LC devices have restricted their use in real-time coherent optical computing.

Two types of LC structure are of interest; the nematic phase which is characterized by 1-D ordering with the molecules aligned with their long axes parallel, and the cholesteric phase which consists of a series of nematic like planes with the direction of alignment of the molecules progressively changing from one plane to the next tracing out a spiral in a direction normal to the planes. The large anisotropy (difference in orthogonal components of the refractive index n or dielectric constant ε) exhibited by these materials is responsible for the dramatic changes in optical properties that occur when this ordering is changed. Nematics do not exhibit memory and their use is thus limited, while a 10% cholesteric-90% nematic compound possesses memory.

As noted earlier, field effects involve tilting and turning molecules which can be accomplished in tenths of msec, whereas current effects require setting a viscous liquid into motion and typically take msec. The use of field effects will thus result in devices with faster rise times. The lifetime of LC devices has always been questioned. The use of ac rather than dc fields has however increased lifetime [43]. Most long life cells will probably be operated in reflection. Writing will generally occur from one side at one wavelength and reading from the other side at a different wavelength. The PC must be chosen to minimize interaction of these two operations, often blocking electrodes are used for this purpose to prevent dc current flow in the LC. This and a careful choice of the materials used for various layers can greatly reduce electrochemical chemical deterioration. It appears that contrast will always be low in these devices.

The basic cell structure consists of this LC-PC layers sandwiched between two electrodes (one of which may be reflective). Polarizers and reflecting dielectric and other thin film layers may also be present, and numerous other structures are possible. Surface orientation effects tend to align the long axes of the LC molecules parallel or normal to the electrode boundaries. These forces compete with the tendency of the molecules to align at angles (negative anistotropy) or parallel (positive anistropy) to an electric field applied between electrodes.

In the twisted nematic effect [44], known also as tunable optical activity, the LC is sandwiched between two electrodes. Parallel microgrooves are formed on each electrode, the grooves on one electrode are normal or at some angle such as 45° to the grooves on the other electrode. If a nematic LC with positive anisotropy is used, the grooves force the LC molecules to align with their long axes parallel to the direction of the grooves. The molecules in the various LC layers between electrodes thus trace out a helical or twisted path. When an electric field is applied, the molecules try to align parallel to the field, thus changing the pitch of the helix and widening it. Light propagating parallel to the axis of the helix experiences a dichroism. The polarization of input light, initially linearly polarized in the direction of the microgrooves, is thus rotated through the twist angle of the helix as it passed through the material. As the applied voltage between electrodes increases, the rotary power of the device can be reduced to practically zero (from its initial 90° or 45° value). By properly orienting the polarizers and adjusting the applied voltage, the element can be opened or closed or intermediate levels obtained. Contrast reversal and level slicing of images are special operations that are possible. Because the rotary power cannot be made completely zero electronically, contrast will not be extremely high.

The voltage controlled birefringence effect [45] is closest to the conventional Kerr and Pockels effects. The cell structure is similar to that used before. Parallel microgrooves are again formed on the electrodes and coated with an ultra-thin surfactant. If a nematic LC with negative anisotropy is used, all LC molecules will tilt in a direction determined by the direction of the microgrooves. The frequency of the applied ac field is chosen to be high enough to prevent dynamic scattering. Phase modulation of the incident light results. The effective birefringence of the LC layer is a function of the normal and extraordinary indices of refraction, the applied voltage V and material constants. The transmitted light intensity is related to V by

$$I = I_o \sin^2 [K_1(1-\frac{K_2}{V} \tanh\frac{V}{K_3})]$$

where the K are constants. This non-classical dependence on V must be considered in devices where linearity is needed. The twisted nematic effect is thus preferable on this basis.

These simplified explanations of these two field effects in LCs assumed uniform voltages on each electrode. Spatial light modulators can of course be fabricated by using a matrix of stripe electrodes (EALM) or placing a PC layer on the LC. In this latter case, the applied field is initially across the PC because of its high resistivity. When illuminated by light, the resistance of the PC will vary locally in accord with the illuminating pattern. This causes the voltage across the LC to vary spatially in agreement with the input pattern.

The structure of a new optically addressed LC spatial light modulator [46] using a hybrid field effect is shown in Fig. 7. The device uses the twisted nematic effect with a 45° rotation to determine the off state and the birefringence effect to determine the on state. It exhibits no storage and has only a moderate 60 line /mm resolution at present. It uses noncoherent Xenon arc lamp write illumination from one side and 632.8/nm coherent readout in reflection from the other side. The dielectric mirror and isolation electrode enable simultaneous write and read, prevent dc current flow in the LC, and should result in a long lifetime device. Contrast is 100:1 and resolution over 100 lines/mm. However at 50% modulation only 60 lines/mm resolution and a far lower 3:1 contrast result. The LC layer is only 2 μm thick, which may explain the fast 10 msec rise and 15 msec decay times observed. The applied voltage is 6 Vrms at 10kHz. Contrast reversal and level slicing of images have been demonstrated on this device.

Residual images are generally expected in LC devices and their contrast and resolution will probably always be less than for other spatial light modulators. But they may be far cheaper. Devices such as this (Fig. 7) are one method of increasing the lifetime of LC devices. Optical and cosmetic device quality are other areas needing attention. Trapped air bubbles, inclusions in the thin film layers, misaligned regions, surface contamination, and internal stresses in the thin films are areas still requiring attention to allow ease of fabrication of reproducible devices. LC modulators may find their largest use in combination with other spatial light modulators rather than alone.

The guest-host LC devices [51] may also have use in certain cases. In these devices a plechroic dye is combined with a nematic host. The alignment of the host LC molecules with an applied field affects the orientation of the dye molecules. The effect is a device that absorbs light with no field but not with a field present. The absorption spectrum is that of the dye and can be electronically switched on and off.

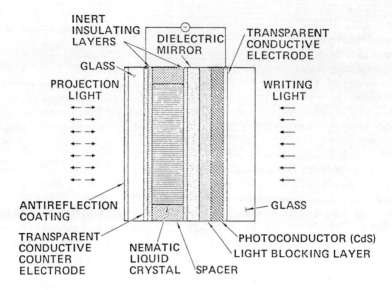

Fig. 7 Schematic structure of hybrid field effect liquid crystal OALM
Courtesy Hughes Res. Lab.

Various electroded LC devices have been developed [33] for use as displays [47] and as page composers [48]. The display LC has 100x100 elements and uses MOSFETs integrated into each element of the structure for storage. The page composer has 1024 elements. For these applications, such resolution and the contrast obtainable may be acceptable, while most coherent optical processing applications require better performance. Two electron beam addressed LC devices [49-50] were also fabricated for feasibility purposes. They were used for projection display only and not in coherent light. Target destruction and cathode poisoning were apparently overcome but resultant resolution was low (150 x 150) and ease of fabrication unclear.

ACOUSTO - OPTIC DEVICES

The interaction of a sound wave with a light beam has been of considerable interest in many applications [52-53]. The Raman - Nath effect describes the effect of an acoustic wave on a plane wave parallel to the acoustic wavefronts at frequencies below 100 Mhz. The Bragg effect is of more importance however. The devices are inherently 1-D modulators and fabricating parallel cells in sufficient number is a severe problem. These devices seem to be most appropriate for special purposes such as deflectors, or in radar [52] or speech [54] processing, or other 1-D signal processing applications. They have recently been used in a clever, multiplexing scheme [55] as a page composer. The high bandwidths possible with these devices are their major advantage. They are thus not given further attention in this present survey.

MAGNETO - OPTIC DEVICES

Considerable attention has been given to point-by-point storage using thermomagnetic effects for writing and the Kerr (reflection) or Faraday (transmission) magneto-optic effect for readout [56-57]. The laser is used to locally heat areas to destroy their magnetization or in combination with electric fields. Curie point writing has been demonstrated in MnBi at room temperature and EuO at cryogenic temperature. GdIG allows for compensation temperature writing, while Ni-Fe has been used with a conductor matrix. High resolution (500 lines/mm) and nsec response but a low SNR ≈ 8 and η of only a few percent result. Precise control of the intensity and duration of the laser pulse is the major problem and prevents practical holographic storage on these devices [57] although holograms have been recorded. Material sensitivity and diffraction efficiency must be improved in room temperature materials and the present requirement [47] for reproducible and accurate 10^3 W peak nonosecond laser pulses must be overcome before holographic recording (with larger recording area requirements and the need for far shorter laser pulses to avoid reduction of the thermal modulation depth by diffusion) will be practical. These devices will thus not be given further consideration in a survey such as this.

OPTICAL DENSITY CHANGE DEVICES

The most common examples of materials utilizing this effect for optical storage or modulation are the alkalai halides. These materials are characterized by the presence of color centers (defect centers intentionally introduced into these materials). These defect centers exhibit <u>absorption</u> in various wavelength bands. These materials also have the ability to switch color centers under control of optical irradiation (e.g. by transferring an electron from one type of color center to another). To use these materials for storage, the information is optically imaged onto the material at λ_1. This changes the crystal's absorption (point-by-point) at another λ_2 where the recorded image is read.

The materials are of interest because of their inherent unlimited resolution and their ability to record multiple absorption volume holograms. Reflection holograms with a fringe spacing of 0.1×10^{-12} m have been formed [36] on $SrTiO_3$:Fe, Mo. Considerable work in the Soviet Union [58] and Czechoslovakia [59] on spiropyran has involved successful cyclic recording and erasure of holographic information. These results should be reported on elsewhere in this text.

Attention in the United States has recently focused on the use of the photodichroic nature [60] of the higher order centers in KCl, NaCl, and NaF to obtain polarization dependent modulation. NaF is of special interest and many recent results are of importance. The material operates at room temperature. Lithium doped NaF samples with low OH^- content seem to be resistant to fatigue which has plagued many photochromic materials. Ion implantation promises quite useful thin layers. Multiple holograms have been recorded and reconstructed in certain materials. Additive coloration has greatly improved the optical density obtainable [61]. The material's sensitivity is still high by comparison to others ($150 mJ/cm^2$). Although the use of dispersion [62] and extinction writing techniques [63] promise improvements, this limitation will always exist. Another limitation that will persist is the low 1-3% diffraction efficiency obtainable [36].

Fig. 8. Reconstruction of a suppressively written image in NaF using only one wavelength of coherent light.

Another recent development [64] has overcome an objection common to many OALMs: the necessity for two different wavelengths of light for writing and reading. In coherent optical processing, this causes considerable difficulty. We have recently demonstrated a technique for writing NaF where only one wavelength of coherent light is needed. This is of extreme importance in holographic recording and correlation. By simultaneously illuminating the sample with broadband radiation from a Hg arc lamp and a coherent 508 nm Argon laser (with the input data placed in the Argon beam), the ionization of M centers can be inhibited in those regions of the sample illuminated by both beams (bright areas of the image). M-center ionization occurs in dark regions of the input, the dipole moment of the centers located in those regions switches and modulation between crossed polarizers results. The suppressively written image in Fig. 8 was obtained on 40 μm of an NaF sample. Image reversal is possible. Recent developments such as these make these materials more attractive and preferable to photochromics and cathodochromics. No voltages, layers, photoconductors, etc. are needed so operation and fabrication are simple and inexpensive.

ELECTRO-OPTIC OALMS

The most fruitful area for research seems to be devices utilizing the electro-optic effect. Various separate categories of these materials (most are OALMs) follow.

POCKELS EFFECT DEVICES

Various versions [65-68] of an electron-beam addressed light valve using a DKDP target crystal have been well described in the literature so that only summary data need be presented here for unification purposes. This is probably the one electron-beam addressed device with no <u>inherent</u> cycle time, erase time, or lifetime limitations. Its only disadvantages are that the target material must be cooled to $-60°C$ (although this is possible by Peltier Cells) and that the target and support electronics are sophisticated but within present technology. The device operates by depositing charge on the DKDP crystal target by a scanning electron beam modulated with the input data. The spatial field developed across the crystal produces a point-by-point phase modulation that is detected by crossed polarizers. Erasure is by secondary emission using a second flood electron gun in one system [66]. Several other writing and erasing schemes are possible. This erase mechanism allows for very rapid (msec) erasure. Useful storage of over one hour is also possible. $10^3 \times 10^3$ point resolution and operation (complete write/read/erase cycle) at 30 frames/sec with a 60:1 contrast ratio has been demonstrated. The target crystal in our system [66] is

2" x 2" x 0.010". A sealed system has operated for over 2 years and DKDP crystals flat to $\lambda/5$ can be fabricated. Thus lifetime and optical quality are not problems.

The device has seen extensive use in many areas of coherent optical processing. Holograms have been reconstructed in real-time using the device [67,69], real-time radar processing [70,71], pattern recognition, area correlation and text correlation [66] have been demonstrated.

The structure of an optically addressed DKDP device [72] is shown in Fig. 9a. A Se photoconductor and 170 μm DKDP crystal are used. Voltage applied between the electrodes divides between the PC and DKDP, most across the PC. When exposed to write light, charges are created and drift to the crystal-PC interface, causing a field to develop across the crystal (30 μsec write time). Readout is in reflection by the Pockels effect. When the applied voltage is reduced to zero and the PC is flooded, the trapped charge migrates (3000 μsec erase) to the electrodes and the device is erased. If the applied voltage (80V) is reversed or intermediate voltages used, image reversal, subraction and level slicing on both devices shown in Fig. 9 is possible. Phototitus has the same sensitivity for both polarities of applied voltage, whereas the Prom does not and image subtraction is more complex. 40 line/mm resolution at 30:1 contrast has been shown. Contrast of over 100:1 is possible but is a function of the cone angle of the readout light since DKDP is uniaxial. Useful storage is one hour and 100 erg/cm^2 sensitivity results.

An available and promising OALM using a thin (600-900 μm) $Bi_{12}SiO_{20}$ crystal [73] (which is both photoconductive and electro-

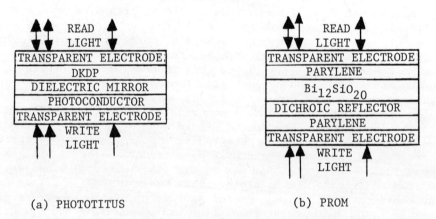

(a) PHOTOTITUS　　　　　　　　(b) PROM

Fig. 9. Structure of two Pockels Effect OALMs.

optic) is shown structurally in Fig. 9b. A separate PC layer is not necessary as in photo-DKDP, but large DKDP crystals can presently be more easily fabricated. The thin (3 μm) Parylene layer provides the voltage division needed for operation with no PC layer. The device requires two wavelengths for writing (blue light) and reading (red He-Ne) from opposite sides. Both devices operate similarly except 2000V is needed in the Prom, although operation is at room temperature as opposed to $-50°C$ for Phototitus. The Prom has a sensitivity of 50 erg/cm^2, can be fabricated flat to $\lambda/5$ and has 10 line/mm resolution at a much higher 2000:1 contrast (100 line/mm resolution with baseline subtraction) and η = 10%. The contrast measurements were performed differently for both devices however. The origin of the increased Prom resolution is not certain, but appears to be due to a nonuniform exposure over the crystal's thickness; most of the write light is absorbed at the input of the crystal, while the back region appears as a thin crystal (better resolution) [20].

Care must be taken in these devices that resolution is not quoted at the visible limit (MTF = 0.05) and contrast elsewhere. Both devices have seen considerable use in coherent light and correlation. The dual in-line image correlation scheme (Fig. 2) seems most appropriate with the comparatively (by holographic standards) low device resolution.

FERROELECTRIC MATERIALS

This second division of electro-optic EO devices is characterized by materials that can record high efficiency multiple volume phase holograms. The basic storage mechanism involves generation of photocarriers in the bulk of the material by absorption of the input write light. These carriers then diffuse or drift to regions of lower light intensity where they are trapped. The fields created by these carriers correspond to the input illumination and locally modulate the material's index of refraction causing modulation of linearly polarized light. Erasure is by thermal ($300°C$) or optical excitation. Stored images must be fixed or they will be erased after repeated readout (10-20 reads in $LiNbO_3$).

Undoped $LiNbO_3$ can be fixed by heating to $100°C$ for 30 min, cooling, and reexposing. $LiNbO_3$:Fe can be fixed by recording (20 sec-2 min) on it at elevated temperatures ($150°C$) and cooling it before heating for another exposure. SNB can be fixed by application of a field for 0.5 sec after recording [74].

Sensitivity has been a problem in these materials. Remarkable progress [75] has been made in iron doped $LiNbO_3$. Sensitivity has been decreased from 100J/cm^2 (undoped) to 200 mJ/cm^2. The primary purpose of research efforts on these materials has been multiple hologram storage. This has been achieved in undoped $LiNbO_3$ with a field present (10 recorded holograms with η = 30% and 100 with η > 1%). If the field is removed, erasure requires an energy equal to that used for recording. With the field present, 10-20 readouts are possible. In doped $LiNbO_3$, η depends on the writing energy used. For 1.25 J/cm^2, η = 10%, while it decreases to 1% for 150 mJ/cm^2 write energy.

Strontium barium niobate SBN [76] is operated similar to $LiNbO_3$ but its sensitivity (200 mJ/cm^2) is much lower and can be reduced by applying an electric field (3 mJ/cm^2 at 3 kv/cm field). Image storage and resistance to readout depend on the input exposure. Images recorded with 100 J/cm^2 resist readout for minutes while those recorded with low exposures are destroyed in seconds. Efficiency is also increased [77] by the presence of a field (1.6 kV/cm). SBN can be electrically fixed whereas $LiNbO_3$ requires heating. Both materials have been used for hologram storage.

Many tradeoffs in write illumination, storage, efficiency, fixing, resistance to readout, operating complexity, etc. exist. Decisions on materials, dopants, and operation should be made for each intended application. Some cases require long storage, others do not, for in some cases sensitivity is more important than η, etc. Other materials are candidates for volume storage, but these two and the photodichroics seem the best.

FERROELECTRIC-PHOTOCONDUCTOR DEVICES

The final devices to be discussed are sandwich structures of a ferroelectric FE (such as $Bi_4Ti_3O_{12}$) or ceramic (PLZT) and PC between electrodes. Both form thin phase holograms. Bismuth titanate [78] has been used for holographic recording, 90 line/mm resolution and 100:1 contrast (possibly not simultaneously obtainable since η = 0.01% [36]) has been demonstrated. Sensitivity is higher than in the Pockels OALM devices (10^4 erg/cm) and requires 300-400 V present during writing. Readout is nondestructive (if the field is removed). The erase process is the same as writing in both materials. In both materials the writing light does not directly produce the storage as in SBN and $LiNbO_3$, rather carriers produced in the PC by the write light increase the conductivity of the PC and switch the applied voltage across the FE. Information is stored in bismuth titanate as domain patterns in zero field. $Bi_4Ti_3O_{12}$ is difficult to fabricate but is included here since thin film devices would make it quite attractive.

PLZT has received considerable attention as a display device [79] and for use as flash goggles. Electroded devices [80] have also been used extensively as page composers. Their resolution is too low for the uses considered here however. Holograms have recently been recorded on PLZT (9/65/35) using the Kerr effect and a 6.5 kV/cm dc bias field during writing and a -10 kV/cm field during reading [81]. The presence of these fields increased η from 3 to 35%; a 1 J/cm^2 exposure was used however. Device resolution is typically 50 lines/mm. Fatigue [82] is a problem in PLZT. Contrast will apparently always be low. The scattering mode is not usable as noted before. Strain-biased mode devices are hard to fabricate and breakage results. Major areas needing attention are the nonuniformity of results: large variations in chemical, electrical, electromechanical, and electro-optic responses make device consistency a major problem [24].

Although the cost of the FE-PC devices are far less, the Pockels effect OALMs of comparable resolution, sensitivity, etc. seem more promising if their costs can be reduced.

SUMMARY

Recent advances in real-time materials and devices have been noted in a rather comprehensive survey, in which remarkable progress in many areas has been noted. Device selection should bear in mind the specific application and the need for lifetime, storage, efficiency, etc. must be assessed for each case. Often more of a systems approach to real-time optical computing will be needed to produce viable systems; several devices may have to be combined in one system. Attention should also be paid to the source available in OALM applications: low sensitivity is needed for CRT imaging, while deflector/scanner specifications should also be addressed in point-by-point OALMs. Above all, only more cooperation between researchers in this area may produce viable systems. In many of the devices described, more critical comments were possible only because the device has seen considerable research. This fact should be borne in mind in considering new devices and comparing those reported on in this chapter. All are also in various stages of development or research.

ACKNOWLEDGMENT

The author wishes to thank the Office of Naval Research for continued support of his work included here on Contract NR048-600, the AFOSR, Dept. of Defense, NASA, the U.S. Army Missile Command, and Battelle Research Labs. for other related support. The assistance of many persons and corporations made this survey possible. Of special note is the assistance of Hughes, Ampex, Itek, Radiation (Harris Semiconductor), RCA, ERIM, and others including my own research group.

REFERENCES

1. G. W. Stroke, *An Introduction to Coherent Optics and Holography*, Second Ed., N. Y.: Academic Press, 1969.

2. J. W. Goodman, *Introduction to Fourier Optics*, N. Y.: McGraw Hill, 1968.

3. K. Preston, Jr., *Coherent Optical Computers*, N. Y.: McGraw Hill, 1972.

4. A. Vander Lugt, *Opt. Acta*, 15, 1 (1968).

5. G. W. Stroke, *IEEE Spectrum*, 9, 22 (1972).

6. A. Vander Lugt, *Proc. IEEE*, 62, 1300 (1974).

7. G. W. Stroke, *Proc. IEEE*, 63, 829 (1975).

8. *Optical Comput. Symp. Digest*, Darien, IEEE Cat. No. 72CH0687-4-C

9. *Digest Internat. Opt. Comput. Conf.*, Zurich (1974), IEEE Cat. No. 74CH0862-3C.

10. "   "   " "   " , Wash. D. C. (1975).

11. *Appl. Opt.* Special Issue on Optical Storage of Digital Data, 13, April 1974.

12. *J. Soc. for Info. Disp.*, 15, third quarter 1974.

13. *Opt. Engr.*, Spec. Issue on Digital and Optical Image Proc., May 1974.

14. *IEEE - Trans. Comput.*, Special Issue on Opt. Comput., April 1975.

15. *Proc. IEEE*, Special Issue on Opt. Comput., Nov. 1976 (tentative).

16. *IEEE - Trans. Elec. Dev.*, Special Issue on Display Devices, ED-20, Nov. 1973.

17. *Liquid Crystal Bibliography,* Eastman Kodak Co., Publication JJ-193.

18. B. Ellis and A. Walton, Royal Aircraft Estab. Tech. Rept. 71009.

19. A. Agajanian, *J. Soc. for Info. Disp.*, 14/2, 76 (1973).

20. S. Lipson, Advances in Holography, Dekker Pub. (to be published).

21. R. J. Collier, C. B. Burckhardt, and L. H. Lin, Optical Holography, N. Y.: Academic Press, 1971.

22. E. Ramberg, RCA Rev., 33, 5, March 1972.

23. T. C. Lee, Appl. Opt., 13, 888 (1974).

24. Harris Semic. Div., Final Report on NAS 8-26360.

25. J. Urbach and R. Meier, Appl. Opt., 5, 666 (1966).

26. R. Doyle and W. Glenn, Appl. Opt., 11, 1261 (1972)

27. Environ. Res. Inst. Mich., Final Report on AFAL-TR-73-88.

28. General Elec. Co., Tech. Info. Series, R75ELS-12.

29. T. Turpin, Ref. 9, p. 34.

30. N. Sheridon, IEEE - ED, 19, 1003 (1972).

31. J. Bordogna, S. Keneman and J. Amodei, RCA Rev., 33, 227 (1972).

32. K. Preston, Jr., IEEE - AES, 6, 458 (1970).

33. D. Casasent, J. Soc. Info. Disp., 15/3, 131 (1974).

34. D. Casasent, IEEE Spectrum, to appear.

35. O. Tufte and D. Chen, IEEE Spectrum, 10, 26 (1973).

36. J. Bordogna, et al., RCA Rev., 33, 227 (1972).

37. J. Urbach and R. Meier, Appl. Opt., 8, 2269 (1969)

38. F. Reizman, Proc. EOSD Conf., N. Y., Sept. 1969.

39. J. Flannery, IEEE - ED, 20, 941 (1973).

40. Eastman Kodak & Co., Public JJ-193.

41. G. Heilmeir, et al., Proc. IEEE, 56, 1162 (1968).

42. R. Bartolino, et al., Appl. Opt., 12, 2917 (1973).

43. T. D. Beard, et al., Appl. Phys, Lett., 22, 90 (1973).

44. A. Boller, et al., Proc. IEEE, 60, 1002 (1972)

45. R. Sorel and M. Rafuse, J. Appl. Phys., 43, 2029 (1972).

46. J. Grinberg, A. Jacobson, et al., Opt. Engr., to be pub.

47. M. Ernstoff, et al., Dig. IEDM, Wash. D. C., 548 (1973).

48. G. Labrunie, J. Robert, and J. Borel, Appl. Opt., 13, 1355 (1974).

49. J. Van Raalte, Proc. IEEE, 56, 2146 (1968).

50. J. Hansen and R. Schneeberger, IEEE - ED, 15, 896 (1968).

51. J. Heilmeir and L. Zanoni, Appl. Phys. Lett., 13, 91 (1968).

52. W. Maloney, IEEE Spectrum, 6, 40 (1969).

53. A. Korpel, Proc. SPIE, 38, 3 (1973).

54. Yu, IEEE Spectrum, Feb. 1975.

55. A. Bardos, Appl. Opt., 13, 832 (1974). See p. 841 also.

56. B. Brown, Appl. Opt., 13, 761 (1974).

57. D. Chen, Appl. Opt. 13, 767 (1974).

58. A. Mikaeliane, et al., IEEE - QE, 4, 757 (1968).

59. M. Lescinsky and M. Miller, Opt. Commun., 1, 417 (1970).

60. I. Schneider, et al., Appl. Opt., 9, 1163 (1970).

61. W. Collins, et al., Appl. Phys. Lett., 24, 403 (1974).

62. I. Schneider, Phys. Rev. Lett., 32, 412 (1974).

63. I. Schneider, et al., Appl. Phys. Lett., (1974).

64. D. Casasent and F. Caimi, to be published.

65. G. Marie and J. Donjon, Proc. IEEE, 61, 942 (1973).

66. D. Casasent and W. Sterling, IEEE - TC, 24, 348 (1975).

67. G. Goetz, Appl. Phys. Lett., 17, 63 (1970).

68. W. J. Poppelbaum, p. 368 in Pict. Pattern Recog., Wash.: Thompson, 1968.

69. G. Groh and G. Marie, Opt. Commun., 2, 133 (1970)

70. D. Casasent and F. Casasayas, IEEE - AES, 11, 65 (1975).

71. D. Casasent and F. Casasayas, Appl. Opt., June 1975.

72. J. Donjon, et al., IEEE - ED, 20, 1037 (1973).

73. P. Vohl, et al., IEEE - ED, 20, 1032 (1973).

74. L. Anderson, Ferroelectrics, 7, 55 (1974).

75. D. Staebler and W. Phillips, Appl. Opt., 13, 788 (1974).

76. T. Inagaki, et al., Appl. Opt., 13, 814 (1974).

77. J. Thaxter and M. Kestigian, Appl. Opt., 13, 913 (1974).

78. S. Cummins and T. Luke, IEEE - ED, 18, 761 (1971).

79. C. E. Land, Ferroelectrics, 7, (1974).

80. H. Roberts, Appl. Opt., 11, 397 (1972).

81. F. Micheron and G. Bismuth, Topical Meeting on Opt. Storage (March 1973).

82. W. Stewart and L. Cosentino, Ferroelectrics, 3, (1971).

Comments by D. Casasent

<u>Paper by D. Casasent.</u>

Page 5, lines 13-14, should read (e.g., photochromics and film) not (e.g., film and photochromics).

On page 21, line 15, the image size should be listed as 200 μm, not 40 μm. The resolution obtainable on PC and PD materials is limited by the optical system, not the material. The presence of a 10^{-12} m fringe spacing [36] can thus be detected, but an image of such resolution could not be viewed.

Examples of imagery [46] obtained on the LC structure of Figure 7 are shown in Figures A-1,a, and A-1,b. The images were noncoherently imaged onto the LC from a transparency by a tungsten lamp whose output was centered at 527.5nm with a 50% bandwidth of 23.3nm. The projected readout image in Figure A-1, a, was formed with a 632.8nm He-Ne laser with 5.5V rms at 10 kHz applied to the crystal. The contrast reversed image of Figure A-1, a, was obtained by changing the frequency of the bias voltage to 2 KHz. This alters the threshold and slope of the LC response. As intermediate frequencies are used, different portions of the gray scale distribution can be examined. Contrast reversal and intermediate results can also be achieved by varying the voltage. Image subtraction and baseline line subtraction are clearly possible.

Figure A-1, a, - Projected Image Written on LC Structure of Figure 7 Using 5.5V rms Bias Voltage at 10 KHz.

Figure A-1, b, - Contrast Reversed Version of Figure A-1, a, Obtained by Changing the Frequency of the Bias Voltage to 2 KHz.

Paper by R. Landauer

A large NASA Program [Goddard SFC Report X-943-75-14 January 1975] to implement a parallel 1024 x 1024 point non-coherent optical logic processor should be noted. The system (Called a TSE Computer) uses fiber optics bundles to transport the parallel data. Logical AND, OR, and NOT devices operating in parallel on 1024 x 1024 point data are in the early development stages. Present components consist of 16,384 elements in a 128 x 128 matrix and have 5msec response times for a 3×10^6 bit per sec. rate and a fan out of 10. Both thin film and silicon technologies are being pursued.

The alkali halide memory system noted need not require volume holography, although such an approach would offer a far larger packing density than a planar memory. Thin 10μm layers have been fabricated by ion implantation although the OD of such structures is presently low. Low energy acceleration particles can also be used to effectively color only thin areas of the material. This later approach seems more feasible when single rather than multiple volume holograms are desired.

INFORMATION PROCESSING IN OPTICAL SYSTEMS

OF HOLOGRAPHIC MEMORY DEVICES

I.S. Gibin, P.E. Tverdokhleb

Institute of Automation and Electrometry
Novosibirsk, USSR

ABSTRACT

A class of integral transforms that can be performed in holographic memory devices (HMDs) with a matrix memory organization has been determined. The structures of the optical systems realizing these transforms have been considered. The possibilities for data processing in the HMDs have been illustrated by the examples of an associative retrieval and a spectral analysis of images.

I. INTRODUCTION

For bulk data storage and processing of (analysis and transform of images, information retrieval in data banks, multichannel processing of signals, etc.) in increasing frequency are used optical techniques providing the information storage density up to 10^8 bit/cm^2 and high processing rate mainly achieved due to high capacity of optical systems (the number of independent channels can be 10^6 and more). These possibilities are realized in special-purpose processors and optical memory devices.

As a rule optical processors are analog devices with a parallel mode of operation. They are well adapted for performing integral transforms of the images of information files.

Optical memory devices provide information recording, storage and reading, and in the main perform the functions of the computer external memory devices.

Among the optical memory devices the devices realizing information recording and storage by a holographic technique are of special interest. The memory devices of this type have high noise immunity for the defects of an information carrier and also provide the possibility of storage and read-out of information by pages of up to 10^4 and more bits. According to numerous opinions the holographic memory devices (HMDs) with a page-by-page recording of information and with a matrix organization of the memory are most promising for recording and storage of bulk data [1-5]. The capacity of the HMD can achieve $10^8 - 10^9$ bits, and the arbitrary access time equals to 1-10 μ s.

At present a tendency exists to develop and produce memory devices in which the functions of information storage are brought into coincidence with its processing functions. This makes it possible to increase the computer efficiency and to decrease the information content transmitted over the memory device-computer transmission lines.

The above tendency manifests itself in developing the HMD as well. Here the attempts are known to realize an associative retrieval [6,7], logical and arithmetic operations [8] and functional transforms by the table arithmetics methods [9]. The investigations in this direction are particular to their type and refer to various structures of the HMD. Thus, the data processing is based on using an HMD with a linear [6,8] and a matrix memory organization [7,9], the latter organization is advantageous in its memory size as compared with the former one .

Therefore the necessity arises in a more complete investigation of the possibilities of data processing in the HMD with a unified perspective memory organization. The urgency of this investigation is dictated by the fact that the work on the development of the HMDs is actively carried out in many laboratories of the world. It is expected that in the near future the commercial versions of such devices will be produced. Therefore the possibility of multifunctional use of the HMDs assumes special importance.

The present paper is devoted to the investigation of a class of transforms in the optical systems of the HMD with a matrix memory organization. The versions of the systems have been described capable of processing the information pages, and the possibilities have been shown for their application in solving the problems of associative retrieval and image spectral analysis. The problems of information processing in the HMD opto--electronic elements e.g. in a photomatrix with memory and developed parallel logics have not been touched here.

II. TRANSFORMS IN HMDs

The characteristic feature of the HMDs to be considered is that in such devices data are recorded in pages on separate holograms in the form of a rectangular (most commonly a square) matrix. Holograms are small in size (about 1 mm and less) and their number in the matrix achieves the values of $10^3 - 10^4$ and more. In this connection the HMD optical system can be considered as a linear non-invariant data processing one. In this case the input pulse action is represented by a narrow light beam directed to one of the stored holograms and the response to this action is an information page image reconstructed in the photomatrix plane. The pulse response type is determined by the hologram content and varies from hologram to hologram.

Let us show that in this system general type linear integral transforms can be realized. For this purpose let the scheme represented in Figure 1 be considered. Here 1 is a plane of the image under recording, 2 is a plane of the hologram matrix, 3 is a plane of the image reconstructed, 4 is the output frequency plane. Planes 1-2, 2-3, 3-4 are interconnected by a Fourier transform performed with the help of the objectives O_1, O_2, O_3. Let the coordinates in planes 1 and 3 be designated by the symbols x, y and in planes 2 and 4 by the symbols ξ, η. The angular spatial frequencies u and v in planes 2 and 4 are related with the linear coordinates ξ and η by the relations $u = 2\pi\xi/\lambda f$, $v = 2\pi\eta/\lambda f$ where f is a focal distance of the objective realizing a Fourier transform, λ is a light wavelength.

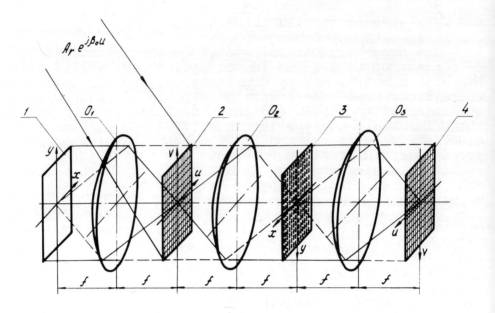

Figure 1. General Structure of Holographic Memory Device with Data Processing Orientation.

If the hologram matrix is placed in plane 2 and all the holograms of this matrix is illuminated by a coherent reference beam $Ar\exp(ju\beta_0)$, then in plane 2 in the direction of the first diffraction order to the accuracy of a constant factor the amplitude distribution will be obtained described by the expression

$$M_h(u,v) = \sum_{m=-\frac{M}{2}}^{\frac{M}{2}} \sum_{n=-\frac{M}{2}}^{\frac{M}{2}} F_{mn}(u+mG, v+nG)\,\mathrm{rect}\left(\frac{u+mG}{H}\right)\mathrm{rect}\left(\frac{v+nG}{H}\right),$$

(1)

$$m, n = 0, \pm 1, \pm 2, \ldots, \pm \frac{M}{2},$$

where $F_{mn}(u,v) = \iint_\Omega f_{mn}(x,y)\exp[j(ux+vy)]\,dx\,dy$
is a spatial-frequency spectrum of the image of a page of information $f_{mn}(x,y)$ stored in the hologram;

HOLOGRAPHIC MEMORY DEVICES

function rect $\left(\frac{u}{H}\right) = \begin{cases} 1, & |u| \leq \frac{H}{2} \\ 0, & |u| \leq \frac{H}{2} \end{cases}$,

$H = 2\pi h/\lambda f$ is an angular size of the hologram (h is a linear size); $G = 2\pi g/\lambda f$ is an angular step of the hologram matrix (g is a linear step); $M+1$ is the number of holograms in the matrix row (column); Ω is an area of the function $f_{mn}(x,y)$.

Since the spatial-frequency spectrum of the function $f_{mn}(x,y)$ is limited by the hologram size H, according to the theorem of samples [10] this function can be represented by separate samples at regular intervals $\Delta x = \Delta y = 2\pi/H$. If the spatial frequencies $|u| = |v| > H/2$ give a fair contribution to the image then, provided that $f_{mn}(x,y)$ is limited by the area $\Omega = L_r \times L_r$, the number of samples $N_x = N_y = L_r H/2\pi$ and the function $f_{mn}(x,y)$ is expanded into a finite series

$$f_{mn}(x,y) = \sum_{k=-N/2}^{N/2} \sum_{\ell=-N/2}^{N/2} f_{mn}^{k\ell} \frac{\sin[\frac{H}{2}(x+\frac{2\pi k}{H})]\sin[\frac{H}{2}(y+\frac{2\pi \ell}{H})]}{(x+\frac{2\pi k}{H})(y+\frac{2\pi \ell}{H})}, \quad (2)$$

$k, \ell = 0, \pm 1, \pm 2, \ldots, \pm \frac{N}{2}$.

Here $f_{mn}^{k\ell}$ are samples of the function $f_{mn}(x,y)$ at the points $x = 2\pi k/H$, $y = 2\pi \ell/H$.

Here and in what follows the indices mn are assumed to refer to the elements of the rows and columns of hologram matrix (1) and the indices $k\ell$ to the elements of the rows and columns of sample matrix (2), describing the image.

With the account of (2) the spatial-frequency spectrum of the image $f_{mn}(x,y)$ to the accuracy of a constant factor can be represented as the sum

$$F_{mn}(u,v) = \sum_{k=-N/2}^{N/2} \sum_{\ell=-N/2}^{N/2} f_{mn}^{k\ell} \exp[\frac{2\pi}{H} j(uk + v\ell)] \text{rect}(\frac{u}{H}) \text{rect}(\frac{v}{H}) \quad (3)$$

and Expression (1) takes the form

$$M_h(u,v) = \sum_{m=-\frac{M}{2}}^{M/2} \sum_{n=-\frac{M}{2}}^{M/2} \sum_{k=-\frac{N}{2}}^{N/2} \sum_{l=-\frac{N}{2}}^{N/2} f_{mn}^{kl} \exp\left[\frac{2\pi}{H}\{(u+mG)k+(v+nG)l\}\right] \times$$
$$\times rect\left(\frac{u+mG}{H}\right) rect\left(\frac{v+nG}{H}\right). \quad (4)$$

The reconstruction of images from the hologram matrix in the HMD optical system under consideration is equivalent to the realization of a Fourier transform operation for expression (4). Then in the plane of the image reconstructed we obtain the amplitude light distribution

$$M_r(x,y) = \mathcal{F}\{M_h(u,v)\} =$$
$$= \sum_{m=-\frac{M}{2}}^{M/2} \sum_{n=-\frac{M}{2}}^{M/2} \sum_{k=-\frac{N}{2}}^{N/2} \sum_{l=-\frac{N}{2}}^{N/2} f_{mn}^{kl} \frac{\sin\left[\frac{H}{2}(x+\frac{2\pi k}{H})\right]\sin\left[\frac{H}{2}(y+\frac{2\pi l}{H})\right]}{(x+\frac{2\pi k}{H})(y+\frac{2\pi l}{H})} \exp[jG(mx+ny)]$$

(5)

Here the symbol $\mathcal{F}\{\cdot\}$ designates the Fourier transform operation. Let us designate

$$f^{kl}(uv) = \sum_{m=-\frac{M}{2}}^{M/2} \sum_{n=-\frac{M}{2}}^{M/2} f_{mn}^{kl} \exp\left[\frac{2\pi G}{H}j(mk+nl)\right] rect\left(\frac{u+mG}{H}\right) rect\left(\frac{v+nG}{H}\right)$$

(6)

Having substituted (6) and (2) into (4) and (5), respectively, we obtain that the amplitude light distribution in the plane (u,v) can be described by the expression

$$M_h(u,v) = \sum_{k=-\frac{N}{2}}^{N/2} \sum_{l=-\frac{N}{2}}^{N/2} f^{kl}(u,v) \exp\left[\frac{2\pi}{H}j(uk+vl)\right], \quad (7)$$

and in the plane (x,y) by the expression

$$M_r(x,y) = \sum_{m=-\frac{M}{2}}^{M/2} \sum_{n=-\frac{M}{2}}^{M/2} f_{mn}(x,y) \exp[jG(mx+ny)]$$

(8)

From (7) it follows that the hologram matrix can be represented as a sum of $(N+1) \times (N+1)$ images $f^{k\ell}(u,v)$ and in this case each element $k\ell$ in the plane (x,y) corresponds to its own image $f^{k\ell}(u,v)$ in the plane (u,v). The phase multiplier $\exp[\frac{2\pi}{H}j(uk+v\ell)]$ determines the angle at which the element $k\ell$ is seen from the hologram matrix plane.

In a similar way from (8) it follows that when reconstructing images from the hologram matrix in the plane (x,y), a sum of the images $f_{mn}(x,y)$ is formed, in this case each of these images corresponds to the element mn in the hologram matrix plane. The images $f_{mn}(x,y)$ are projected at angles determined by the phase multipliers $\exp[jG(mx+ny)]$

The observation or photoelectric reading of light distributions (7) and (8) is possible only at the level of intencities determined as a square of the amplitude distribution module.

$$I_h(u,v) = M_h(u,v) M_h^*(u,v) = \sum_{k=-\frac{N}{2}}^{N/2} \sum_{\ell=-\frac{N}{2}}^{N/2} |f^{k\ell}(u,v)|^2 +$$

$$+2 \sum_{k=-\frac{N}{2}}^{\frac{N}{2}-1} \sum_{p=k+1}^{\frac{N}{2}} \sum_{\ell=-\frac{N}{2}}^{\frac{N}{2}-1} \sum_{q=\ell+1}^{\frac{N}{2}} f^{k\ell}(u,v) f^{pq}(u,v) \cos\left\{\frac{2\pi}{H}[u(k-p)+v(\ell-q)]\right\},$$

(9)

$$I_r(x,y) = M_r(x,y) M_r^*(x,y) = \sum_{m=-\frac{M}{2}}^{M/2} \sum_{n=-\frac{M}{2}}^{M/2} |f_{mn}(x,y)|^2 +$$

$$+2 \sum_{m=-\frac{M}{2}}^{\frac{M}{2}-1} \sum_{s=m+1}^{\frac{M}{2}} \sum_{n=-\frac{M}{2}}^{\frac{M}{2}-1} \sum_{t=n+1}^{\frac{M}{2}} f_{mn}(x,y) f_{st}(x,y) \cos\{G[x(m-s)+y(n-t)]\}.$$

(10)

Expressions (9) and (10) comprise high-frequency interference terms determined by coherent addition of images from the adjacent elements hl (or mn). If, however, each image element kl in recording or the element mn in hologram reconstructing is given a random phase shift [11] or the holograms are reconstructed by a partially-coherent reference source so that the wave fronts from the adjacent holograms are non-coherent, the interference terms in expressions (9) and (10) are averaged and can be neglected. Then

$$I_h(u,v) = \sum_{k=-\frac{N}{2}}^{N/2} \sum_{l=-\frac{N}{2}}^{N/2} |f^{kl}(u,v)|^2, \qquad (11)$$

$$I_r(x,y) = \sum_{m=-\frac{M}{2}}^{M/2} \sum_{n=-\frac{M}{2}}^{M/2} |f^{mn}(x,y)|^2 \qquad (12)$$

Taking into account that the number of the elements M and N is large, let the discrete independent variables m, n, k, l be replaced by the continuous variables u, v, x, y and let us change in (11) and (12) from sums to integrals. As a result we obtain

$$I_h(u,v) = \iint_{-L_r/2}^{L_r/2} K(x,y;u,v)\,dx\,dy, \qquad (13)$$

$$I_r(x,y) = \iint_{-L_h/2}^{L_h/2} K(u,v;x,y)\,du\,dv, \qquad (14)$$

where $L_r = 2\pi(N+1)/H$ - is a linear dimension of the image restored, $L_h = (M+1)G$ - is an angular size of the hologram matrix and

$$K(x,y;u,v) = |f(x,y;u,v)|^2; \quad K(u,v;x,y) = |f(u,v;x,y)|^2.$$

Generally speaking, this change is not mathematically rigorous and is necessary only for better understanding of the nature of transforms being performed in the HMD.

Let in the reconstructed image plane (see Fig. 1) a transparency with its amplitude transmission $\varphi^{1/2}(x,y)$ be placed, then the intensity distribution in frequency plane 4 equivalent to hologram plane 2 will be described by the expression

$$A_h(u,v) = \int\int_{-L_{r/2}}^{L_{r/2}} \varphi(x,y)\, K(x,y;u,v)\, dx\, dy\, . \tag{15}$$

In an analogous way, if the transparency with an amplitude transmission $\varphi^{1/2}(u,v)$ is placed in hologram matrix plane 2 then in plane 3 the intensity will be distributed according to the law

$$A_r(x,y) = \int\int_{-L_{h/2}}^{L_{h/2}} \varphi(u,v)\, K(u,v;x,y)\, du\, dv \tag{16}$$

Expressions (15) and (16) represent a pair of linear two-dimensional integral transforms with a general-type kernel. The kernel $K(x,y;u,v)$ is determined by the hologram content and is real and normalizable. The kernel formation is performed via recording the image which intensity distribution is proportional to $K_{mn}(x,y)$ into each Fourier-hologram with the coordinates (u_m, v_n). In this case each point $x = x_k, y = y_\ell$ in the reconstructed image plane corresponds to the intensity distribution in the hologram matrix plane $K^{k\ell}(u,v)$.

Due to the finite dimensions of the holograms and thus to spatial discretization of the images under processing in the HMD optical systems in fact not integral transforms (15), (16) are realized but their discrete analogs that can readily be represented in a matrix form. For example, transform (15) is equivalent to the multiplication of a rectangular matrix by a matrix-column

$$\begin{Vmatrix} a_1 \\ a_2 \\ \vdots \\ a_J \end{Vmatrix} = \begin{Vmatrix} K_{11} & K_{12} & \ldots & K_{1I} \\ K_{21} & K_{22} & \ldots & K_{2I} \\ \vdots & & & \vdots \\ K_{J1} & K_{J2} & \ldots & K_{JI} \end{Vmatrix} \times \begin{Vmatrix} \varphi_1 \\ \varphi_2 \\ \vdots \\ \varphi_I \end{Vmatrix}, \tag{17}$$

where the matrix-column components $\{a_j\}$ are calculated according to the formula

$$a_j = \sum_{i=1}^{I} K_{ji} \varphi_i$$

and transform (16) is equivalent to the multiplication of the matrix-column by a rectangular matrix

$$\|a_1\ a_2 \ldots a_I\| = \|\varphi_1\ \varphi_2 \ldots \varphi_J\| \times \begin{Vmatrix} K_{11} & K_{12} & \ldots & K_{1I} \\ K_{21} & K_{22} & \ldots & K_{2I} \\ \cdot & \cdot & \cdot & \cdot \\ K_{J1} & K_{J2} & \ldots & K_{JI} \end{Vmatrix} \quad (18)$$

where the matrix-row components $[a_i]$ are calculated according to the formula

$$a_i = \sum_{j=1}^{J} \varphi_j K_{ji}$$

In expressions (17) and (18) the values φ_i and φ_j are samples of the image to be processed; a_j and a_i are samples of the image resulting from the processing; K_{ji} are elements of the transform kernel forming a rectangular matrix of the $J \times I$ size. Here $J = (M+1)^2$ and $I = (N+1)^2$, that corresponds to the continuous numeration of the input and output image elements. In this case each row of the rectangular matrix describes the content of one of the memory holograms.

III. WIDENING THE CLASS OF TRANSFORMS

The class of transforms realized in the HMD optical systems can be widened via changing the reconstructing beam incident angle to the hologram. Thus, if the hologram matrix is illuminated by a series of tilted plane waves

$$\sum_{\mu=-G/2}^{G/2} \sum_{\gamma=-G/2}^{G/2} \exp[j\alpha(u\mu + v\gamma)],$$

where $\mu = 0, \pm 1, \pm 2, \ldots, \pm G/2$; $\gamma = 0, \pm 1, \pm 2, \pm G/2$; α is a factor determining the angle of light

HOLOGRAPHIC MEMORY DEVICES

beam rotation when their numbers μ, ν increase, then directly behind the hologram matrix plane we obtain the wave front

$$\bar{M}_h(u,v) = \sum_{\mu=-\delta/2}^{\delta/2} \sum_{\nu=-\delta/2}^{\delta/2} M_h(u,v) \exp[j\alpha(u\mu + v\nu)],$$

and in the reconstructed image plane the front

$$\bar{M}_r(x,y) = \sum_{\mu=-\delta/2}^{\delta/2} \sum_{\nu=-\delta/2}^{\delta/2} M_r(x+\alpha\mu, y+\alpha\nu). \tag{19}$$

This is equivalent to the multiplication of amplitude distributions (8) and respectively to intensity distribution (14). Since the transform kernel in this case is

$$\bar{K}(u,v;x,y) = \sum_{\mu=-\delta/2}^{\delta/2} \sum_{\nu=-\delta/2}^{\delta/2} K(u,v;x+\alpha\mu, y+\alpha\nu),$$

then provided that integration is carried out independently over the areas $\Omega_{\mu\nu}$ where the images $\varphi_{\mu\nu}(x,y)$ are preset expression (15) is transformed to

$$\bar{A}_h(u,v) = \sum_{\mu=-\delta/2}^{\delta/2} \sum_{\nu=-\delta/2}^{\delta/2} A_{\mu\nu}(u+u_\mu, v+v_\nu) =$$

$$= \sum_{\mu=-\delta/2}^{\delta/2} \sum_{\nu=-\delta/2}^{\delta/2} \iint_{\Omega_{\mu\nu}} \varphi_{\mu\nu}(x+\alpha\mu, y+\alpha\nu) K(u+u_\mu, v+v_\nu; x+\alpha\mu, y+\alpha\nu) \, dx \, dy, \tag{20}$$

where $u_\mu = \dfrac{2\pi\alpha\mu}{\lambda f}$, $v_\nu = \dfrac{2\pi\alpha\nu}{\lambda f}$.

From (20) it follows that in the system output frequency plane the result of the integral transform of the images $\varphi_{\mu\nu}(x,y)$ with a kernel $K(u,v;x,y)$ is obtained.

Thus, when reconstructing the hologram matrix by light waves with a various tilt of the phase front, simultaneous processing of several images can be done. The respective expression (20) is equivalent to the matrix equation

$$\begin{Vmatrix} a_{11} & a_{12} & \ldots & a_{1P} \\ a_{21} & a_{22} & \ldots & a_{2P} \\ \cdot & \cdot & \cdot & \cdot \\ a_{J1} & a_{J2} & \ldots & a_{JP} \end{Vmatrix} = \begin{Vmatrix} K_{11} & K_{12} & \ldots & K_{1I} \\ K_{21} & K_{22} & \ldots & K_{2I} \\ \cdot & \cdot & \cdot & \cdot \\ K_{J1} & K_{J2} & \ldots & K_{JI} \end{Vmatrix} \times \begin{Vmatrix} \varphi_{11} & \varphi_{12} & \ldots & \varphi_{1P} \\ \varphi_{21} & \varphi_{22} & \ldots & \varphi_{2P} \\ \cdot & \cdot & \cdot & \cdot \\ \varphi_{I1} & \varphi_{I2} & \ldots & \varphi_{IP} \end{Vmatrix} \quad (21)$$

where $a_{jp} = \sum_{i=1}^{I} K_{ji} \varphi_{ip}$, $j = 1, 2, \ldots, J$ is the result of one of the p images transform.

The size of the matrices under multiplication in (17) and (18) is determined by the data recording density and the HMD memory capacity. With real technical limitations the HMD memory capacity achieves 10^8 and more bits. In this case the number of holograms and elements in the image reconstructed can be of the order of 10^4. So the matrix size is $J \times I = 10^4 \times 10^4$ and the number of elements in the matrix-row $[\varphi_i]$ or in the matrix-column $\{\varphi_j\}$ is $J = I = 10^4$, respectively.

The size of the matrices entering into expression (21) is determined by the admissible variation in the angle of the reconstructing beam incidence $\mathcal{E}$ to the hologram matrix. For the multiplication of square matrices $J = I = P = (\mathcal{E} \ell_h / \lambda)^{2/3}$ where ℓ_h is a linear dimension of the hologram matrix side. The angle $\mathcal{E}$ variation area is limited by the admissible deformations and fall in brightness of the image reconstructed. It can be shown that with a relative deformation of about 1.5% and fall in the image brightness by 10% the angle $\mathcal{E}$ value should not exceed $\mathcal{E} \approx \pm 2° = 0.35$ rad. If $\ell_h = 100$ mm, $\lambda = 0.63 \cdot 10^{-3}$ mm and provided that square matrices are multiplied their size equals to $J \times I = I \times P = 4 \cdot 10^2 \times 4 \cdot 10^2$.

IV. CALCULATION OF CORRELATION FUNCTIONS

Since the correlation functions result from realization of the linear integral transform with a difference kernel, they can be calculated in the HMD optical systems realizing transforms (17), (18), and (21). In this case in the Fourier holograms images should be recorded with the same intensity distribution but shifted in the output plane per value proportional to the hologram coordinate. It is evident that in this case the HMD memory capacity is used inefficiently. At the same time another more simple method of the correlation function calculation exists.

It is known [12] that the Fourier hologram equation comprises a term proportional to the complex-conjugated spatial-frequency image spectrum $f(x,y)$. If the light distribution equivalent to the spatial-frequency spectrum $\Phi(U,V)$ of the image $\varphi(x,y)$ is directed to such a hologram then as a result of realization of the inverse Fourier transform in the reconstructed image plane we obtain the light amplitude distribution proportional to the image cross-correlation function. It follows from the relation

$$F^*(u,v)\, \Phi(u,v) \xrightarrow{\mathcal{F}} \Psi(x,y), \qquad (22)$$

where $\Psi(x,y) = \iint\limits_{-\infty}^{\infty} f(x'+x, y'+y)\, \varphi(x',y')\, dx'dy'$

is a function of the cross-correlation.

The calculation of the cross-correlation functions in the HMD optical systems can be performed both successively and in parallel.

In the first case the transparency with the image $\varphi(x,y)$ located in plane 1 of the system which scheme is shown in Fig. 1 is successively illuminated by plane waves with a different tilt of the phase front. Behind hologram matrix 2 the products $\Phi(u+mG, v+nG)F_{mn}(u+mG, v+nG)$ are successively obtained which in the reconstructed image plane form light distributions equivalent to the correlation functions. The location of these functions is independent of the hologram coordinates.

In the second case the transparency with an image is illuminated by a series of plane tilted waves as a result of which in the hologram matrix plane the spatial-frequency spectrum $\varPhi(u,v)$ multiplication occurs.

If behind the hologram matrix a lens array is positioned then in its back focal plane we obtain the light amplitude distribution

$$\overline{\Psi}(x,y) = \sum_{m=-\frac{M}{2}}^{M/2} \sum_{n=-\frac{M}{2}}^{M/2} \Psi_{mn}(x+mg, y+ng) \qquad (23)$$

equivalent to the matrix of correlation functions. The number of simultaneously calculated functions is determined by the number of holograms in the matrix.

V. OPTICAL SYSTEM VERSIONS

In the past to determine the class of transforms realized in the HMD optical systems a general structural scheme of such a system was considered. Now we restrict ourselves to separate versions of the optical systems that can be used as the basis for an HMD with data processing functions.

The scheme of the optical system for performing transforms of type (15) equivalent to multiplication of a rectangular matrix by matrix-column (17) is given in Figure 2. It comprises a hologram matrix 1 determining the transform kernel $K(x,y;u,v)$; a Fourier transform objective O_1 ; a transparency with the image $\varphi(x,y)$ placed in the reconstructed image plane 2; a Fourier transform objective O_2 ; an output frequency plane or a transform plane 3.

Images reconstructed by a reference beam from the total of holograms are projected to the same place of plane 2 where an optical multiplication of each reconstructed image $f_{mn}(x,y)$ by the image under transform $\varphi^{1/2}(x,y)$ takes place. The result of each multiplication is integrated by the objective O_2 and is transferred to plane 3 to the point with the coordinates (U_m, V_n) . The light intensity at this point is proportional to

$$A_h(U_m, V_n) = \int\int_{-L_{r/2}}^{L_{r/2}} \varphi(x,y) K(x,y;u,v) dx dy.$$

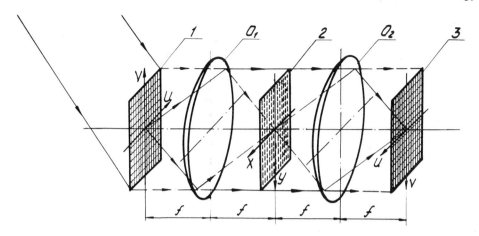

Figure 2. Scheme of Optical System for Matrix by Matrix-Column Multiplication.

Hence the intensity distribution in plane 3 corresponds to Ex. (15).

In this system to introduce an image "passive" transparencies should be used operating for transmission (or reflection). The number of the image $\varphi(x,y)$ elements is determined by one hologram capacity and the number of the resulting distribution $A_h(u,v)$ elements is determined by that of holograms.

The transform of Type (16) equivalent to multiplication of a rectangular matrix by a matrix-row (18) in performed with the help of the system which scheme is shown in Figure 3. It comprises a transparency with the image $\varphi(u,v)$ under transform - 1; a hologram matrix 2, determining the transform kernel $K(u,v;x,y)$; an objective O_1 ; a transform plane 3 coincident with the reconstructed image plane.

The light beam reconstructed illuminates hologram matrix 2 through transparency 1 with the image under transform. An optical multiplication occurs of the ho-

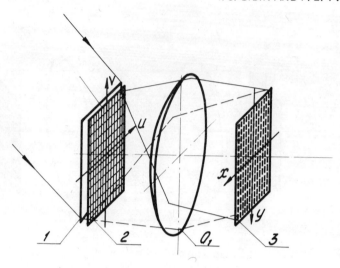

Figure 3. Scheme of Optical System for Matrix-Row by Matrix Multiplication.

logram matrix transmission function by the amplitude image $\varphi^{1/2}(u,v)$. In this case each of the images $f^{k\ell}(u,v)$ is multiplied by $\varphi^{1/2}(u,v)$. As a result the light intensity at the point with the coordinates (x_k, y_ℓ) of plane 3 is proportional to

$$A_r(x_k, y_\ell) = \int\int_{-L_{h/2}}^{L_{h/2}} \varphi(u,v) K(u,v; x_k, y_\ell) \, du \, dv$$

Thus the intensity distribution in plane 3 corresponds to Ex. (17). The characteristic feature of this system is that the image under transform is projected directly to the hologram matrix. Therefore for the introduction of images not only "passive" but also "active", transparencies can be used, e.g. in the form of a semiconductor laser matrix. The number of the image $\varphi(u,v)$ elements is determined by that of matrix holograms and the number of the distribution $A_r(x,y)$ elements is determined by their capacity.

HOLOGRAPHIC MEMORY DEVICES

The optical system scheme applicable to performing transform (20) equivalent to rectangular matrix multiplication (21) is given in Figure 4, where 1 is a hologram matrix plane; 2 is a Fourier transform objective; 3 is a reconstructed image plane; 4 is a lens array performing multichannel integration over the areas $\Omega_{\mu\nu}$ where the images $\varphi_{\mu\nu}(x,y)$ are preset; 5 is an output frequency plane.

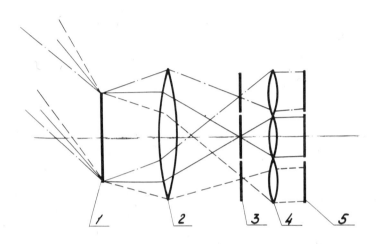

Figure 4. Scheme of Optical System for Rectangular Matrices Multiplication.

The tilted light beams reconstruct from hologram matrix 1 the multiplied sums of the images

$$\sum_m \sum_n f_{mn}(x+\alpha\mu, y+\alpha\nu) \exp[jG(mx+ny)].$$

Each light distribution equivalent to $\sum_m \sum_n f_{mn}(x+\alpha\mu, y+\alpha\nu) \exp[jG(mx+ny)]$ is optically multiplied by the corresponding image $\varphi_{\mu\nu}^{1/2}(x+\alpha\mu, y+\alpha\nu)$. The result of multiplication is transformed by lens array 4 and in its back focal plane the intensity distribution is formed proportional to Ex. (20).

Multichannel calculation of the cross-correlation functions is realized in the optical system shown in Figure 5 where 1 is a transparency with the image $\varphi(x,y)$, 2 is a Fourier transform objective, 3 is a hologram matrix, 4 is a lens array and 5 is an output correlation plane.

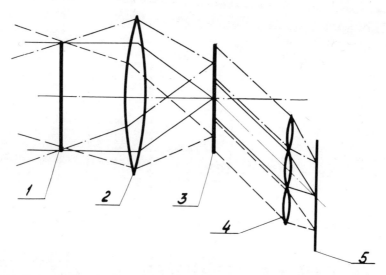

Figure 5. Scheme of Optical System for Multichannel Calculating of the Correlation Functions.

When illuminating the transparency by a set of plane tilted waves the multiplied image of the spectrum $\Phi(u,v)$ is projected to each holograms in plane 3. As a result of this light beams diffracted on the holograms in the reference beam direction comprise information on the product of Fourier spectra of the images being compared. Then lens array 4 transforms these light beams into the matrix of correlation functions.

Now let some practically important applications of the systems considered be pointed out.

HOLOGRAPHIC MEMORY DEVICES

VI. ASSOCIATIVE DATA RETRIEVAL

This type of retrieval is made according to the interogator structure composed of coded graphical, alphabetic, numerical and other chacacters. The content of the interrogator is compared with that of the memory according to then closeness criterion after which either necessary information or its address is given. The information and key words in the HMD with data arrayr oriented recording are preset in the form of the corresponding images, therefore in retrieval as a closeness criterion of the images being compared correlation functions and scalar products can be used.

The correlation functions are calculated utilizing the optical system which scheme is shown in Figure 5. As it was shown in [13] the same system can be used to calculate difference functionals as well, in case the images being compared represent one-dimensional contour curves. For this purpose a special mask is applied to the correlation function image and integration of the light distribution obtained is performed. As an example of the associative HMD using the above closeness criteria the data-retrieval device with a holographic memory can be given described in [14].

Scalar products of the images are calculated by using the optical systems shown in Fig. 2 and 3. Such criteria are efficient in the case when the key word images are of the form of a regular spatial matrix of light points. In this case the problem of comparing images of the interrogators and key words consists in calculation of the matrix-column $\{a_j\}$ elements according to the matrix logical equation

$$\begin{Vmatrix} a_1 \\ a_2 \\ \vdots \\ a_J \end{Vmatrix} = \begin{Vmatrix} K_{11} \bar{K}_{11} K_{12} \bar{K}_{12} \dots K_{1I} \bar{K}_{1I} \\ K_{21} \bar{K}_{21} K_{22} \bar{K}_{22} \dots K_{2I} \bar{K}_{2I} \\ \vdots \\ K_{J1} \bar{K}_{J1} K_{J2} \bar{K}_{J2} \dots K_{JI} \bar{K}_{JI} \end{Vmatrix} \times \begin{Vmatrix} \bar{\varphi}_1 \\ \varphi_1 \\ \bar{\varphi}_2 \\ \varphi_2 \\ \vdots \\ \bar{\varphi}_I \\ \varphi_I \end{Vmatrix} \quad (24)$$

where $(K_{j1} \bar{K}_{j1}, K_{j2} \bar{K}_{j2}, \ldots, K_{jI} \bar{K}_{jI})$, $j = 1, 2, \ldots, J$,

are bit key words each bit of which is represented in a paraphase code, i.e. "1" is recorded as 10 and "0" as 01; $(\bar{\varphi}_1 \varphi_1, \bar{\varphi}_2 \varphi_2, \ldots, \bar{\varphi}_I \varphi_I)$ is an interrogator which binary bits are represented respectively in the inverse paraphase code so that "1" is recorded as 01 and "0" as 10.

From equation (24) it follows that the element a_j of the matrix-column $\{a_j\}$ is equal to

$$a_j = \sum_{i=1}^{I} (K_{ji} \bar{\varphi}_i + \bar{K}_{ji} \varphi_i).$$

In the case of coincidence between the key word with the indexs j and the interrogator we obtain $a_j = 0$; in the case of incoincidence $a_j = 1$; if the physical parameter modelling this variable takes the value exceeding the threshold a_0. But if in the calculation the comparison with a threshold is not carried out, then the case $a_j > 0$ corresponds to the number of incoincident zeroes and unities in the words being compared.

The image comparison of the interrogator and the key words can also be made according to the matrix logical equation

$$\|a_1 a_2 \ldots a_J\| = \|\bar{\varphi}_1 \varphi_1 \bar{\varphi}_2 \varphi_2 \ldots \bar{\varphi}_J \varphi_J\| \times \begin{Vmatrix} K_{11} & K_{12} & \ldots & K_{1I} \\ \bar{K}_{11} & \bar{K}_{12} & \ldots & \bar{K}_{1I} \\ K_{21} & K_{22} & \ldots & K_{2I} \\ \bar{K}_{21} & \bar{K}_{22} & \ldots & \bar{K}_{2I} \\ \cdot & \cdot & \cdot & \cdot \\ K_{J1} & K_{J2} & \ldots & K_{JI} \\ \bar{K}_{J1} & \bar{K}_{J2} & \ldots & \bar{K}_{JI} \end{Vmatrix} \quad (25)$$

where

$$a_i = \sum_{j=1}^{I} (\bar{\varphi}_j K_{ji} + \varphi_j \bar{K}_{ji}).$$

Experimental devices with the associative data retrieval were described in [15]. They are based on the optical systems shown in Fig. 2 and 3. The coincidence between the key word and interrogator is concluded as a result of performing transforms of the type (24) and (25).

For the data retrieval problems more general transforms can be used described by the product of two matrices in the form of (21). In this case the first matrix describes the memory hologram content and the second the content of p interrogators. The parallel retrieval according to several interrogators can be realized with the optical system shown in Fig. 4.

VII. IMAGE SPECTRAL ANALYSIS ON ARBITRARY BASIS

The necessity of image expansion into a generalized Fourier series arises in image coding problems [16], choosing criteria in pattern recognition 17, data retrieval, etc. In this case the values of the spectral components

$$a_{mn} = \iint_\Omega \varphi(x,y) \cdot K_{mn}(x,y) \, dx \, dy,$$

are to be determined, where $K_{mn}(x,y)$ are functions of the chosen system expansion.

The analysis process is described by the matrix equations equivalent to equations (17) and (18), i.e.

$$\{a_j\} = \|K_{ji}\| \times \{\varphi_i\}, \qquad (26)$$

$$[a_i] = [\varphi_j] \times \|K_{ji}\|. \qquad (27)$$

The functions on which basis the analysis is made can be represented either by the rows $[K_{ji}]$ or by the columns $\{K_{ji}\}$ of the matrix $\|K_{ji}\|$. In the first case memory holograms store images of the expansion functions $[K_{ji}]$ and the analysis of the image described by the matrix-column $\{\varphi_i\}$ is performed according to equation (26). In the second case the holograms store the "cross-sections" $\{K_{ji}\}$ of the system of expansion functions, therefore the analysis of the image described by the matrix-row $[\varphi_j]$ is made according to equation (27). In both cases the matrix-column $\{a_j\}$ or the matrix-row $[a_i]$ describes a generalized spectrum of the image under analysis.

In the analysis process not only absolute values of the spectral components but also their signs should be determined. A sign can be determined provided that the analysis is made separately according to positive and negative parts of the expansion function and then the values obtained are algebraically summarized. This method of determination of the signs and absolute values of the spectral components results from the identity

$$\iint_\Omega K^+_{mn}(x,y)\varphi(x,y)\,dx\,dy - \iint_\Omega K^-_{mn}(x,y)\varphi(x,y)\,dx\,dy =$$
$$= \iint_\Omega \varphi(x,y)[K^+_{mn}(x,y) - K^-_{mn}(x,y)]\,dx\,dy = \iint_\Omega K_{mn}(x,y)\varphi(x,y)\,dx\,dy.$$

The information on the positive and negative components of the expansion functions can be recorded either in different holograms as a result of which equation (26) is transform to the form

$$\begin{Vmatrix} a_1^+ \\ a_1^- \\ a_2^+ \\ a_2^- \\ \vdots \\ a_J^+ \\ a_J^- \end{Vmatrix} = \begin{Vmatrix} K^+_{11} & K^+_{12} & \cdots & K^+_{1I} \\ K^-_{11} & K^-_{12} & \cdots & K^-_{1I} \\ K^+_{21} & K^+_{22} & \cdots & K^+_{2I} \\ K^-_{21} & K^-_{22} & \cdots & K^-_{2I} \\ \vdots & \vdots & & \vdots \\ K^+_{J1} & K^+_{J2} & \cdots & K^+_{JI} \\ K^-_{J1} & K^-_{J2} & \cdots & K^-_{JI} \end{Vmatrix} \times \begin{Vmatrix} \varphi_1 \\ \varphi_2 \\ \vdots \\ \varphi_I \end{Vmatrix}$$

or in one hologram as a consequence of which equation (27) is transformed to the form

$$\begin{Vmatrix} a_1^+ \bar{a}_1 & a_2^+ a_2^- & \cdots & a_I^+ a_I^- \end{Vmatrix} = \begin{Vmatrix} \varphi_1, \varphi_2, \ldots, \varphi_J \end{Vmatrix} \times \begin{Vmatrix} K^+_{11} K^-_{11} K^+_{12} K^-_{12} \cdots K^+_{1I} K^-_{1I} \\ K^+_{21} K^-_{21} K^+_{22} K^-_{22} \cdots K^+_{2I} K^-_{2I} \\ \vdots \\ K^+_{J1} K^-_{J1} K^+_{J2} K^-_{J2} \cdots K^+_{JI} K^-_{JI} \end{Vmatrix}$$

In both cases the number of the calculated values twice exceeds that of the spectral components. Real values of the components are determined according to the formula $a_i = a_i^+ - a_i^-$.

In photoelectric reading of the light distribution which intensity is proportional to the values a_i^+ and a_i^- this difference is found by subtracting the electrical signals obtained from each of the i-th pair of photoreceivers.

The results of investigation of the optical systems shown in Fig. 2 and 3 under the modes of the image Walsh-basis analysis have been given in [18,19].

In a general case the hologram matrix being a component of the above optical systems is obtained by using in-turn recording of the images of two-dimensional functions to each hologram. It is evident that the preparation and holographic recording of a great number of images require great labour and time expenditures and results in the necessity to increase the quality of the optical system, components. At the same time in the image spectral analysis according to the system of functions with separated variables the hologram matrix formation process can reasonably be simplified. Thus, in [20] it has been shown that the basis of expansion of $N \times N$ two-dimensional functions can be formed by using combined holographic recording of $2N$ one dimensional functions.

VIII. CONCLUSION

The transforms realized in the HMD optical systems refer to the class of general type linear integral transforms. They are equivalent to the operations of multiplication of a rectangular matrix by a matrix-column and a matrix-row by a rectangular one. The size of the rectangular matrix determining the transform kernel may achieve the value of the order of $10^4 \times 10^4$. A possibility exists for the simultaneous linear processing of several images. It can be made due to a controlled shift of the images reconstructed from holograms. The transforms performed in this case are equivalent to the operations of multiplication of rectangular matrices which size is of the order of $4 \cdot 10^2 \times 4 \cdot 10^2$.

The transform kernel depends on the hologram matrix content and can be varied by information re-recording in the case of the operative HMD.

The above transforms represent the basis for a data associative retrieval, an image spectral analysis according to an arbitrary basis, correlation calculations and other types of processing.

REFERENCES

1. F.M. Smits, L.E. Gallaher. Design Considerations for a Semipermanent Optical Memory. - The Bell System Technical J., 1967, 46, 6, p. 1267-1278.

2. L.K. Anderson. Holographic Optical Memory for Bulk Data Storage. - Bell Laboratories Record, 1968, 46, 10, p. 318-325.

3. A.L. Mikaelyan, V.I. Bobrinev, S.M. Naumov, L.Z. Sokolova. Vozmozhnosti primeneniya metodov golographii dlya sozdaniya novykh tipov zapominayushchikh ustroistv. - Radiotekhnika i elektronika, 1969, I, s. 115-123.

4. J.A. Rajchman. Promise of Optical Memories. - J. of Applied Physics, 1970, 41, 3, p. 1376-1383.

5. A.L. Mikaelyan, V.I. Bobrinev, A.A. Akselrod, S.M. Naumov, M.M. Koblova, E.A. Zasovin, K.I. Kushtanin, V.V. Kharitonov. Golographicheskie zapominayushchie ustroistva s zapisyu informatsii massivami. - Kvantovaya elektronika, 1971, 1, s. 79-84.

6. M. Sacaguhi, N. Nishida, T. Nemoto. A New Associative Memory System Utilizing Holography. - IEEE Trans. on Computers, 1970, C - 19, 12, p.1174-1181.

7. G.R. Knight. Page - Oriented Associative Holographic Memory. - Applied Optics, 1974, 13, 4, p. 904-912.

8. S.M. Maiorov, Li Si Ken. Ob odnom metode vypolneniya ariphmeticheskikh i drugikh operatsii na golographicheskikh ustroistvakh. - Izvestiya vuzov, Priborostroenie, 1974, 17, 2, s. 53-55.

9. I.I. Korshever, G.G. Matushkin, P.E. Tverdokhleb. Tsifrovye funktsionalnye preobrazovaniya na osnove opticheskikh zapominayushchikh ustroistv. - Avtomatria, 1974, 1, s. 9-15.

10. A.A. Feldbaum et al. Teoreticheskie osnovy svyazi i upravleniya. M. "Phizmatgiz", 1963.

11. C.B. Burckhardt. Use of a Random Phase Mask for the Recording of Fourier Transform Holograms of Data Masks. - Applied Optics, 1970, 9, 3, p. 695-700.

12. A. Vander Lugt. Signal Detection by Complex Spatial Filtering. - IEEE Trans. on Information Theory, 1964, IT - 10, 2, p. 139-145.

13. E.S. Nezhevenko. Opredelenie blizosti funktsii v koherenthykh opticheskikh vychislitelnykh ustroistvakh. - Avtometria, 1971, 6, s. 81-86.

14. G.A. Voskoboinik, I.S. Gibin, E.S. Nezhevenko, P.E. Tverdokhleb. Primenenie kogerentnykh opticheskikh vychislitelnykh ustroistv dlya resheniya zadach informatsionnogo poiska. - Avtometria, 1971, 1, s. 77-81.

15. I.S. Gibin, M.A. Gofman, E.F. Pen, P.E. Tverdokhleb. Assotsiativnaya vyborka informatsii v gologrammnykh zapominayushchikh ustroistvakh, - Avtometria, 1973, 5, s. 12-18.

16. P.A. Wintz. Transform Picture Coding. - Digital Picture Processing. Proceedings of th IEEE, 1972, v. 60, N 7, special issue.

17. I.T. Turbovich, E.F. Yurkov, V.G. Gitis. Approksimatsiya i normirovanie opisaniya obrazov. (pod red. I.T. Turbovicha), "Nauka", M., 1968.

18. I.S. Gibin, E.S. Nezhevenko, O.I. Potaturkin, P.E. Tverdokhleb. Kogerentno-opticheskie ustroistva dlya obobshchennogo spektralnogo analiza izobrazhenii. - Avtometria, 1972, 5, s. 15-21.

19. E.S. Nezhevenko, O.I. Potaturkin, P.E. Tverdokhleb. Lineinye opticheskie sistemy s impulsnoi redaktsiei obshchego vida. - Avtometria, 1973, 6, s. 88-90.

20. I.S. Gibin, M.A. Gofman, Yu.V. Chugui. Obobshchennyi spektralnyi analiz izobrazhenii s ispolzovaniem golographicheskogo metoda formirovaniya kodiruyushchei plastiny. - Avtometria, 1975, 3.

HOLOGRAPHIC MEMORY DEVICES

QUESTIONS TO THE REPORT BY I. GIBIN

1. Dr. A. Korpel. Why is a two-dimensional image represented not with a two-dimensional matrix but with a one-dimensional one?

Dr. I. Gibin. In this case it is more convenient to represent the image in a vector form, that is, as a matrix-row or a matrix-column. Then the transforms realized in the optical systems are described by well-known matrix equations. If the image is represented as a two-dimensional matrix the transforms being realized should be described by a tensor equation. It is less convenient.

2. Dr. J. Goodman. It is not clear in what way a transform is realized in the system whose scheme is given in Fig. 2.

Dr. I. Gibin. Images reconstructed by a reading beam from the whole of holograms are projected to the same place of plane 2 where the optical multiplication of each reconstructed image $f_{mn}(x,y)$ by the image under reconstruction $\varphi(x,y)$ occurs. The result of each multiplication is integrated by the objective O_2 and transmitted to the output plane point with the coordinates (U_m, V_n). The light intensity here is proportional to

$$A_h(U_m, V_n) = \iint_{\Omega_{;r}} \varphi(x,y) K(x,y; u_m, v_n)\, dx\, dy$$

3. Dr. A. Vander Lugt. What logical transforms are realized in the systems considered?

Dr. I. Gibin. In the considered optical systems of holographic memories the logical operations "and", "or", "exclusive or" are performed. These operations can be performed in a multichannel mode, that is, simultaneously over several data files.

4. Dr. A. Vander Lugt. What practical applications can the transforms realized in the optical systems of holographic memories be used for?

Dr. I. Gibin. The transforms considered form the basis for associative retrieval, image spectral analysis, correlation calculations, and so on. We believe that one of the main practical applications consists in the realization of associative retrieval in holographic memories and in the development of information-retrieval systems on their basis.

5. Dr. S. Lee. Earlier we proposed a method for matrix multiplication based on the application of holograms (R.A.Heinz, I.O.Artman, S.H.Lee, Matrix Multiplication by Optical Methods. Applied Optics, No. 9). Have you been familiar with it?

Dr. I. Gibin. Yes, we have. Your method suggests a holographic complex-matched filter to be used. We believe that a correlation mechanism of calculation forming a basis for such filtering does not enable us to realize multiplication of large-size matrices to a good accuracy. In the systems considered matrix multiplication is performed by using scalar methods.

Dr. S. Lee. Yes, I agree with you, scalar methods are more accurate.

Dr. I. Gibin. In addition, our laboratory has proposed a specialized optical system for matrix multiplication described in the paper by E.S.Nezhevenko, P.E.Tverdokhleb "Matrix Multiplication by Optical Method", Avtometria No. 6, 1972. This system also uses a scalar method.

Dr. S. Lee. It will be of interest to be familiar with it.

GENERATION OF ASYMMETRIC INTERFERENCE FRINGES IN REFLECTED LIGHT

N.D. Goldina, Yu.V. Troitsky

Institute of Automation and Electrometry
Novosibirsk, USSR

ABSTRACT

A possibility has been considered to generate asymmetrical interference fringes in light reflected by a two-mirror interferometer. The front mirror has high absorption and represents a combination of a very thin $\ll \lambda$ absorbing film and a dielectric multilayer coating. By varying the imaginary part of the film surface admittance and the parameters of the dielectric layers, fringe asymmetry can be controlled over a wide range. The suitable formulas have been given for a mirror with a dielectric multilayer looking inside the interferometer. The curves have been plotted illustrating a fringe shape calculated for one particular case. The experimental results supporting the asymmetry control principle and a method for its calculation have been described.

At present in holography interest arises [1] for using an interference pattern with asymmetrical fringes. The purpose of the present communication is to describe the possibilities for the achievement of this effect by means of a reflecting two-mirror interferometer.

It is known that the fringe shape in a conventional transmitting Fabry-Perot interferometer is determined only by a reflectance of mirrors: these fringes are always symmetrical and their sharpness is uniquely interrelated with contrast. Radically new possibilities arise when a front mirror of the interferometer has high absorption and in addition an interference pattern is observed in the reflected light. A general formula for a fringe shape of this interferometer was first obtained by M. Hamy [2] in 1906. Later on it became clear that this interferometer posessed potentiality to change a fringe profile, since it depends on 3 parameters in contrast to 1 in a transmitting Fabry-Perot interferometer. For example, one may independently vary sharpness, contrast and asymmetry of the fringes. In the past, however, possibilities of such interferometers were used to a very small extent. It might be explained as follows. First, on the opinion of many researchers, the interference patterns of this interferometer should be of low brightness because of high absorption (e.g., 90%) in the front mirror. Second, no methods existed to make the front mirror so that it provided the predetermined finesse and fringe profile.

A method to overcome these difficulties was proposed in [3]. As an absorption component of the front mirror high-absorbing metallic film was proposed with a thickness much below a wave length. When using such a metallic film, the brightness of the reflected interference pattern is not worse than in the transmitting Fabry-Perot interferometer; for example, the maximal reflectance for normal incidence is equal to unity. In addition, it was suggested to combine an absorbing film with dielectric multilayer coatings. When the front mirror is made of a metallic film alone, a set of the available profiles will be very limited. For example, an attempt to increase sharpness results in drastic decrease in contrast and uncontrolled variation of asymmetry; the substitution of one metal for another does not give an appreciable result. The presence of dielectric layers which number and thickness can readily be varied allows the fringe width, contrast and asymmetry to be independently changed over a wide range. The thus-fabricated interferometer is superior to a Fabry-Perot one to a certain extent. In particular, an ideal contrast can be obtained independently of sharpness. The width of light fringes against a dark background may achieve, e.g. 1/70 of the fringe spacing [4,5].

For the design of the above interferometer it is convenient to use well-developed methods of the transmission-lines theory based on a known analogy between the electromagnetic wave propagation in space and the propagation of current and voltage waves in a transmission line. This procedure is significantly facilitated by the fact that a very thin metallic film can be considered in the visible light region as a surface having a complex surface admittance [6,7]. It corresponds to lumped parallel admittance $\xi = \xi' + i\xi''$ in the line. The interferometer parameters calculation in a general form is presented in [8].

At the top of Figure 1 a scheme of the interferometer is represented. Here n_1 is a refractive index of the incident medium, n_2 is the same for the medium between two mirrors. A combination of metallic absorbing film M and a dielectric multilayer stack TR (indices of dielectrics are n_a, n_3, n_4) forms a front mirror. The back mirror MR has a reflectance equal to unity (it presents no special problems to consider a more general case [8]).

Figure 1 shows a multilayer situated to the right of the film. In this case dielectric layers do not influence on contrast of the reflected interference pattern and affect only sharpness and asymmetry of the fringes. If contrast is to be varied parameters of the absorbing film should be changed or dielectric layers should be introduced on the light source side of the film [8].

An equivalent scheme of the system of transmission lines is represented at the bottom of Figure 1. Here wave admittances of each line are shown. The dotted arrows show the correspondence with the interferometer components. The reflected fringe intensity distribution of the interferometer is given by

$$\widetilde{R} = \frac{(n_1 - \xi' - y')^2 + (\xi'' + y'')^2}{(n_1 + \xi' + y')^2 + (\xi'' + y'')^2} \qquad (1)$$

where $y = y' + iy''$ is an input admittance of the system situated to the right of the film M measured in the film plane. Provided that losses in the dielectrics are negligible $y' = 0$.

The film surface admittance $\xi' + i\xi''$ of Eq. (1) can be found by measurements in a traveling wave. The

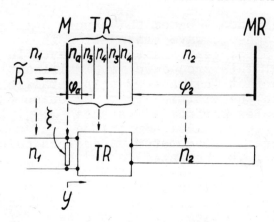

Figure 1. Scheme of Two-Mirror Interferometer: at the top - an optical scheme, at the bottom - the equivalent network of transmission lines.

relation ξ''/ξ' depends on thickness, kind of metal and absorbing film technology. If $\xi'' \neq 0$, the interference fringes in the reflected light are asymmetrical even in the absence of dielectric layers. The dielectric layers, however, allows us to change the fringe asymmetry in the direction desired.

The dependencies of y'' on φ_2 may be numerous. Consider the case for a dielectric layer adjacent to the metal film with the index n_a and phase thickness φ_a. The remainder of the dielectric stack is composed of k - quarter-wave layers of two dielectrics n_3 and n_4 (k is the even number, in Figure 1 $k = 4$). Then

$$y'' = n_a \frac{y_2 + i n_a \tg \varphi_a}{n_a + i y_2 \tg \varphi_a} \qquad (2)$$

where $y_2 = -i(n_3/n_4)^4 \mathrm{ctg}\, \varphi_2$, φ_2 is a phase length of the interferometer.

Figure 2 shows the results of $\tilde{R}(\varphi_2)$ calculation at $n_1 = 1.46$ (fused quartz), $n_2 = 1$, $n_4 = n_a = 2.3$, $n_3 = 1.34$, $k=4$, $\xi = 1.46 + i\, 0.7$; here the value of ξ'' corresponds to our experiments with Ni-films.

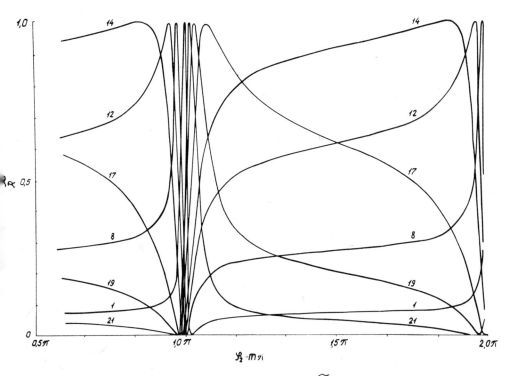

Figure 2. Calculated Dependence $\tilde{R}(\varphi_2)$ for various values of φ_a : Curve 1 - $\varphi_a = \pi/30$, 8 - $8\pi/30$, 12 - $12\pi/30$, 14 - $14\pi/30$, 17 - $17\pi/30$, 19 - $19\pi/30$, 21 - $21\pi/30$. Admittance of Ni- film $\xi = 1.46 + i\, 0.7$.

For all the curves of Figure 2 $\tilde{R}_{max} = 1$. The asymmetry is extremely high, on one side from the maximum $\tilde{R}$ sharply decreases and on the other side it decreases smoothly. In several cases the fringe shape is close to a saw-tooth one. All the curves of Figure 2 have maximum-possible contrast since $\xi' = n_1$. Thus, by changing the thickness of one of the dielectric layers, it is possible to achieve the required fringe shape and asymmetry sign. If another fringe profiles are necessary not entering into the family shown in Figure 2 the values of n_3, n_4, ξ', ξ'', k can be changed. At $\widetilde{R}_{max} = 1$ the fringe shape is determined by three independent parameters one of which simultaneously affects contrast.

Provided that the necessary intensity distribution is predetermined, it can be matched to the available fringe profiles by means of a computer. The calculated values for each of three parameters allow a physical structure of the front mirror to be synthesized.

Formulas (1) and (2) give the intensity distribution of the interference pattern of constant-thickness fringes. It is possible to observe the interference fringes of constant inclination.

The above method for calculation of $\widetilde{R}(\varphi_2)$ provides good results being in accordance with an experiment. The only noticible difference consists in that in practice $\widetilde{R}_{max}$ is slightly below unity (see [4,5]).

Figure 3a demonstrates the experimentally measured ($\lambda = 633$ nm) dependence of $\widetilde{R}(\varphi_2)$ at normal light incidence to the interferometer on its length. The front mirror was supported onto a quartz substrate and consisted of an Ni-film and 5 layers of zinc sulfide and cryolite, the top four layers having a quarter-wave thickness. Thus the interferometer parameters should correspond to one of the curves from the family of Figure 2. The second mirror of the interferometer was a multilayer laser one with low transmittance. The length of interferometer was equal to 4 mm, for the scanning of the length piezoceramics was used. The horisontal line on the photograph of Figure 3a equals to $\widetilde{R} = 0$, the maximum value of $\widetilde{R}$ for this sample is about 0.9. Comparing the obtained dependence of with the calculated curves (to avoid overcrowding Figure 2 represents only a small part of the calculated curves of the family given) it may be stated that the dependency measured is very close to that calculated for $\varphi_a = 0.6\pi$. In Figure 3b the dependence of $\widetilde{R}(\varphi_2)$ corresponds to the calculated curve for

ASYMMETRIC INTERFERENCE FRINGES

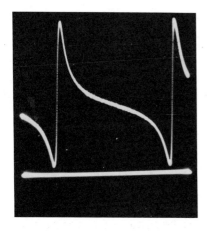

Figure 3. Experimental Dependencies of $\tilde{R}(\varphi_2)$ for Different Structures of Front Mirror:

a) $MH'LHLH$ where M is a Ni-film with $\xi = 1.43 + i\,1.17$, H' is a ZnS - layer with $\varphi_a = 0.6\pi$, and L, H is a $\frac{\lambda}{4}$-layer cryolite and ZnS, respectively.

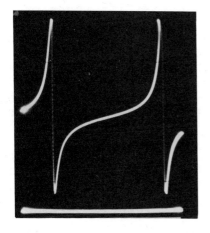

b) $MH'LHLH$ where $\varphi_a = 0.3\pi$.

c) Ni- film without dielectric coatings.

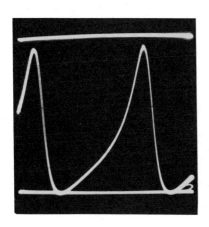

$\varphi_a = 0.3\pi$. There is no doubt that it is possible to obtain other curves of Figure 2 as well.

In those cases when extremely high asymmetry and interference fringe sharpness are not necessary dielectric coatings can be omitted, an absorbing film can be used alone.

Figure 3c gives a photograph of $\widetilde{R}(\varphi_2)$ for the cases when the front mirror is made of the same Ni-film as in Figure 3a but without dielectric layers. In this case $y = -i n_2 \operatorname{ctg} \varphi_2$ and the asymmetry of $\widetilde{R}(\varphi_2)$ results only from reactance of the film itself. In several cases the value ξ'' at predetermined ξ' may be varied due to the material and method of preparation of the film. The maximum asymmetry is demonstrated by relatively thick films of well-reflecting metals Ag, Al, Au for which the difference of squares of the refraction and absorption indices is high. As usual for such films $\xi'' < 0$ (inductive admittance) and a more sharp decrease in $\widetilde{R}$ is obtained on that side of the maximum which corresponds to the decrease in φ_2. Dielectric layers allow us to obtain any sign of the asymmetry. To carry out the experiment described in [1] an Al-film was used. The dependence of $\widetilde{R}(\varphi_2)$ for it is given in Fig. 11a there.

Conclusion. The method of generation of asymmetrical interference fringes in the reflected light has been described. This method utilizes a two-mirror interferometer, one of the mirrors being a combination of a thin absorbing film and a dielectric multilayer coating. The interference fringe profile is determined by three independent parameters that may be varied so that to obtain a dependence close to that required. The brightness of interference pattern is high. The fringe shape calculation method based on the use of a model of the complex-admittance surface is confirmed by the experiment with a scanning interferometer.

REFERENCES

1. V.P. Koronkevich, G.A. Lenkova, I.A. Mikhaltzova, V.G. Remesnik, V.A. Fateev, V.G. Tsukerman, in this issue.

2. M. Hamy, Journal de Physique, 5, 789, 1906.

3. Yu.V. Troitsky. Pis'ma JETF, 11, 281, 1970.

4. N.D. Goldina, Yu.V. Troitsky, "Optika i Spektroscopia", 31, 147, 1971.

5. N.D. Goldina, M.I. Zakharov, Yu.V. Troitsky. Avtometria No.3, 107, 1975.

6. Yu.V. Troitsky, Avtometria, No.6, 91, 1972.

7. Yu.V. Troitsky. In "Dispergirovannie metallicheskie plenki", Kiev, 1972, s. 214.

8. Yu.V. Troitsky, "Optika i spektroscopia, 30, 544, 1971.

NOISE IN COHERENT OPTICAL INFORMATION PROCESSING

J.W. Goodman

Department of Electrical Engineering
Stanford University
Stanford, California 94305

INTRODUCTION

The noise appearing at the output of an information processing system poses one important limit to the accuracy with which the desired processing operations can be performed. The detailed statistical character of the noise, as well as its degree of severity, depend on many factors, including: (1) the quality of the input data to be processed; (2) the character of the mathematical data processing operations to be performed; (3) the physical embodiment of the data processing system (e.g., electronic, optical, digital, analog, etc.); and (4) the degree of accuracy with which the ideal processing operations are realized physically.

In this paper we consider only coherent optical information processing[1,2,3]. The assumed geometry is shown in Fig. 1. Coherent light illuminates the input data transparency, which may or may not be photographic in character. The Fourier transform of the transmitted light amplitude distribution appears in plane P_2, where a "filter" with amplitude transmittance t_f is inserted. Finally, lens L_2 Fourier transforms the light amplitude distribution transmitted by the filter, yielding a processed output in plane P_3.

What are the sources of noise in this system, and what is the statistical character of the output fluctuations they generate? To this and related questions we address our attention in the sections that follow.

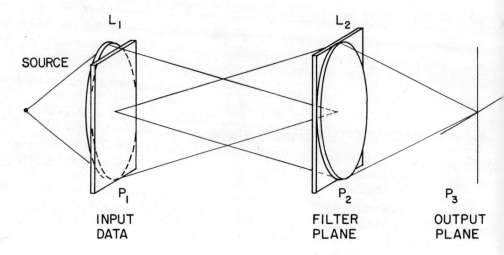

Fig. 1 Coherent optical processor

NOISE STATISTICS UNDER INCOHERENT AND COHERENT ADDITION

There are fundamental differences in the statistical character of noise processes that arise from incoherent and coherent addition of optical fields. We discuss these differences here in very general terms, without specifying the source of noise or the location of the observation plane in the optical system.

Incoherent Addition

Consider first the case when the observable quantity, intensity measured at a point, arises from the sum of many uncorrelated intensity contributions which add to yield the final measured value. Implicit here is the assumption that the underlying field contributions are totally uncorrelated. The result is a measured intensity I_M that can be expressed as the sum of K independent intensity contributions i_k, $k = 1, 2, \ldots K$,

$$I_M = \sum_{k=1}^{K} i_k . \qquad (1)$$

Typically, the elementary contributions i_k are random variables; although it is not absolutely necessary to do so, we assume that they are identically distributed, with means $\bar{i}$ and standard deviations $\sigma_i \triangleq \overline{(i - \bar{i})^2}$.

The statistical properties of the measured intensity I_M are of prime interest. From equation (1) above,

$$\overline{I_M} = K \overline{i} \quad , \quad \overline{I_M^2} = K \overline{\sigma_i^2} + K^2 \overline{i}^2$$

$$\sigma_M^2 = \overline{(I_M - \overline{I_M})^2} = K \sigma_i^2 \quad . \tag{2}$$

The ratio of the mean measured value $\overline{I_M}$ to the standard deviation σ_M can often be called the "rms signal-to-noise ratio" associated with the measurement. Clearly, under incoherent addition,

$$\left(\frac{S}{N}\right)_{rms} = \frac{\overline{I_M}}{\sigma_M} = \sqrt{K} \frac{\overline{i}}{\sigma_i} \quad . \tag{3}$$

Thus <u>the rms signal-to-noise ratio increases in proportion to the square root of the number of elementary contributions when incoherent addition is involved</u>.

If we know that the intensity contributions are <u>independent</u> rather than simply uncorrelated, then the statistics of I_M can be specified in greater detail. In particular, if the number K of contributions is extremely large, as is often the case in practice, then under rather general conditions the Central Limit Theorem[4] can be invoked to conclude that I_M is approximately a gaussian random variable, with probability density function

$$p(I_M) \cong \frac{1}{\sqrt{2\pi}\,\sigma_M} \exp\left\{-\frac{(I-\overline{I_M})^2}{2\sigma_M^2}\right\} \quad . \tag{4}$$

It should be noted that incoherent illumination of an object in an optical system does not guarantee incoherent addition of noise contributions. After leaving the object plane, the light takes on some degree of partial coherence. Hence if the origin of the noise lies in a plane following the object plane, the noise contributions may add with partial correlation, and the above conclusions may not be valid. Nonetheless, the ideal case of incoherent addition of noise is an important limiting situation, and is closely approached when the noise process arises in an incoherently illuminated object plane.

Coherent Addition

The case of greater interest to us here is that of <u>coherent</u> addition of noise fields. Note that, if the object illumination is coherent, and if there are no time-varying phenomena within the optical system, then the noise contributions add coherently regardless of where they originate in the optical system.

If the noise contributions are coherent, then they must be added on an <u>amplitude</u> basis rather than an intensity basis. Thus we write

$$\underline{U}_M = \sum_{k=1}^{K} \underline{u}_k \tag{5}$$

where $\underline{u}_k$ represents the complex field contributed at the measurement point by the kth noise scatterer, and $\underline{U}_M$ represents the complex amplitude of the resultant field. An underbar is used to indicate complex-valued quantities.

To understand the statistical properties of the observed noise intensity I_M, it is necessary to first understand the properties of the underlaying complex field $\underline{U}_M$. To this end, Eq.(5) is represented pictorially in Fig. 2, and we see that the field $\underline{U}_M$ may be regarded as the result of a random <u>walk</u> in the complex plane.

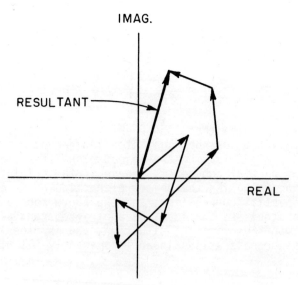

Fig. 2 Random walk in the complex plane

The real and imaginary parts of the noise field $\underline{U}_M$ may be expressed as sums of independent contributions,

$$\mathcal{R}_M \stackrel{\Delta}{=} \text{Re}\{\underline{U}_M\} = \sum_{k=1}^{M} |\underline{u}_k| \cos \phi_k$$

$$\mathcal{I}_M \stackrel{\Delta}{=} \text{Im}\{\underline{U}_M\} = \sum_{k=1}^{M} |\underline{u}_k| \sin \phi_k \tag{6}$$

where $\phi_k = \arg\{\underline{u}_k\}$. In the most common case, the phases ϕ_k of the elementary contributions are independent of each other, independent of the $|\underline{u}_k|$, and are uniformly distributed on $(0, 2\pi)$, with the result that

$$\overline{\mathcal{R}_M} = \overline{\mathcal{I}_M} = 0,$$

$$\sigma_\mathcal{R}^2 = \sigma_\mathcal{I}^2 = \sigma^2, \tag{7}$$

$$\overline{\mathcal{R}_M \mathcal{I}_M} = 0.$$

Thus $\mathcal{R}_M$ and $\mathcal{I}_M$ are zero mean random variables, uncorrelated, and have identical variances. Now if in addition the number K of noise contributions grows large, we can again invoke the Central Limit Theorem to conclude that $\mathcal{R}_M$ and $\mathcal{I}_M$ are gaussian random variables,

$$p(\mathcal{R}_M) \cong \frac{1}{\sqrt{2\pi}\,\sigma} \exp\left\{-\frac{\mathcal{R}_M^2}{2\sigma^2}\right\}$$

$$p(\mathcal{I}_M) \cong \frac{1}{\sqrt{2\pi}\,\sigma} \exp\left\{-\frac{\mathcal{I}_M^2}{2\sigma^2}\right\} \tag{8}$$

Of course, the optical fields are not directly measurable, and we must now inquire as to the statistics of the measured intensity,

$$I_M = \mathcal{R}_M^2 + \mathcal{I}_M^2. \tag{9}$$

A simple transformation[5] shows that I_M obeys negative exponential statistics,

$$p(I_M) = \begin{cases} \dfrac{1}{\overline{I_M}} \exp\{- I_M/\overline{I_M}\} & I_M \geq 0 \\ 0 & \text{otherwise} \end{cases} \quad (10)$$

where $\overline{I_M} = 2\sigma^2$. An important property of this type of statistical distribution is that the standard deviation σ_M is identically equal to the mean $\overline{I_M}$. Hence, in the case of coherent addition, we find that

$$\left(\dfrac{S}{N}\right)_{rms} = \dfrac{\overline{I_M}}{\sigma_M} = 1 , \quad (11)$$

and this result is quite independent of how large K may be, provided only that K is larger than about 10. Thus the important difference between incoherent and coherent addition of noise contributions is that, in the former case the r.m.s. signal-to-noise ratio improves in proportion to $\sqrt{K}$, while in the latter case it is asymptotic to unity for large K.

It should be noted that there exist a number of possible ways in which the assumptions implicit in the above argument may be violated. First, if there is some degree of depolarization introduced by the scatterers, the intensity statistics will depart from negative exponential. Second, if the number of independent contributions is less than about ten, departures from negative exponential statistics are again observable. Finally, if the phases of the individual field contributions are not uniformly distributed on $(0, 2\pi)$, different intensity statistics result. This latter case is sometimes encountered when the noise originates in the object plane or input plane.

NOISE INHERENT IN THE DATA TO BE PROCESSED

It is useful to distinguish between noise that is inherent in the data to be processed, and noise that is introduced by the data processing system itself. In this section we concentrate on the former type of noise, the latter type being treated later.

The data to be processed is never perfect, i.e., there always exist departures from what we would consider to be ideal data. Such noise will, of course, degrade any data processing operation, regardless of the physical means for carrying it out. Our examples will be drawn from the field of picture restoration, but similar considerations arise in most data processing problems. We discuss a number of sources of these data imperfections.

Background Noise

If the data to be processed represents a picture of some object of interest, that object may be accompanied by a background. The background may arise from radiation sources within the field-of-view of the original imaging system, or it may arise due to some form of electronic interference during the detection process.

Generally speaking, background that coincides with the object of interest is of greatest concern, especially if the general location of the object is known. Of course, a constant or uniform background can in principle be subtracted out, but any noise associated with the detection of the background remains. This problem is especially severe when the image is formed at low light levels, in which case quantum fluctuations associated with the detection of the background light can be a severe source of noise.

There is little more that can be said about background noise in a general sense.

Speckle Noise

If the data to be processed consists of an image that was formed with fully coherent light, that image will suffer from so-called <u>speckle noise</u>. Such noise arises due to the coherent addition of field contributions from a large number of independent scatterers lying within any single resolution element on the object. If those scatterers are randomly phased, fluctuations of the recorded images result, those fluctuations having precisely the statistics described previously for coherent addition of noise components. Thus the standard deviation of the image intensity at any given point is precisely equal to the mean intensity at that point.

Speckle noise can be said to be a truly multiplicative noise in the following sense. The measured image intensity can be written as

$$I_M(x,y) = \overline{I_M}(x,y) n(x,y) \qquad (12)$$

where $\overline{I_M}(x,y)$ generally is the ideal desired image intensity and $n(x,y)$ is a noise process. The significant fact is that, when defined in this multiplicative form, $n(x,y)$ is a <u>stationary</u> random process, independent of the ideal intensity, and with the further properties

$$\overline{n} = 1 \quad , \quad \sigma_n = 1 \quad . \qquad (13)$$

Quantum Noise

The most fundamental and, in many respects, the most interesting type of data noise arises from quantum fluctuations in the interaction of the incident image intensity distribution and the detectors that sense this radiation. In this case we may regard the data to be processed as a spatial array of counts, N_i ($i = 1,2,\ldots,M$), one for each of the M detectors in the image plane.

If the radiation incident on the detector array is thermal in origin, and (as is usually the case) has a low degeneracy parameter[6], then the counts will obey Poisson statistics

$$P(n_i) = \frac{(\overline{N_i})^{N_i}}{N_i!} \exp(-\overline{N_i}) \qquad (N_i = 0,1,\ldots.) \qquad (14)$$

where $\overline{N_i}$ represents the mean number of counts from the ith detector. $\overline{N_i}$ is given explicitly by

$$\overline{N_i} = \frac{\eta \tau_c \iint_A I(x,y)\,dx\,dy}{h \overline{\nu}} \qquad (15)$$

where η is the quantum efficiency of the detector, τ_c is the counting time, A is the detector area, h is Planck's constant, $\overline{\nu}$ is the mean frequency of the incident radiation, and $I(x,y)$ is the image intensity distribution.

It is, of course, well known that the Poisson distribution has the property that the standard deviation σ_i of the counts is equal to the square root of the mean number of counts,

$$\sigma_i = \sqrt{\overline{N_i}} \ . \qquad (16)$$

In this case we can see that the noise is not truly multiplicative, for if we write

$$N_i = \overline{N_i}\, n_i \ , \qquad (17)$$

where n_i is the "noise", we can easily show that for Poisson statistics,

$$\overline{n_i} = 1 \ , \quad \sigma_{n_i} = \frac{1}{\sqrt{\overline{N_i}}} \qquad (18)$$

and hence the "noise" n_i depends on the "signal" $\overline{N_i}$. We can thus only say that Poisson noise is "signal dependent".

Some extremely interesting questions arise when we consider the restoration of images that are limited by Poisson noise. Two examples are given in the following. Suppose that we wish to record an image of a distant faint object which is moving with constant velocity v. Further suppose that, due to certain practical constraints, we are unable to track the object with the imaging system. The question then arises as to how long we should leave the camera shutter open. With a very short exposure, the image suffers very little blur, but very little light flux is collected and the image is extremely noisy. On the other hand, if the shutter is left open longer, more flux is collected, less image noise is present, but the image is blurred due to motion of the object. If we are willing to process the recorded image to restore resolution, what is the optimum exposure time?

A second closely related problem comes from the field of nuclear medicine. We wish to form an image of an object emitting gamma rays. Since lenses do not exist in this region of the spectrum, we are forced to use a pinhole for imaging. Now for a fixed measurement time, the smaller the pinhole is made the sharper the resulting image, but also the fewer gamma rays collected and the noisier the image. The larger the pinhole, the more flux collected, the higher the image signal-to-noise ratio, but the more blurred the image. If post-detection restoration of the image is to be attempted, what is the optimum pinhole size?

We make a short digression to consider the first of the above examples in more detail; the second can be treated in a similar manner. We wish to construct a measure of the "restorability" of the image of a moving object as a function of exposure time. Since noise poses the fundamental limit to restorability, our measure must take into account the signal-to-noise ratio present in the image at all spatial frequencies. The MTF associated with the blur is unfortunately not a good measure of restorability, for MTF's are by convention always normalized to unity at the origin, and do not take account of the amount of transmitted light flux. However, if we multiply the MTF by a number proportional to the square root of the flux reaching the image, we have a quantity that is at least useful for comparing the distributions of signal-to-noise ratios vs. spatial frequency for different exposure times.[7]

For the case of an object moving with constant velocity v in the X-direction, the MTF is given by

$$H(\nu_X, \nu_Y) = \mathrm{sinc}(vT\nu_X) \qquad (19)$$

where T is the exposure time. In order to make comparisons of image restorability for different exposure times, we multiply the MTF by $\sqrt{vT}$, which, for fixed v, is in direct proportion to the square root of the collected light flux. Thus our comparisons are based on the quantity

$$\rho(\nu_X, \nu_Y) = \sqrt{vT}\, \text{sinc}(vT\nu_X) \qquad . \tag{20}$$

Fig. 3 shows plots of $\rho/\sqrt{10}$ vs. spatial frequency ν_X for various values of vT. If we consider the frequency range 0 to 1 on this rather arbitrary scale, we see that, of the four exposure times shown, the curve for vT = 1 probably represents the highest degree of restorability. Both longer exposure times and shorter exposure times lead to poorer signal-to-noise ratios over most of the frequency range of interest. It should be noted, however, that while the curves drawn are accurate representations of <u>relative</u> restorability, they do not yield much information about <u>absolute</u> restorability. To specify absolute restorability, it is necessary to know the brightness of the object as well as its spatial frequency power spectrum.

We shall not pursue this subject further here. Walkup[8] has studied similar questions from a somewhat different point of view. Further consideration would lead us quite naturally to the study of optimum "shutter modulation"[9] and optimum "coded apertures".[10] It suffices to say here that a full understanding of questions such as these is essential to a full understanding of the data processing limitations posed by noise inherent in the data to be processed.

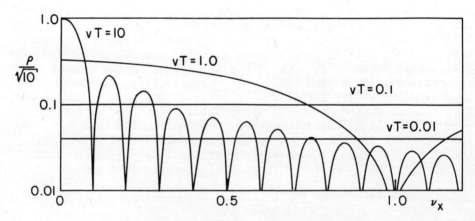

Fig. 3 Image restorability vs. spatial frequency for a moving object photographed with different exposure times

NOISE INTRODUCED BY THE OPTICAL PROCESSOR

No data processing system is capable of performing the desired data processing operations with an infinite degree of precision. Digital computations are limited in accuracy by quantization noise due to finite work lengths. So too are optical data processors limited in accuracy, but generally by a greater diversity of physical phenomena than encountered with digital computations. Here we shall discuss a few of the sources of output noise generated by the coherent optical data processing system itself.

Noise Associated with the Input Transparency

Because coherent optical data processing systems operate on complex field, any noise associated with either the amplitude or the phase of light transmitted by the input transparency will degrade the output of the processor. Such effects are particularly noticeable in image deblurring experiments, for in such experiments most of the light transmitted by the input transparency is thrown away in the process of attenuating the low frequency components of the blurred image. A small scatterer, such as a dust speck, can produce a very significant response in the output plane, for most of the light scattered will pass through the highly transmissive portion of the filter corresponding to high spatial frequencies. The same is true for small phase defects, such as tiny bubbles in the emulsion or the base of the input transparency.

Small defects such as these produce a characteristic response in the output plane of the processor, namely they each produce an impulse response of the deblurring filter. Such effects are seen in many image deblurring experiments, including those performed with digital computers. We illustrate here with an example obtained by Tichenor[11] using a coherent optical processor. Fig. 4 shows photographs of (a) the blur function associated with a severe focusing error, (b) the intensity of the impulse response of a deblurring filter designed to compress the blur function by about 10 to 1 in radius, and (c) the intensity distribution in the deblurred image of a point source. The deblurring filter is made after the method of Ragnarsson[12], as modified by Tichenor[13], and properly controls the amplitude transmittance over a dynamic range of 100 to 1 in the frequency plane.

When a more general blurred image is entered into the processing system, the effects of impulsive noise become evident. Fig. 5 illustrates this point. Part (a) shows the original object before blurring; part (b) shows the object blurred by a severe focusing error; part (c) shows the restored image produced by the deblurring filter. Note in particular the multitude of circular

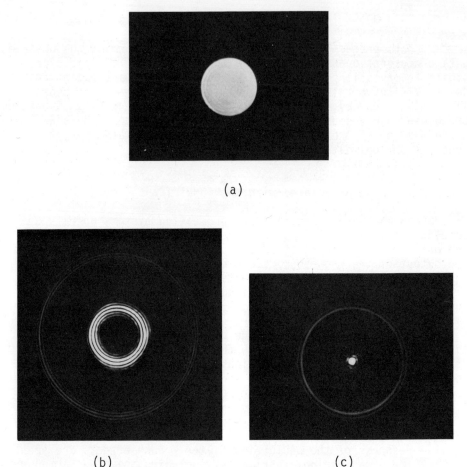

Fig. 4 (a) blur function for severe defocusing;
(b) intensity impulse response of the deblurring filter, (c) intensity distribution in the deblurred image of a point

noise pulses which overlap and obscure the restored image. Each such pulse is an impulse response of the deblurring filter. In general, noise such as this becomes more and more objectionable as the degree of attempted restoration increases.

Noise Associated with the Filter Transparency

Another source of noise in the optical processor is the frequency plane transparency. Usually, this transparency is recorded

NOISE IN COHERENT OPTICAL INFORMATION PROCESSING

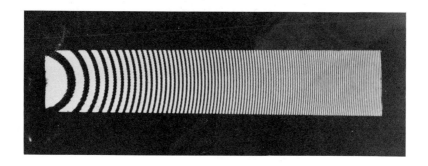

(a)

(b)

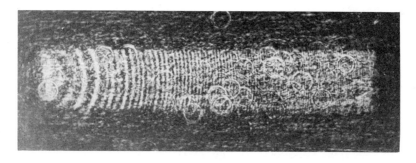

(c)

Fig. 5 (a) original object; (b) blurred object; (c) deblurred image

on very-high-resolution plates, and as a consequence we might assume that the amount of noise it introduces is quite negligible. In most applications this assumption is justified. An exception occurs in the application to image deblurring, for in this case the amount of light transmitted by the filter is often so small that scattered flux from the filter cannot be ignored. As we discuss below, the zero-frequency region of the filter plane may often be the prime source of noise.

The picture to be processed is usually represented in the processor as a positive and real-valued field distribution, corresponding to a positive and real-valued irradiance distribution in the blurred image. As a consequence of the blurring operation, the power spectral density of the blurred image is generally much more intense at low frequencies than at high frequencies. In addition, another phenomenon occurs which makes the relative intensity of the zero-frequency region of the spectrum depend on the size of the image field that is to be deblurred. Imagine that, at the input plane of the processor, we have a diaphragm which controls the extent of blurred image field. As we open this diaphragm, gradually enlarging the input field, the zero-frequency component of the image spectrum, which arises from the positive bias of the picture, increases in intensity as A^2, where A is the open area of the diaphragm. On the other hand, the fluctuations of the image about its positive bias may be regarded as a random process with some correlation area A_c. If $A \gg A_c$, then increasing the diaphragm area further will result in an increase in the intensity of the image spectrum (other than the zero-frequency component) in proportion to the product $A \cdot A_c$. Hence, as the size of the image field to be deblurred increases, the dominance of the zero-frequency component of the spectrum likewise increases as A/A_c.

For bleached holographic deblurring filters of the type introduced by Ragnarsson[12], the noise scattering efficiency of the filter is expected to be relatively constant across the frequency plane, and hence the vast majority of the scattered flux arises from the very strong zero-frequency component of the spectrum. In the case of the "sandwich" type deblurring filters, as pioneered by Stroke[14], the situation is somewhat more complex. Such filters attenuate the light by absorption, rather than diffraction, and hence are highly absorbing at the origin of the frequency plane. Since the noise scattering efficiency of an emulsion is a function of the average transmission, the tendency of the emulsion to scatter light varies across the frequency plane. We expect this type of filter to scatter less light than the bleached holographic filters, but this point has not yet been confirmed experimentally in the image deblurring context. Again, as the size of the image field increases, eventually the zero-frequency component will be the dominant source of scatter.

The experimentally observed effects of Fourier plane scattering are illustrated in Fig. 6. In this case the object was subjected to only a slight blur. Part (a) shows the deblurred image obtained with only a small input field; effects of Fourier plane scattering are not evident. However, in part (b) the size of the input field is increased, and the Fourier-plane scattering begins to appear in the background. The coarse structure of this noise indicates that it arose primarily from a small region in the frequency plane near zero frequency. These experiments were performed by Tichenor.

For bleached holographic filters, Fourier plane scattering becomes more and more important the larger the dynamic range of the filter in the frequency plane. The reason is quite simple: for a fixed image field, the amount of scattered noise flux is essentially independent of dynamic range, while the amount of transmitted "signal" light flux decreases with the square of the dynamic range of filter amplitude transmittance. Tichenor[13] has shown that for this type of filter, Fourier plane noise imposes a limit on the product of image size and the square of dynamic range of amplitude transmittance.

Computer-generated holograms are sometimes used as Fourier-plane filters, and these too give rise to Fourier plane scatter. In the case of binary filters, Fourier domain quantization noise is usually much more important than grain noise. The fact that most scattered flux arises from the intense zero-frequency component of the spectrum suggests that much finer quantization should be used in this region of the filter plane than in other regions.

Other Comments on System-Induced Noise

No discussion of noise in coherent optical processing systems would be complete without mention of a few additional facts. First, it is found experimentally that system-induced noise is strongest at very low spatial frequencies. Low frequency variations of the thickness of film base and of the phase fronts transmitted by liquid gates are no doubt partially responsible. This fact points out the advantage of using carrier-frequency filters, which deflect the system output away from the low frequency components of the Fourier plane scatter.

A second point worth emphasis is the fact that in a coherent optical processing system, the signal and noise amplitudes _interfere_ to produce the output intensity distribution. Thus, it is possible for a rather weak noise field to produce relatively strong noise fluctuations upon interference with a strong signal wave. As a consequence, the signal-to-noise ratio at the output of the processor is proportional to only the _square root_ of the ratio of the intensity

(a)

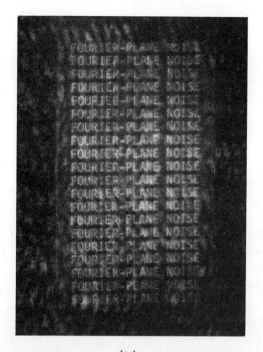

(b)

Fig. 6 (a) deblurred image of a small input field, (b) deblurred image of a large input field

of the signal field to the intensity of the noise field. This fact is largely responsible for the very significant role played by system noise in coherent optical data processing.

Finally, we mention that an additional source of noise is dust specks and dirt on various optical elements in the processor. A variety of clever techniques have been devised to suppress this type of noise. For example, if the input film and output film are moved in synchronism through the processor, all noise arising beyond the input plane will be at least partially suppressed.

CONCLUDING REMARKS

These days it is quite common to find discussions of the relative merits of digital computations, coherent optical computations, and even incoherent optical computations. There can be no doubt that the tremendous advantages of parallelism in optical data processing are gained only at the price of decreased accuracy. One of the reasons for this loss of accuracy is the system noise associated with optical processors. To arrive at a rational choice of data processing technology best suited for any given problem, it is necessary to have a solid understanding of the noise limitations associated with each technology. Because of the diversity of noise sources encountered in coherent optical processing, and because of their complicated dependence on various system parameters, it is not yet a simple matter for a potential user to assess the limitations he will encounter in practice.

ACKNOWLEDGMENTS

Partial support of the Office of Naval Research and the Stanford Joint Services Electronics Program is gratefully acknowledged.

REFERENCES

1. K. Preston, Coherent Optical Computers, McGraw-Hill Book Co., New York, N.Y. 1972.

2. G.W. Stroke, I.E.E.E. Spectrum, pp.24-40, December 1972.

3. A.B. Vander Lugt, Proceedings of the I.E.E.E. 62, 1300 (1974).

4. D. Middleton, Introduction to Statistical Communication Theory, McGraw-Hill Book Co., New York, N.Y. 1960.

5. J.W. Goodman, "Statistical Properties of Laser Speckle Patterns", in <u>Laser Speckle and Related Phenomena</u> (J.C. Dainty, ed.) Springer-Verlag, Heidelberg (in press).

6. L. Mandel and E. Wolf, <u>Rev.Mod.Phys</u>. <u>37</u>, 231 (1965).

7. F.D. Russell and J.W. Goodman, <u>J.Opt.Soc.Am</u>. <u>61</u>, 182 (1971).

8. J.F. Walkup, <u>Limitations in Interferometric Measurements and Image Restoration at Low Light Levels</u>, Ph.D. Dissertation, Department of Electrical Engineering, Stanford University, 1971 (University Microfilm Incorporated, Order No. 72-11685).

9. O. Bryngdahl and A.W. Lohmann, <u>J.Opt.Soc.Am</u>. <u>59</u>, 1175 (1969).

10. H.H. Barrett et al. "Fresnel Zone Plate Imaging in Radiology and Nuclear Medicine", <u>Opt.Eng</u>. <u>12</u>, 8 (1973)

11. D.A. Tichenor and J.W. Goodman, <u>Proceedings of the International Optical Computing Conference</u>, 82-84, April 1975 (I.E.E.E. Catalog No. 75 CH0941-5C).

12. S.I. Ragnarsson, <u>Physica Scripta</u> <u>2</u>, 145 (1970).

13. D.A. Tichenor, <u>Extended Range Spatial Filters for Image Deblurring</u>, Ph.D. Dissertation, Department of Electrical Engineering, Stanford University, 1974 (University Microfilm Incorporated, Order No. 74-27128).

14. G.W. Stroke and M. Halioua, <u>Phys.Lett</u>. <u>39A</u>, 269 (1972).

QUESTIONS AND COMMENTS

Dr. P. Tverdokhleb (question to Dr. J. Goodman)

a) Have you taken into account the noise statistics in filtering multiplicative noises?

Dr. J. Goodman. No, we have not.

b) Your method of filtering multiplicative noises needs realization of the image logarithmation. In order to take into account the noise statistics in matched filtering one should know in what way the initial statistics of these noises vary after logarithmation. Can you show how one or another law of distribution of noise probabilities vary after logarithmation?

Dr. J. Goodman. No, I cannot. It is a mathematical problem.

c) You have mentioned about the method of fabrication of a reconstructing filter in the process of one exposure. Has this method been proposed by you?

Dr. J. Goodman. No, it has not. The method mentioned was described by S. I. Ragnarsson in "Physica Scripta", 2, 145 (1970).

INFORMATIONAL CAPACITY OF COHERENT OPTICAL PROCESSING SYSTEMS

S.B. Gurevich

A.F.Ioffe Physical-Technical Institute

Leningrad 194021, USSR

Let us consider an image optical filtering system which is one of the coherent optical processing systems. Fig.1 shows that the monochromatic point-source light is collimated by the lens L_k, forming the plane wavefront in the input of the system on the plane P_1. An image to be filtered is put on this plane. The Fourrier-transform lens L_1 is placed at a distance "f" from the plane P_1. In the so-called frequency plane P_2 the spatial image spectrum is distributed with an accuracy to some constant. This spectrum can be modified by a filter placed at the same plane. The lens L_2 produces the second Fourrier-transform of the filtered spectrum and generates a processed image at the output plane P_3.

Our purpose is to evaluate the information capacity of the system as well as the effect of different sources of information loss.

It should be noted that the coherent information processing systems have a number of features in common with non-coherent lens image forming systems and holographic systems.

The following features of these three systems are qualitatively compared in Table 1:
1) character of the light source,
2) character of the transform performed by the system,
3) presence or absence of the lens,
4) presence or absence of the intermediate recording,

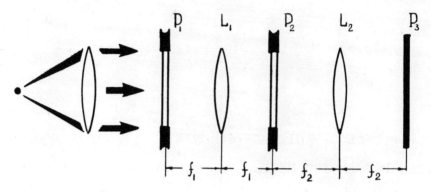

Fig.1 The optical image filtering scheme.

5) time of the object image transmission,
6) information contained in the modulated beam,
7) character of light modulation.

The features listed in the table are arbitrary to some extent. It should be noted that the presence of both the lens and holographic systems in coherent optical information processing system leads to the simultanious action of the sources of information loss inherent in both systems, and therefore leads to a larger loss of information. However, if this fault were not compensated by other advantages (for example, a possibility to increase the volume of transmitted useful information at the expense of unnecessary information, and the reduction of redudancy etc), these optical processing systems would not have been appreciated as highly as they are now.

Let us consider the information characteristics of the lens systems (1).

If every element of a group of objects cannot reflect (pass) light flux over a certain value, one can calculate the ability of an objective to pass information, i.e. the maximum of information passed through an objective per time unit provided the number of object characteristics is known. If the information passing time is limited by the time when the object is in a stable state, by the exposure time, by the frame time of the TV tube or by some other time, one can calculate the information capacity of the objective in this

COHERENT OPTICAL PROCESSING SYSTEMS

TABLE 1

	Light Source	Lens	Intermediate Recording	Time of Object Image Transmission	Information contained in the modulated beam in the plane P_2	Modulation of the beam in the plane P_2
Noncoherent lens image systems	Noncoherent	+	−	Real time	+	−
Holography systems	Coherent	−	+	Nonreal time	−	+
Optical Processing systems	Coherent	+	+	Real time	+	+

system. This value would indicate the maximum volume of the information that the objective is capable to transfer to the input of the next system link in a given time under the conditions of the limited light energy in the objective aperture.

The lens-produced information loss affects two components of information. The lens aperture passes the limited light flux through the system thus reducing the volume of the gradational component of the information; the reflection and scattering in lenses produce the same effect. On the other hand, the aperture limitation and similar attending effects result in the loss of spatial information depending on reduction of the bandwidth of the spatial frequencies to be transferred. This effect increases in the direction from the centre to the periphery of the image field. The spatial information loss and the irregularity of its transmission increase due to various aberrations in an objective. When the influence of factors leading to the frequency characteristic irregularity is negligible, the objective information capacity is determined by:

$$I_c = S I'_c = \int_0^{N'_p} \log_2 [m(\Omega)+1] d\Omega = S \int_0^{N'_p} \log_2 [m_0 \delta(\Omega)+1] d\Omega,$$

where S is the area of the objective image field, I'_c = the information capacity for the square unit, N'_p = the maximum number of elements, in which the image field square unit can be decomposed, $m(\Omega)$ = the dependence of the number of resolvable gradations on the two-dimensional object image spatial frequency Ω. $\delta(\Omega) = m(\Omega)/m(0)$ = is the gradational-frequency characteristic, this depends on the contrast-frequency

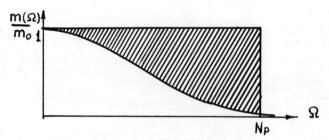

Fig.2 The information loss due to the character of contrast variation. The upper part is the loss.

characteristic and also on the variation of signal-to-noise ratio with frequency.

The spatial information loss H_{aps} is determined by the following relation:

$$H_{aps} = S \int_{N_p'}^{0} \log_2 [\delta(\Omega) + 1/m_0] \, d\Omega .$$

Fig.2 shows the information loss due to the nature of the dependence $\delta(\Omega)$.

In practice, when the intensity and frequency-contrast characteristics depend on the distance to the image field centre, the value of "m" happens to depend not only on the spatial frequency Ω but also on the distance to the field centre. The objective information capacity is then determined by the following:

$$I_c = 2\pi \int_0^R \tau \, d\tau \int_0^{N'} \log_2 [m_0(\tau) \, \delta(\Omega,\tau) + 1] \, d\Omega .$$

The characteristics $m_0(\tau)$ and $\delta(\Omega,\tau)$ can be calculated or found experimentally, and then the information capacity of the objective forming an image can be evaluated.

In the spectral plane P_2 of the scheme of Figure 1 the modulated light passes through a filter that can be a hologram. So it is of interest to consider the process of passing of information through a holographic system (2, 3). There is a phase and amplitude distribution recorded in a hologram. When the reconstructive wave has passed through the hologram an intensity distribution appears in the output of the image plane which varies with the coordinate of this plane and with direction in which the wave of the given intensity passes this plane.

If we take the value of intensity for all the directions and elements to refer to one of the equipossible "m" intervals, the informational capacity I_c of the holographic system is determined by the following expression:

$$I_c = N_\ell N_\alpha \log_2 (m+1) \simeq N_\ell N_\alpha \log_2 m ,$$

where N_ℓ is the number of the spatial elements (along the axes X and Y), N_α = the number of angle directions from one element, in which intensities are distinguishable.

The information contained in the output image is qualitatively more abundant than that contained in a photographic image because it includes not only the planary spatial and gradational information but also the directionally spatial information. Naturally, this does not mean that the hologram recorded on the same photographic material in amplitude record alone should contain more information than the photographic plane image. The limit of information capacity of amplitude-recording material should not depend on the way of recording (holographic or lens), while depending only on the type of the contrast-frequency characteristics, the signal-to-noise ratio and the dynamic range of the material. Both recording material and lens may limit the volume of information transmitted by the system if the contrast-frequency characteristics are not good enough. However, it should be noted that new recording materials having an increased information capacity are being produced. For example, it is shown that the resolution of chalcogenide amorphous semiconductor films is 10000 lines/mm which corresponds to the value I_c at m = 1 being more than 10^{10} bit/cm^2 for binary record and almost to 10^{11} bit/cm^2 for half-tone record generally, the resolution can be more than that but we cannot use it because of the information capacity of light flow being less. One can show that if the gradation information is not taken into account, the light flow can transfer information which does not exceed the value of 10^9 bit/cm considerably. In holography this value is still less because of information loss due to the presence of many orders for difracted beams. If we consider now the light field in the frequency plane of Fig.1 we can notice that information about the object is contained only in a certain region of the field, the most abundant information being contained in the central region of the field. The peculiarity of the light field is that in the frequency plane it is axially symmetrical, hence only a half of the field signal transferring contains information (Fig.3).

The lens aperture limitation leads to the establishing of the maximum object field sizes, where all the object spatial frequencies are transferred. Let us suppose the object size in one of the dimensions (for example, X) to be equal l_x. The number of resolvable elements (of size Δx) in this direction is $N_x = N'_x l_x$, where N'_x is the number of elements on the object length unit, which is equal to the double maximum spatial frequency in this direction $N'_x = 2 \nu_{x \, lim}$. The limit angle including these frequencies is determined by the expression: $\sin \alpha_{lim} = \lambda \nu_{lim}$. If the object is centered with respect to the lens axis, the lens aperture is "D", and the object is placed in the lens frontal focal plane, then we have:

$$(D - l)/2f = tg \alpha_{lim} \approx \sin \alpha_{lim} = \lambda \nu_{lim},$$

hence, $D - l \approx 2f \lambda \nu_{lim}$.

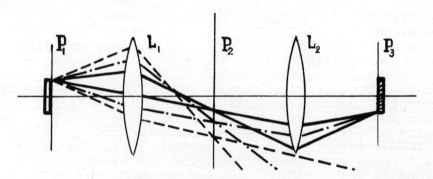

Fig.3 The passage of different spatial object frequencies through an optical channel.

COHERENT OPTICAL PROCESSING SYSTEMS

An ideal lens (without aberrations) transferring all the object frequencies must have an aperture size larger than that of the object by $2f\lambda\nu_{lim}$. For example, if we have to transfer $\nu_{lim} = 250\frac{lines}{cm}$ and $f = 30$ cm, $\lambda = 6{,}3 \cdot 10^{-5}$ cm, we have $D - 1 \approx 1$ cm.

The object size limitation on the value of "1" leads to the fact that the resolvable value $\Delta\nu$ cannot be less than $\frac{1}{2} l_x$ because $N_x = 2\nu_{x\,lim} l_x$. Therefore, if N_x and $\nu_{x\,lim}$ are given, l_x cannot be less than $l_x = \frac{N}{2\nu_{x\,lim}}$. Hence, if we suppose N_x to be 1000 elements/line, $l_x = \frac{1000}{2 \cdot 250} = 2$ cm, and the required lens size is $D = 3$ cm. The given volume of information of the object to be transferred is:

$$I_{c_1} = N_x N_y \log_2(m+1) = 4\nu_{x_1 lim}\nu_{y_1 lim} l_{1x} l_{1y} \log_2(m+1) =$$
$$= 2\Omega_{lim} S_1 \log_2(m+1).$$

Here we suppose the gradation transmission over the whole field to be regular $m(x, y) = $ const, and

$$\nu_{x\,lim}\,\nu_{y\,lim} = \frac{\Omega_{lim}}{2} \quad \text{and} \quad l_x l_y = S.$$

Provided the loss is absent the same information volume will take place in the frequency plane:

$$I_{c_2} = N_x N_y \log_2(m+1) = l_{x_2} l_{y_2} 4\nu_{x_2 lim}\nu_{y_2 lim} \log_2(m+1).$$

The first pair of multipliers:

$$l_{x_2} l_{y_2} = \lambda^2 f^2 \nu_{x_1 lim}\nu_{y_1 lim}.$$

The second pair of multipliers is expressed as:

$$\nu_{x_2 lim}\nu_{y_2 lim} = \frac{N'_{x_2}}{2} \cdot \frac{N'_{y_2}}{2} = \frac{1}{4} \cdot \frac{1}{\Delta x_2} \cdot \frac{1}{\Delta y_2} = \frac{1}{4\lambda^2 f^2 \Delta\nu_{x_1}\Delta\nu_{y_1}} = \frac{l_{x_1} l_{y_1}}{\lambda^2 f^2}$$

because $\Delta\nu_{x_1} = \frac{2}{l_{x_1}}$, and $\Delta\nu_{y_1} = \frac{2}{l_{y_1}}$.

Therefore:

$$I_{c_2} = \nu_{x_1 lim}\nu_{y_1 lim} = 4 l_{x_1} l_{y_1} \log_2(m+1) = I_{c_1}.$$

So the information capacity is invariable, but the components interchange: the object field size determines the bandwidth of the part of Fourrier-image in the frequency plane, and the object spectrum determines the size of the field in this plane. And although the value of transferred information is invariable, the density of information in the frequency plane differs appreciably from that in the object plane: if m = I in the second case the density of information is

$$I'_{c_1} = N'_x N'_y = 4 \nu_{x,lim} \nu_{y,lim},$$

but in the first case $I'_{c_2} = \ell_{x_1} \ell_{y_1} / \lambda^2 f^2$. Therefore the density of information depends on the ratio of the bandwidth and the object size, and also on the chosen scale (coefficient λf).

The limitations on the volume of the transferred information are introduced not only by the lenses but also by the spatial filter which is situated in the frequency plane. The information capacity of the filter also gives the limitations for the whole system. But one should remember that the additional loss of information could be introduced by the lack of information component accordance, in particular multipliers, which determine the geometrical field sizes and the frequency bandwidth. The filter material has a certain frequency-gradation characteristics and limited sizes which have to be made in accordance with the appropriate field parameters. Fig.4a shows the following cases: 1 - the density of the field binary information for both components (X & Y) is less than the density of the binary information that could be recorded in the filter in this case the filter is in a certain excess: in this system its capacity is not completely full. 2 - the density of the field binary information is more than that could be recorded in the filter, the filter lets only a part of the lower frequency information contained in the light field to pass, while the remaining information is lost. Fig.4b shows the case of the total uniformity of density of the field information and the filter material, but X- and Y- components have different densities: X-density component in the field is larger than that in the filter, and Y-component of the density in the filter is larger than that in the field. Therefore, the part of X-component information is lost, and the part of the Y-component filter is not occupied by the information. This occurs if sizes of the field in the field in X- and Y-directions are different, and this

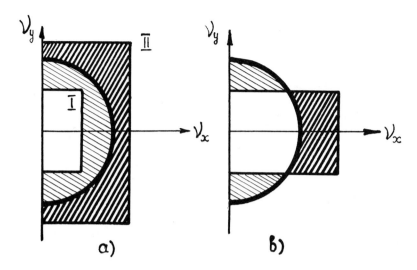

Fig.4 The match of the spatial filter characteristics with the characteristics of the light field in the frequency plane.

produces the difference in the density of the information in the field in X-and Y-directions while the recorded information density in the filter material is, generally, anisotropic.

The limitation on the information capacity can be also introduced by the limitations on the filter size, though, as a rool, it is easier to avoid such limitations than the frequency limitations. In addition to the frequency and the field limitations of the information capacity of the system, which are related to the characteristics of the lens and the filter of the system, there can be considerable limitations caused by the noise, and especially by the filter noise.

One of the features of the noise generation in this system is the illumination of the noising material by the coherent light. It is known that if the illumination is coherent, the signal and the noise are added by an amplitude but not by the intensity. Hence, the average intensity in the image plane is the sum of the average values of the signal and the scattered background:
$$\overline{J}_i = \overline{J}_{si} + \overline{J}_{bi}$$

If the fluctuations of I_{si} could be neglected as compared with the fluctuations of I_{bi}, one can write (4) an expression for the signal-to-noise ratio ψ :

$$\psi_i = \frac{\bar{J}_{si}}{\bar{J}_{Ni}} = \frac{\bar{J}_{si}}{\bar{J}_{bi}\sqrt{1 + 2\bar{J}_{si}/\bar{J}_{bi}}} .$$

Hence, $\psi_i = \bar{J}_{si}/\bar{J}_{bi}$, if $\bar{J}_{si} \ll \bar{J}_{bi}$ and if $\bar{J}_{si} \gg \bar{J}_{bi}$, $\psi_i = \sqrt{\frac{\bar{J}_{si}}{2\bar{J}_{bi}}}$.

The value of I_{bi} can be both calculated and found experimentally. If one introduces the φ function - the Wiener spectrum, $\bar{J}_b = J_{il} \, \varphi \, S_f \, (\lambda^2 f^2)^{-1}$, where I_{il} is the input intensity (the intensity of the field in the frequency plane, where the filter is also situated), S_f = the area of the filter. But, on the other hand,

$$\bar{J}_s = J_{il} \, S_\varphi \, \eta / S_{im} ,$$

where η is the filter diffraction effectivity, S_{im} = the area of the image. Therefore:

$$\psi = \frac{1}{\sqrt{2}} \sqrt{\frac{\bar{J}_s}{\bar{J}_b}} = \sqrt{\frac{\eta \lambda^2 f^2}{2 S_{im} \varphi}} .$$

If $\varphi = 10^{-8} mm^2$, $\eta = 10^{-2}$, $\lambda = 6,3 \cdot 10^{-4} mm$, $f = 150$ mm, $S_{im} = 30 \times 30 \approx 10^3 mm^2$, we shall get the value $\psi \simeq 2,2$. It is readily seen that if we use a higher - noise material with a lower diffraction effectivity or if we take a larger number of elements in the output image (hence in the input also), the signal-to-noise ratio will become so small that the details of the given size will not be resolvable. In this case we shall have the loss of both spatial and gradation information but the gradation information loss (though to a lesser degree) will take place also when the signal-to-noise ratio remains above "1".

It is worth mentioning that we have not taken into account the effect of nonlinearity of the material and the effect of the phase irregularities of the filters used. The examples considered show the importance of the matching the filter information characteristics with the requirements for the problem in question and with the parameters of the optical filtering system used. The same evaluation can be made for other optical processing systems.

References

1. Gurevich S.B. "Effectivity and sensibility of T V systems" Moscow-Leningrad, "Energy", 1964.

2. Gabor D. "Progress in Optics", 1961, 1, p.109.

3. Gurevich S.B., Sokolov V.K. Limited information capacity of holographic systems. JTP (USSR), 1973, v.43, p.675.

4. Goodman J.W. Film-grain noise in wavefront-reconstruction imaging. JOSA, 1967, v.57, 4, p.493-502.

EXTENSIONS OF SYNTHETIC APERTURE RADAR INFORMATION PROCESSING

Winston E. Kock
Visiting Professor and Director
The Herman Schneider Laboratory
The University of Cincinnati

ABSTRACT

Various extensions of the holographic synthetic aperture concept are discussed, including the use of recently developed electronic devices to permit real-time operation of synthetic aperture systems and the use of circular polarization whereby shorter wavelength waves can be employed without serious masking of the desired target echoes by rain reflections.

INTRODUCTION:

One of the earliest practical applications of hologram concepts occurred in radar, and the success of this technique depends heavily upon optical information processing. We here discuss several new developments in this special radar field.

In a synthetic aperture radar system (1), an aircraft moving along a very straight path continually emits successive microwave pulses. The frequency of the microwave signal is very constant (the signal remains coherent with itself for very long periods). During these periods the aircraft may have traveled several thousand feet, but because the signals are coherent, all the many echoes which return during this period can be processed as though a single antenna as long as the flight path had been used. The effective antenna length is thus quite large and this large "synthetic" aperture provides records having extremely fine detail.

The photographic record of the echoes received by such a coherent radar is a form of hologram with the microwave generator which provides the illuminating signal also providing a reference wave. The

reflected signals received along the flight path are made to interfac
with this reference signal (by synchronous detection) and the complex
interference pattern thereby generated and photographed is a form of
hologram. A detailed description of this technique and of several
recent developments can be found in reference (2).

The optically-processed hologram yields photos of extremely good
detail. During 1971 and 1972, the entire Amazon River basin was
recorded (2); Fig. 1 is an example of a (reconstructed) record made
during this mapping project.

When this radar concept was first described, many could not unde
stand how its high resolution could be maintained for all objects.

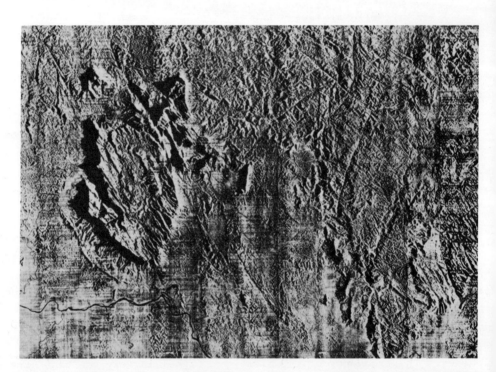

Fig. 1 One of many reconstructed synthetic aperture radar records
made during a 1971 radar survey of the entire Amazon River
basin, (3 million square kilometers), conducted by the
Aero-Service Corporation of Litton Industries under the
supervision of Homer Jensen. It shows the area near the
town of Esmerelda in the state of Amazonas in Venezuela,
the river in the lower left hand corner being the upper
Orinoco. The radar, was a goodyear Aerospace X-band (10
GHz) one, and the word used for the photo is "Mosaico de
Radar." (Courtesy Homer Jensen).

SYNTHETIC APERTURE RADAR INFORMATION PROCESSING

Because of the great length of the synthetic aperture antenna, reflecting objects recorded by it are in its near field, not in the distant, or Fraunhofer region, where most radars and sonars operate. For a near-field reflecting object, radars or sonars having large aperture receiving arrays can only achieve maximum efficiency and resolution when the time delays inserted in the receiving element circuits are adjusted so as to cause the element positions to correspond to being arrayed along the arc of a circle centered on that object. Yet for other near-field objects, at other ranges, all of these delays would have to be changed so as to correspond to arcs of other circles, centered on these other points. How the optical processor could cause all points in the near field, at any distance and at any angle, to be in sharp focus, was beyond comprehension to many. When it is recognized, however, that the radar record is a hologram, the answer becomes quite clear. In holography, each small light-reflecting point generates its own zone plate, and each zone plate later causes coherent laser light to be reconstructed exactly at the point in space from whence the light emanated. Similarly, a synthetic aperture radar captures photographically the curved wave fronts emanating from reflecting points by combining them with a reference wave and thereby generating (one dimensional) zone plates. Later, as in an optical hologram, the coherent light used in the reconstruction process acquires, through diffraction by the zone plates, the properly curved wave fronts to concentrate the light at points corresponding to the reflecting points in the original landscape. Just as an optical hologram causes each point of a three-dimensional scene to be brought into sharp focus no matter what its distance from the photographic plate, so the synthetic aperture hologram record is responsible for the good focus of all of its reconstructed points.

PARALLEL PROCESSING ASPECTS OF SYNTHETIC APERTURE SYSTEMS

As noted above, the great equivalent length of the synthetic (linear) aperture permits many extremely sharp beams to be generated. In most radars which have such large apertures, it is an extremely complicated procedure to cause all of these beams to be formed simultaneously. Instead, <u>one</u> sharp beam is usually formed and this beam is scanned (serially) over the many possible directions in which it can be pointed. In the synthetic system, all beams are automatically formed, (and information is accepted by all) simultaneously.

Any optical or radar aperture whose dimensions extend over many thousands of wavelengths, has a near field which extends outward to very great distances. For such an aperture to have maximum effectiveness in delineating the large number of objects in its near field, the elements must be phased so as to exhibit a curved wave-front property. This curvature is very great for near objects, is less so for mid-range objects and becomes flat for far field objects. For the standard form of radar, such a phase adjustment procedure, differing for each element of range, would be a very complicated one, and if possible at all, it would surely be done in a serial manner. In the synthetic system, such curvature is automatically introduced through the zone plate action. Information is thus accepted from all range points in a highly efficient manner, and in parallel (3,4).

In certain radar applications it is desirable to have the metric (not the angular) resolution constant for all ranges. For this to occur, a much larger aperture must be utilized for more distant point (thereby providing a much higher angular resolution) than for near points. In an ordinary radar, the procedure to accomplish this is again a very complicated one. In synthetic aperture systems, this change of aperture length for the various ranges also occurs automatically because of the finite (small) aperture size of the actual antenna employed. Its constant angular resolution versus range (usually tens of degrees) causes its metric resolution to be less at great ranges. This provides longer zone plates for the more distant objects and shorter ones for the nearby objects. The result is a metric resolution which is independent of range. Parallel acceptance of information is thereby automatically provided, thereby achieving the desired result.

Finally, in the accepting function, it is of interest to note that the beam width of a synthetic aperture is twice as sharp as that of a standard radar of equal aperture.

The processing of the received data in a snythetic aperture system can be considered to be, as in a hologram, the photographic recording procedure, which results in myriads of superimposed zone plates, and in the developing in parallel, of all of the zone plates in that photographic record.

As in the holographic reconstruction of a scene, the display function of a synthetic aperture system is accomplished by the illumination of the photographic record with laser light, causing the parallel displaying of myriads of objects located at many different points in range.

Real-Time Reconstruction

In holography and synthetic aperture systems, there is a time delay because of the requirement to develop and reconstruct the hologram. Several electronic systems have recently been developed in which microwave (or ultrasonic) holograms can be recorded and quickly reconstructed in real-time. An experimental version of one such system is shown in Fig. 2. The equipment consists of a 0.05 cm thick deuterium-KDP (DKDP) crystal, an off-axis scanning electric gun, and associated optics. In operation, the scanning electron beam is modulated with the holographic information, so that a hologram is

Fig. 2. An experimental model of a DKDP reconstructor tube for real-time viewing of acoustic holograms (Bendix Research Laboratories).

written on the crystal in the form of a positive charge pattern. The electric field within the DKDP varies over the crystal according to the holograhic signal, thus modulating its refractive index by the electrooptic effect (5). Coherent light transmitted through the DKDP crystal becomes modulated with, that is, it reconstructs, the holographic information. The hologram is periodically erased by flooding the crystal with electrons with an appropriate potential on a nearby grid. In early models, the real-time capabilities were of the order of 16 frames per second.

Just as the hologram (synthetic aperture) procedure can be considered as a (parallel) optical processing computing procedure, this real-time tube can also be considered an optical computer.

The one-dimensional zone plates of a synthetic aperture record can also cause laser light to be brought to a focus on a plane of reconstructed real images. The focal <u>region</u> of each such zone plate is actually a circle but because of the directionality of the plane-wave, laser, illuminating light, the point of that circle where it intersects the tilted plane is very bright. These focal points correspond exactly to the range and azimuth of the reflecting points which generated the zone plates. Accordingly, when the radar information normally recorded on a hologram photographic plate is recorded (electronically) on the DKDP crystal, the (changing) hologram information can be repeatedly recorded and reconstructed at sufficiently rapid repetition rates (e.g. 16 to 30 times per second) to provide a real-time visual readout.

In this radar procedure, successive real-time views of the terrian would be available to the aircraft pilot even during heavy fog or cloud cover. By aiming radars on both sides of the aircraft toward the forward portion of the normal side-looking view, the all-weather radar views could prove useful in commercial aviation applications.

The use of the synthetic aperture procedure in ultrasonic medical diagnosis has recently proved to be promising, yielding resolutions an order of magnitude better than those of the present day sonar type ultrasonic diagnostic devices, commonly referred to as B-scan devices (7,8). Here again, however, the inability to operate in real-time is a drawback and the real-time technique just described could nullify that disadvantage (9). Fig. 3 shows two binary inputs (two square waves) applied to the optical computer of Fig. 2, and Fig. 4 shows the reconstruction of the corresponding (spatially separated) points. I am indebted to R. Mueller and G. G. Goetz for these two photos.

Fig. 3. Binary (square wave) modulation of the tube of Fig. 2

Fig. 4. Reconstruction of the spatial pattern recorded in Fig. 3

SYNTHETIC APERTURE RADAR INFORMATION PROCESSING   123

Circular Polarization

It has been known for some time that a radar set operating with circular polarization can be used in such a way that the unwanted echoes from symmetrical objects such as raindrops can be suppressed (10, 11). Circular polarizers are also used in optics where they are valuable in reducing specular backscatter.

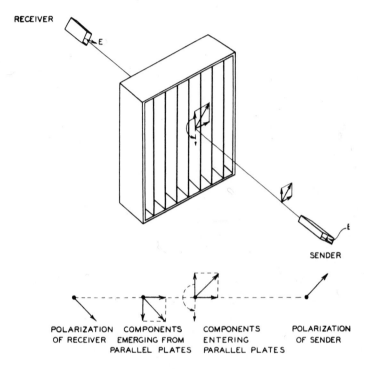

Fig. 5. Rotation of polarization by means of a metal plate waveguide structure.

The rejection of backscattered circularly polarized electromagnetic waves can be described with the help of Fig. 5, showing a metal-plate microwave-waveguide polarization rotating device (12). The vertically polarized component of the transmitted waves experiences a phase velocity which is higher than that of the (unaffected) horizontally polarized component, causing the polarization of the emerging waves to be rotated by $90°$. If the thickness of this half wave plate is reduced by one half, the emerging waves are circularly polarized and upon symmetrical reflection, their second passage through the quarter wave plate will again yield the full 90° rotation shown in the figure. Because these waves now possess a linear polarization which is oriented 90° to the transmitter-receiver polarization, they cannot enter the receiver. Signals reflected from non-symmetrical objects, will, on the other hand, experience, on reflection, varying degrees of depolarization; they therefore will possess a component which can enter the receiver. This effect is shown, for a microwave radar, in Fig. 6.

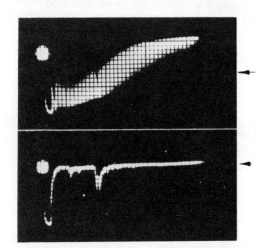

RAIN ECHO

Without λ/4 Plate
  Sweep: - 33 small div./mi

With λ/4 Plate
  Sweep & Gain same as above.

Fig. 6. Suppression of rain echoes in a microwave radar by changing the outgoing signal polarization from linear to circular. The plot is amplitude (vertically downwards) versus range horizontally to the right; top, linear polarization, bottom, circular polarization.

If it is desired to enhance the echoes from raindrops, separate transmitting and receiving antennas can be employed equipped with oppositely oriented quarter wave plates. The symmetrically reflected signals then acquire the polarization needed for reception. In this arrangement, echoes from raindrops are enhanced in strength relative to echoes from non-symmetrical objects.

Fig. 7. Laser radar record, left, of the control tower, right (United Aircraft Research Laboratories).

Figure 7 portrays a laser radar pattern of an airport tower (courtesy of the United Aircraft Research Laboratories). The high detail here recorded, as compared to that obtained by a standard microwave radar, is evident. However, with such optical radars, atmospheric effects (including even light fog) can seriously affect the reflected patterns. If the ratio of the size of fog droplets to rain drops is comparable to the ratio of the size of light wavelengths to radar wavelengths, the use of circular polarization here should yield a result comparable to that shown in Fig. 6. This would permit objects otherwise obscured by fog or mist to be made more visible.

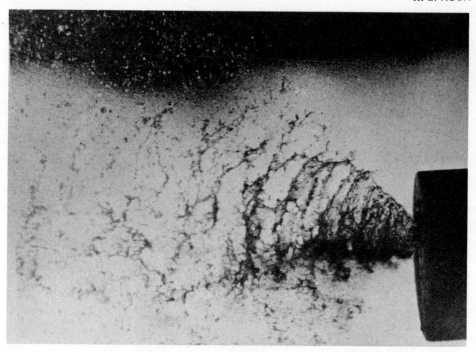

Fig. 8. Reconstructed hologram of a liquid propellant spray in a rocket engine (U.S. Air Force Systems Command).

Fig. 8 is a hologram reconstruction of a spray of droplets (13). It portrays the spray pattern of a liquid propellant in a rocket engine. The size, sphericity, and distribution of propellant droplets are important factors governing the performance of the engine, and they are difficult variables to measure and predict. Since two sets of circularly polarized waves having opposite rotation directions cannot generate an interference pattern, the use for the illumination and reference waves in holography, of circularly polarized waves, can cause symmetrical objects (such as small droplets) in the recorded scene to be either suppressed or accentuated. In either case, those objects which _are_ displayed will appear, as in a normal hologram, in full three dimensions (14).

In the forward-looking, real time, synthetic aperture radar just described, the reduction in rain reflection made possible by circular polarization would permit shorter microwave wavelengths to be used with their accompanying higher resolution.

References

1. L. J. Cutrona, E. N. Leith, L. J. Porcello, W. E. Vivian, "On the Application of Coherent Optical Processing Techniques to Synthetic Aperture Radar," Proceedings of the IEEE, vol. 54, no. 8, pp. 1026-2031, (Aug. 1966).

2. W. E. Kock, "New Forms of Ultrasonic and Radar Imaging," in ULTRASONIC IMAGING and HOLOGRAPHY, (pp. 287-344), Ed. G. W. Stroke, W. E. Kock, Y. Kikuchi, and J. Tsujiuchi (Proceedings of the January 7-13, 1973 U.S.-Japan Science Cooperation Seminar on Pattern Information Processing in Ultrasonic Imaging) Plenum Press, 1974.

3. W. E. Kock, "Optical Computing in Synthetic Aperture Radar," Proc. S.I.D., vol. 15/3 Third Quarter, 1974, pp. 112-118.

4. W. E. Kock, "Parallel Processing Aspects of Synthetic Aperture Hologram Techniques," in NEW CONCEPTS and TECHNOLOGIES in PARALLEL INFORMATION PROCESSING (pp. 51-54), Ed. E. R. Caianiello (Proceedings of the June 17-30, 1973 NATO Advanced Study Institute, Capri, Italy) Noordhoff International Publishing (Leyden), 1975.

5. G. G. Goetz, "Real-time Holographic Reconstruction by Electro-Optic Modulation," Appl. Physics Lett. 17, 63 (1970).

6. W. E. Kock, "A Real Time Parallel Optical Processing Technique," IEEE Transactions on Computers, vol. C-24, No. 4, April, 1975, pp. 407-411.

7. C. B. Burckhardt, P.A. Grandchamp, and H. Hoffmann, "An experimental 2 MHz synthetic aperture sonar system intended for medical use," IEEE Trans. Sonics Ultrason, vol. SU-21, pp. 1-6, Jan. 1974.

8. W. E. Kock, "Synthetic aperture and liquid surface ultrasonic holography," presented at the NATO Advanced Study Inst. Ultrasonics in Medical Diagnostics, Milan, June 1974, Prof. E. Camatini, Ed., to be published in Plenum Press.

9. W. E. Kock, "ENGINEERING APPLICATIONS OF LASERS AND HOLOGRAPHY," Plenum Press, 1975.

10. J. Brown, "Microwave Lenses," Methuen Publishing (London) 1953.

11. W. E. Kock, "Sound Waves and Light Waves, "Doubleday, 1965, pp. 71-72.

12. W. E. Kock, "Related Experiments with Sound Waves and Electromagnetic Waves, "Proc. I.R.E. vol. 47, 1959, pp. 1192-1201.

13. Laser Focus, June 1970, p. 20.

14. W. E. Kock, "Circular Polarization in Certain Laser and Holography Applications," APPLIED OPTICS, July 1975.

CONTROLLED TRANSPARENCIES FOR OPTICAL PROCESSING

I. N. Kompanets, A. A. Vasiliev, A. G. Sobolev

ABSTRACT

Formation and transformation of information represented in digital or analog form by means of controlled transparencies is the basis for construction of memory elements, optical processors and other data processing and computer assemblies. Electro-optic ceramics is a promising material for some controlled transparency applications.

Results on the study of ferro-ceramic materials for creation of the controlled transparencies are reported. Transverse electro-optics effect, fringe effect and asymmetric deformation effect are used in the paper. Controlled transparencies based on nematic liquid crystals and photoconductor-liquid crystal structures are investigated. Application of controlled transparencies in optical coding and recognition of digital and analog data, and in fulfillment of logic operations, is considered.

CONTROLLED TRANSPARENCIES FOR OPTICAL PROCESSING

Optical signal control devices, such as modulators, light deflectors and, particularly, multi-channel controlled transparencies (CT) play an important role in the systems of optical data processing. Being mostly passive-type arrangements, they provide phase or amplitude spatial-time modulation of the light beam.

Formation and transformation of two-dimensional blocks of information in digital or analog form, realized by means of controlled transparencies, constitute the basis for optical memory systems and other data processing and computer asemblies. Fig. I shows various CT applications, such as transparencies controlled both by

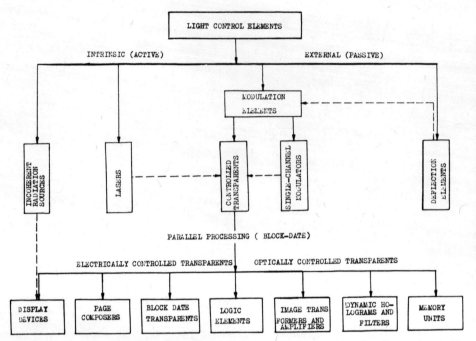

Fig. 1. Controlled transparencies as light control elements and their applications (scheme).

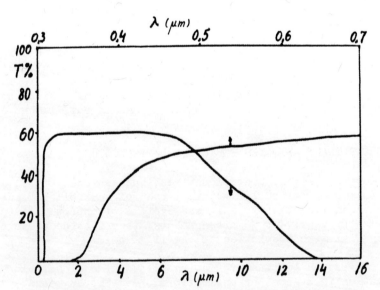

Fig. 2. 100 mcm ceramic sample transmittance versus light wavelength.

electric (CTE) and optical (CTO) signals: display devices (including those of the projection type), page composers, digital information converters, logic elements, coding and decoding image arrangements, transformers (of frequency, coherence) and image amplifiers, dynamic holograms and filters, blocks of optical memory etc.

Work on the creation of controlled transparencies is under consideration in many laboratories. The investigation results have been reported in numerous working including special issues /I-3/ and reviews /4,5/. There the problems are treated on the study of materials for CT, technique of transparency control, construction and technology peculiarities of CT, and optical schemes of connection with the other arrangements. As practice shows, there are some common problems to be solved for various types of CT.

The choice of the working material for CT is the basic problem in the creation of CT. At present no materials are known which satisfy the demands such as high sensitivity to the control signals and resolution, high efficiency and depth of the light modulation, fast operation and existence of essential read-write memory. However, some materials are found now which are attractive candidates for application in CT devices. They include, for example, nematic liquid crystals, electrooptic ceramics and some electrooptic crystals.

The mockups of experimental CTO arrangements are carried out on the basis of nematic liquid crystals applying the thermal effect /6/ and the change of conductivity of the photosensitive semiconductor in contact with a liquid-crystal film /7/. Note some interesting work on the creation of light-valve arrays, individual addressing by means of thin-film transistors /8/ and matrix addressing due to the reduction of the element switching time /9/. Thus, the limitations connected with the relaxation-type memory of liquid crystals and their small fast-operation might be avoided.

We have also studied peculiarities of the effects of electric field on nematic crystals (NLC) for transparencies /IO/. The ways to increase fast-operation of NLC were considered, holographic writing by means of experimental CTE mockups was realized and image transformation with the help of CTO on the basis of the liquid-crystal-photosemiconductor structure was performed. Microseconds of electrooptic switching time in NLC were obtained. Large-sized liquid crystal CTE and CTO systems are under consideration at present.

Electrooptic ceramics are promising materials for application in memory and fast-operation CT systems /II/. In 1974 we initiated the studies of ceramics in the segnetoelectric phase, and began realization of CTE arrangements on this basis /I2/. Some of the results obtained are reported in the first section of this paper. Another part is devoted to the problems of CT application in optical processors for coding, recognition of digital and analog information and for logic elements. We thus emphasize the importance of this prospective field of controlled transparency applications.

Optically transparent segnetoceramics are, at present, the most promising materials to be applied in optical processing technique /II, 13/. These materials in the polycrystalline form comprise a hot pressed solid solution of lanthanum doped lead zirconate titanate (PLZT). Crystal structure of the PLZT ceramics is described in /I4,I5/. This piezoceramic has such distinctive electrooptic properties as induced birefringence and scattering. Dependence of some physical properties on the content and technology of ceramic preparation is reported in /I3, I6-I8/.

In controlled transparencies, one may apply both the effect of birefringence change occuring with a fine grain ceramic (grain size greater than 2 mcm), and the effect of light scattering in a coarse grain ceramic. The study of materials for this purpose and construction of devices is reported in detail in /II, I3/. Much attention has been paid to the longitudinal electrooptic effect applied in a strained segnetoceramic /I9, 20/. This effect is characterized by high contrast ratio (up to 4000:I), high efficiency of the electrooptic switching, and it may be applied in matrix addressing. Unfortunately, a uniform strain bias is not provided over the total area of a ceramic plate, and these difficulties restrict the size of light-valve arrays. Therefore, a page composer sized 128 x 128 elements is manufactured using a so-called fringe effect /2I/.

A photosemiconductor PLZT was placed between transparent electrodes, and operated, in fact, as a transparency controlled by light /I8,20/. This structure has a sufficiently high resolution (up to 50 line/mm) and sensitivity (about MJ/cm^2).

Among various PLZT ceramics the one of 8/65/35 content (8% lanthane, 65% zirconate and 35% lead titanate) is the most attractive It has large values of induced birefringence and scattering, being itself highly transparent. A solid mixture of this content is close to the phase boundary at the diagram of PLZT states. Due to this fact, slight controlled changes in the component ratios lead to a considerable change in ceramic properties.

Below we shall report the results of an investigation of fine grain PLZT-ceramic of the 8/65/35 content. Since in the papers cited above much attention was paid to the study of optically transparent piezoceramics, we shall devote ourselves mainly to the problems that achieved less consideration.

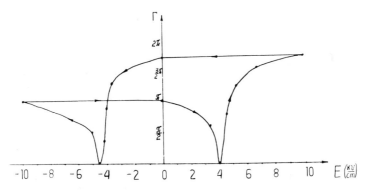

Fig. 3. Phase light delay versus electric field in the Michelson interferometer.

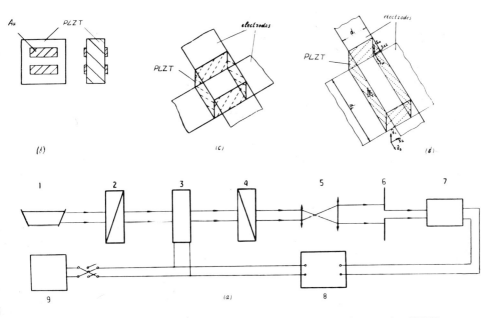

Fig. 4. Experimental study of electro-optic effects in PLZT-ceramics (a) and electrode geometry for the transverse effect (b), fringe effect (c) and asymmetric deformation effect (d): 1- laser, 2 - polarizer, 3 - sample, 4 - analyzer, 5 - objective, 6 - diaphragm, 7 - detector, 8 - recorder, 9 - voltage source.

The ceramic samples are plane-parallel plates, 1.5 x 1.5 cm, with thickness d = 0.1 mm. The optical transparency of these samples is shown in Fig. 2 by the graphs for PLZT ceramic transmission in a broad wavelength band including a visible region. At the wavelength λ = 0.63 μm sample transmission T in the thermally depolarized state was 58 percent. Taking into account the light reflection of 33% (refraction index for ceramics n = 2.6) the losses for light absorption and scattering were 9%.

We have studied the <u>change</u> of the PLZT <u>plate thickness</u> versus the value of the applied electric field. For this purpose, reflecting silver electrodes were deposited on both plate faces. One face of the sample was stuck to the massive substrate, another was used as a mirror in one of the Michelson interferometer arms. The change of plate thickness was measured by an interference fringe shift at the output of the coherent light beam from the interferometer. A single-mode He-Ne laser served as a light source.

Fig. 3 reveals the dependence of the light phase on the electric field voltage. The maximum change of the light phase was 1.75 π at E = I0 kv/cm for 0.I mm sample thickness. This change of phase corresponds to a change of ceramic thickness Δd = 0.276 μm. From the graph it is seen that ceramic has a good memory.

The possibility of an effective light modulation due to the back piezoeffect is worth consideration from the standpoint of ceramic applications in phase controlled transparencies.

Fig. 4a shows the experimental set-up for the study of electrooptic effects. Sample 3 was placed between crossed polarizers 2 and 4. A He-Ne laser I(λ = 0.6328 μm) was the light source. Objective 5 projected an enlarged image of the interelectrode ceramic area onto the diaphragm plane 6 which moved together with the photomultiplier window 7. An electric voltage from the generator 9 was applied to the ceramic electrodes and to the horizontal scanner of the two-coordinate recorder 8. The other coordinate had a signal from a photomultiplier. We did not provide, especially, a preliminary strain bias of samples or switching of the transversal electric field for the vector p orientation of the domain electric polarization in the plane of the ceramic plate.

In the case of the <u>transversal electrooptic effect</u> (electrode geometry is shown in Fig. 4b) the polarizator axis was inclined at 45° angle to the applied electric field direction E. It provided maximum depth of light modulation. The light intensity I recorded by the photomultiplier is associated with the phase difference Γ between ordinary and extraordinary beams by the following expression:

$$I = I_o \sin^2 \Gamma/2 \qquad (1)$$

where I_o is intensity of light outside an analyzer when the optical axes of the polarizer, ceramic plate and analyzer are parallel to each other. The experimental dependence of the normalized intensity $\frac{I}{I_o} \cdot 100\%$ on the electric field obtained for one of the cycles of the ceramic repolarization (time of the cycle is of 6 min) is shown in Fig. 5 (curve I). The normalized intensity was as high as 90% at electric fields of 6 kv/cm-5.5 kv/cm, and the maximum value of I at field values of $\pm$ 9 kv/cm. Maximum contrast was I_{max}/I_{min} = 90, modulation depth m = 0.98.

Using formula (I) and the expression for the phase delay

$$\Gamma = \frac{2\pi d}{\lambda} \cdot \Delta n \qquad (2)$$

one may estimate the dependence of birefringence n on the applied electric field. It is shown by curve I of Fig. 6. Maximum change of the birefringence was $\delta(\Delta n) = 2.2 \times 10^{-3}$; in the state of memory (E = 0) $\delta(\Delta n) = 1.33 \times 10^{-3}$.

Hence in the case of the transversal electrooptic effect we observed considerable changes of birefringence allowing realization of light modulation about 100% in depth. Unfortunately, this effect can hardly be applied to matrix addressed transparents due to a peculiar electrode configuration. Now we are mainly concerned with this problem.

To apply PLZT ceramic material it is of interest to consider the **fringe effect** which is connected with the vector reorientation in the near-to-electrode areas of the ceramic plate due to appearing deformation strain./2I/. In the experiments we used reflecting silver electrodes of 0.I mm width with the distance between them of 0.3 mm. An individual matrix element is shown in Fig. 4c.

The scheme of measurements is demonstrated in Fig. 4a. Dependence of the noramlized light intensity $I/I_o \cdot 100\%$ on the applied electric field for a single cycle of ceramic switching is shown in Fig. 5 (curve 2). Minimum light transmission from the analyzer was achieved when the value of applied electric field was E = $\pm$ 2.4 kv/cm, while at E = $\pm$ 5 kv/cm we observed the maximum intensity $(I/I_o) \cdot 100$-48%. The contrast was equal to 12 which is in agreement with the results of /21/. The modulation depth was m = 0.85.

Dependence of birefringence on the applied electric field has been calculated from formulas (I) and (2). This dependence is represented in Fig. 6 (curve 2). The greatest change of birefringence is $\delta(\Delta n) = 1.2 \times 10^{-3}$; in the memory state (E-0) it was $\delta(\Delta n) = 1.15 \times 10^{-3}$.

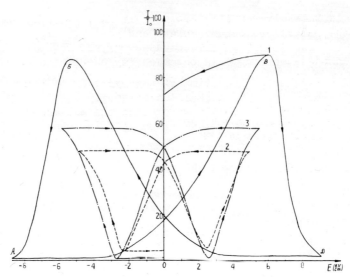

Fig. 5. Normalized light intensity $I/I_o \cdot 100\%$ versus electric field voltage: 1 for the transverse electro-optic effect, 2 - for the fringe effect, 3 - for the asymmetric deformation effect.

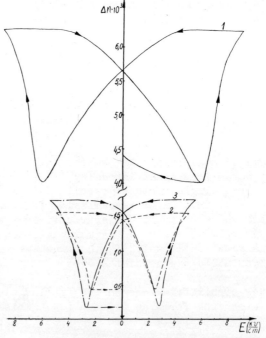

Fig. 6. Ceramic plate birefringence versus electric field: 1 - for the transverse electro-optic effect, 2 - for the fringe effect, 3 - for the asymmetric deformation effect.

Dependence of the switching near-to-electrode area sizes on the ceramic plate thickness /2I/ was studied in our experiments. Moreover, we observed a decrease of this area size with a decrease of the electrode width. For instance, when the electrode width was equal to 75 µm and 500 µm the size of the switching area was 140 µm and 370 µm, respectively, for a plate thickness of 0.1 m (applied electric field was of 5 kv/cm). The fragment of the reconstructed image (Fig. 7a) illustrates switching of transparent elements. In case of light transmittance, 4 bleached ceramic areas near the electrode intersection correspond to a switched element of the matrix.

We measured the speed of PLZT ceramic switching by electric pulses (amplitude U, duration τ). The ceramics were switched from the initial state $I - I_{min}$ to the state where the passing light intensity $I \leq I_{max}$. Here I_{max} means intensity corresponding to the complete polarization of the switched area. This value is reached at the given pulse duration for certain voltage U'.

The dependence $U'(\tau)$ is shown in Fig. 8 with a decrease of pulse duration by 2 µ sec, the switching voltage rapidly grows. Hence, for the given samples of PLZT ceramic the optimum drive pulse duration was about 2 µ sec, since a further increase of fast operation is caused by a sharp growth of the driving voltage. In the range of τ from 2 to 7 µ sec the curve $U'(\tau)$ has a small slope. The limitation of ceramic fast operation $\tau = 0.3$ µ sec was experimentally limited by the duration of the voltage pulse fronts provided by a generator 9. To choose a controlled pulse amplitude for the fringe effect it is necessary to remember that at a certain amplitude U of the driving pulse (for a quasi-stationary switching U = 80 v) the ceramic areas near electrodes undergo strong effects of residual deformation. At a subsequent field switching they participate weakly in repolarization. As a result the contrast grows worse. Note a simplicity of electrode geometry as an advantage of the fringe effect.

We may propose another method for electric control of transparent elements whose physical nature is similar to the described above systems. It is the method of <u>asymmetric deformation</u>. Electrode geometry in this case differs slightly from that at the fringe effect, i.e. the width of an electrode stripe at one of the ceramic planes should be greater than at the other. An individual matrix element is shown in Fig. 4d. To control the light beam we used considerable changes of birefringence occuring not only in the electrode fringes but directly in the interelectrode region of the ceramic plate.

Here is an explanation of this phenomenon: when the electric field is applied along the x_1-axis of the volume (Fig. 4d) we observe an increase of the ceramic thickness due to the vector orientation in some domains along the x_1-axis as in case of the fringe effect. The process is accompanied by ceramic deformation along the x_2- and x_3-direction (axes) with a consequent appearance of mechanical strain components S_{22} and S_{33}. When the ratio of electrode widths is $d_2 > d_1$ in the x_3-direction there appear greater mechanical strains than in case of the x_2-direction, i.e. $|S_{22}|<|S_{33}|$. It leads to a primary orientation of $\bar{P}$ vector in the x_3-direction, while in case of $d_1=d_2$ the mechanical strain components S_{22} and S_{33} are equal, and no birefringence in the interelectrode region is observed.

Quasi-static properties of electrooptic switching for the effect of asymmetric deformation were measured according to the scheme of Fig. 4a. Fig 5 (curve 3) exhibits the dependence of the normalized light intensity on the applied electric field in case of light passing through the electrode intersection, for the ratio of electrode widths 4:1. Maximum intensity was achieved at the field value $E = \pm 6$ kv/cm. Maximum contrast was 58, and in the memory state (E=0) it was equal to 50. Modulation depth was m = 0.97.

Fig. 6 (curve 3) reveals the dependence of PLZT ceramic birefringence change on the applied electric field. The maximum change of the birefringence is $\delta(\Delta n) = 1.55 \times 10^{-3}$, and in the state of memory, $\delta(\Delta n) = 1.4 \times 10^{-3}$.

In the proposed method of birefringence control the ratio of electrode widths is apparently an important point for consideration. Large efficiency of this effect was achieved under the electrode width ratio of $d_2:d_1=4$. Fig. 7b demonstrates the picture of a single switched element having different side sizes. The picture was taken in the transmitted light. The halo around the element is caused by the fringe effect, and bleaching in the interelectrode space by the effect under consideration. Bleaching appears in the center of the element and has an oblong form thus exhibiting development of deformation strains. With increased voltage the bleaching spreads more uniformly over the electrode area. In the experiment we did not observe any worsening of the contrast above electric fields of 15 kv/cm. Despite the fact that electrodes of the non-standard form (neither square nor round) are not always suitable for usage in electric valves, the effect under consideration might be helpful in CT because it allows one to realize a quasi 90°-switching of domains without using mechanical strain of the ceramic plate.

The results we obtained cannot, of course, indicate exactly, what effect should be used in ceramics in order to realize concrete devices controlling the light beam, but still they might be a kind of recommendation. Perhaps, it is worth noting here that the data on the study and application of electrooptic ceramics to be found

CONTROLLED TRANSPARENCIES FOR OPTICAL PROCESSING   139

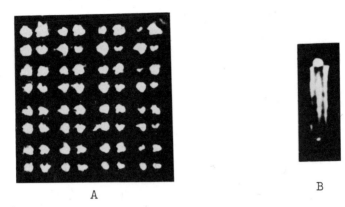

Fig. 7. Photography: a - reconstruction of the holographic transparent image controlled by the fringe effect, b - switched transparent element with various electrode width (300 x 75mcm^2).

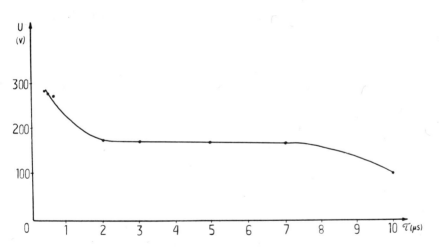

Fig. 8. Pulse amplitude for the ceramic switching versus pulse duration.

in literature are sometimes contradictory. Thus, one should more carefully consider the validity of experimental data for practical purposes. We shall proceed with our work on the development of CTE and CTO systems by applying the segnetoceramic materials. For their successful fulfillment a further progress in fabricating hot pressed high-quality PLZT ceramic is also required.

Application of controlled transparencies for optical processing allows a significant improvement in their productivity, and widens functional possibilities /3/. According to various purposes there are considered various analog and digital optical processors. Then the modulated parameters, such as intensity, phase, light polarization and so forth in controlled transparencies have an uninterrupted number of values (commonly, in CTO) or only two (or several) values.

In the case of binary representation of information it is possible to produce universal computer systems, particularly, arithmetic processors. General principles for the construction of an opto-electronic arithmetic device (OEA) with a speed up to 10^9 oper/sec are reported in /3,22/.

An adder with Boolean algebra using light transmission through controlled transparencies is considered. Moreover, a processor of the table-type has been proposed. It represents, in fact, a long-time holographic memory with the direct access operation results from hologram tables using parameters given by CT /3,23/.

We studied experimentally the possibility of constructing OEAD with controlled transparencies. The main attention was paid to the choice of opto-electronic elements and their common combination in the optical beam line. For simplicity we have chosen a scheme of 2-bit word processors performing two operations, namely addition and multiplication. A single-cycle logic of operation was fully provided.

Schematic representation of the processor is shown in Fig. 9. It consists of optical and electronic parts. He-Ne laser was used as a light source. Output optical signals were registered by the photodiode line. Initial data and output results were realized in binary code. Controlled electric signals corresponding to numbers in the data input block were transferred to CT elements through the diode decoder. The result was at the output of the light table controlled by electric signals from the photodiodes. Both optical and electronic blocks were mounted as a unit assembly, at the input of which the light beam was introduced.

Two variants of transparency arrangement in OEAD (Fig. 10) were tested: series and parallel. In the first case CT are placed one after another, and in the second - in the same plane, in form of the CT matrix. As the working media we used a mixture of two NLC (MBBA:EBBA=I:2). It operated at room temperature; layer thickness was about 20 μm, and the effect of dynamic light scattering was utilized /24/.

The mockup we designed performed operations of addition and multiplication in a satisfactory way. However, it had a complex electric part, because the control block was realized with the help of ordinary electronic elements. It seems preferable to use a combined scheme when a common result is obtained with assistance of an optoelectronic decoder /23/. The drawbacks of the noted schemes are caused by the fact that they apply the logic of digital computers, which restricts the possibilities of CT applications. A further development of mathematical methods for data processing on the basis of bi-dimensional "picture" logic should be realized.

The next step in the experimental tests is supposed to be the construction of the arithmatic table-type processor. It has a simpler scheme and should possess a number of other advantages inherent to holographic methods of data processing.

For the effective operation of systems intended for processing, storage, transformation and transfer of optical data, it is necessary, in some cases, to code bi-dimensional optical signals (images), to compress information contained in them, and to decode it.

By means of coding the following problems may be solved:
- **incre**ase of the channel capacities and productivity of the processing systems due to the decrease of information redundancy /25,26/,
- increase of the holographic memory capacities due to the multiple holographic writing in a single part of the recording medium/27/,
-increase of the noise-stop of data processing and storage units by using special codes to find and eliminate errors and malfunctions /28/,
-formation and separation of feature discovery in character recognition systems /25,26,29/.

The most effective methods of pattern coding are associated with various integral transformers /30/. Linear integral transforms which may be realized in coherent-optical processors (COP) have a total form:

$$\phi(u,v) = \int\int_\Omega f(x,y) \cdot K(x,y,u,v) \cdot dx \cdot dy \qquad (3)$$

where $f(x,y)$, the initial function (original), is a distribution of the light field complex amplitude in the input COP plane (x,y); $\phi(u,v)$ - the transformed function which describes the amplitude distribution in the output plane (u,v); $K(x,y,u,v)$ - the kernel of the transformation, represents the COP impulse response; Ω - integration region defined by an aperture of the COP input pupil.

In particular cases the transformation kernel (3) may be independent of the output plane coordinates or may constitute a difference function (for invariant isoplantic systems /32/):

$$K(x,y,u,v) - K(u-x, v-y) \qquad (4)$$

The Fresnel-transform /33/ and closely connected with it the Giraretransform /35/ are characterized by the kernel of (4) type. Fresneldiffraction lies in the basis of invariant optical systems forming the image which realizes correlation and convolution of optical signals and systems performing spatial filtering of images, holographic systems etc. /32/34/. One must exclude from this row the optical systems realizing Fresnel transform, the kernel of which (K) has a more complex dependence on coordinates in a first approximation /32/34/:

$$K_F \simeq \exp[-i(xu + yv)] \qquad (5)$$

Besides Fourier transforms there exist a lot of internal transformations which are the expansion of an initial function $f(x,y)$ into a series according to a system of orthogonal functions of two variables $K_{ij}(x,y)$. A set of expansion coefficients

$$a_{ij} = \iint_\Omega f(x_1 y) \cdot K_{ij}(x_1 y) \, dx \, dy \qquad (6)$$

is called the general spectrum of the initial function $f(x,y)$ /36/ analogous to the Fourier spectrum. Such systems constitute, for instance, the functions of Haar, Walsh, Legendre /31,37,38/ and others. One should refer also to the intrinsic systems obtained in the result of the Carhunen-Loeve expansion /39/. In work /40/ there is an example of synthesis of a coherent-optical system with a impulse response of the common view (3). Such linear space-invariant systems are applied in the common spectral analysis, plane transformation /31/, associative data search /3,41-43/ etc.

To construct COP systems of high output possessing with functional possibilities for parallel processing of large data blocks in the real time, it is necessary to provide control of impulse

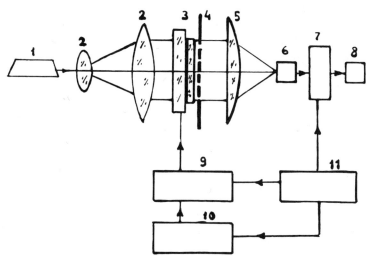

Fig. 9. Block-scheme of the OEAC with the CT matrix. 1 - laser,
2 - collimator lenses, 3 - CT matrix, 4 - mask, 5 - cylinder lens, 6 - liner of photodiodes, 7 - trigger block,
8 - indication block, 9 - d,
10 - block of data introduction, 11 - battery.

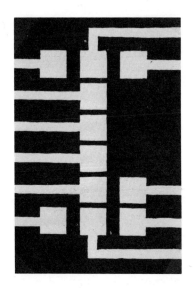
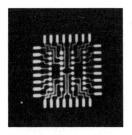

Fig. 10. Electrode geometry in the liquid-crystalline cell:
a - consequent array of cells-transparents (4 cells),
b - parallel array of transparents on the form of CT matrix (single cell).

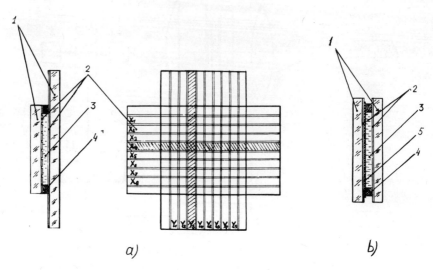

Fig. 11. Construction of DF experimental samples, a - DFE, b - DFO. 1 - glass substrates, 2 - transparent electrodes, 3 - liquid crystal, 4 - 5 - photo-semiconductor layer.

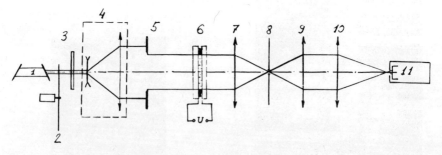

Fig. 12. Amplitude DFO experimental scheme: 1 - laser, 2 - modulator, 3 - inhomogeneous attenuator, 4 - telescope, 5,8 - doaphragms, 6 - liquid-crystalline DFO sample, 7,9,10 - lenses, 11 - multiplier.

response in optical systems in addition to the operative data input. It can also be realized by means of transparencies controlled by electric and optical signals /44/. Note that materials for their creation ought to have improved homogeneity of optical properties and switching characteristics over the total surface. Controlled transparencies applied in optical systems may be called dynamic or retuned filters (DF) by analogy with arrangements controlling pulse reactions of electric systems.

We have studied some electrooptic CT characteristics on the basis of nematic liquid crystals for the purpose of their application as dynamic spatial filters. A schematic representation of DF experimental samples is shown in Fig. 11.

In the filters controlled by electric signals (DFE) we used matrix addressing which simplified sufficiently the DFE element commutation. It also corresponds best of all to the case of filter usage in the systems where Walsh- and Hilbert-transformations are realized /33/. The DF size was 8 x 8 elements, the width of electrode stripes was 2 mm, and the space between them was 50 μm. As electrodes we used a SnO_2 film on polished quartz plates with a resistance 200-300 Ohms per square.

A dynamic filter controlled by optical signals (DFO) was constructed from a structure of photosemiconductor-liquid crystal /45/. A CdS photosensitive layer of 1.5-2 mcm thickness was vacuum deposited on one of two quartz substrates over the SnO_2 transparent electrode film; the layer resistivity was of the order of 10^8 Ohm.cm. To realize homogeneous NLC orientation with a positive dielectric anisotropy, a layer of 70 Å silicon monooxide was vacuum deposited at an angle to the substrate /2/.

Pattern coding by means of Hadamard-Walsh-transformations was considered in /37,38/. Difficulties in realization of this transformation in coherent /36,40/ and, particularly, in incoherent /46,47/ optical processors are associated with the necessity to multiply initial function by -I) when calculating transform coefficients. In the noted papers the authors performed a separate integration over positive and negative Walsh-function parts while substraction was realized electrically at the photodetector output.

In /48/ an optical processor with a nematic liquid crystal DFE was applied to realize the modified Walsh-transform. The effect of light dynamic scattering in NLC was used. We have performed a similar experiment to test the working capacity of a DFE sample /49/. The experimental arrangement is shown in Fig. 12. Laser beam I passes through the modulator 2, the inhomogeneous attenuator 3, which makes the intensity in the beam cross-section uniform, and then the telescope 4. At DF switching any of 64 binary modified Walsh-functions is realized up to the Wal (7.7). The sum power recorded by a

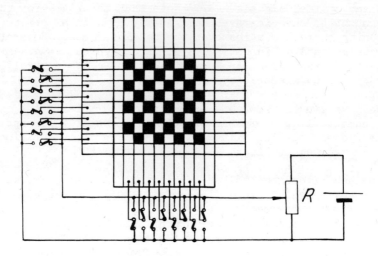

Fig. 13. Commutation scheme of the amplitude liquid-crystalline DFO.

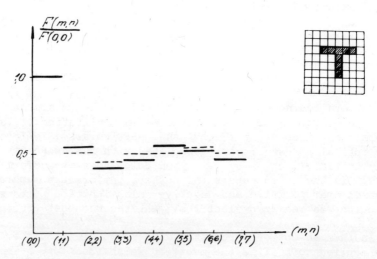

Fig. 14. Measurement (sold line) and calculation (dashed line) result of Walsh diagonal transform coefficients for the letter T image.

photodetector II in the lens focus 10 is proportional to the scalar product of DF-6 transmission function and input transparency (installed after diaphragm 5). It is equal to a corresponding coefficient in the expansion of the input signal into the modified Walsh-Fourier series. To increase the signal-noise ratio we used a spatial filter consisting of lens 7 and diaphragm 8 in its focal plane.

The logic scheme of the liquid crystal DFE commutation is represented in Fig 13. The bi-dimensional modified Walsh-function is realized by means of the operation "exclusive or" over the corresponding one-dimensional functions of rows and columns. Such a condition is fulfilled necessarily in the given filter because dynamic scattering in the crystal layer appears only in those electrode parts where the potential difference is independent of the applied voltage polarity. To this case a "logic zero" corresponds, while in the absence of potential difference and scattering - "logic unit". Thus, the modified Walsh-function differs from the ordinary one by the change of "- unit" to "zero" due to the fact that intensity transmittance cannot be negative. In the filter we used a mixture of nemaic crystals MBBA and EBBA in the weight ratio 2 : I. The mixture resistivity was about 10^7 Ohm-cm, and the thickness of the liquid crystal layer was equal to 10 μm. The nonuniformity of filter transmission was as high as 5%, the uniformity of scattering about 10%, which defined the measurement error. At 20 v voltage by DFE electrodes the contrast ratio of switched on and out elements achieved the value about 100. Comparison of the measurement results on Walsh-coefficient with calculation is represented in Fig. 14.

Unfortunately, substitution of "I" by "0" breaks the transformation orthogonality and decreases significantly the coding efficiency. The same drawback is characteristic of the Golay-transformation /50,51/. One can obtain negative and any complex values of transmission by means of phase dynamic filters. In such filters, for instance, the effects of controlled birefringence in electrooptic segnetoceramics /11, 13/, can be achieved as well as in NLC with a positive dielectric anisotropy /24,52/.

Electro-optic properties of liquid-crystalline dynamic phase filters have been defined by a scheme which includes a Max-Zender interferometer. The DFO sample under study (photosemiconductor-liquid crystal structure) was placed in one of its shoulders. In the experiments we used a mixture with a positive dielectric anisotropy in the nematic phase at room temperature (MBBA+EBBBA+EBAB) with the weight ratio 4:2:I and resistivity 10^6 OhM.cm). The direction of the electric vector of light (read-out wavelength 0.63 μm) was combined with the initial direction of the optical axis in the liquid-crystal layer. Fig 15 demonstrates dependence of the transmitted light phase delay on the applied voltage in the structure. Curve I was taken in the absence of control illumination,

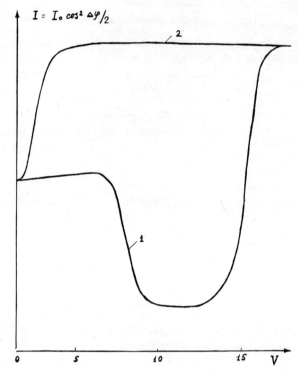

Fig. 15. Phase light delay versus voltage at the liquid-crystal-photosemiconductor structure: 1 - without illumination, 2 - with He-Cd laser illumination.

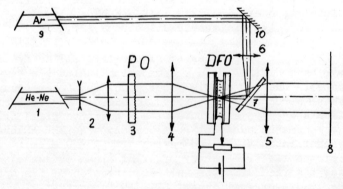

Fig. 16. Optical visualization scheme of the phase contrast using DFO arrangement on the basis of liquid crystal-photosemiconductors.

curve 2 - when the structure was illuminated by He-Cd laser radiation (wavelength of 0.44 μm) with an intensity of about I mw/cm^2. It is seen that a control blue illumination might change the phase of the transmitted red light by π corresponding to 100%-phase modulation.

On the basis of such a structure as a dynamic filter we have realized a scheme of the simplest opto-electronic processor (Fig.16). It is destined for a phase contrast visualization. The complex amplitude distribution in the back focal plane4 is the Fourier-spectrum of the complex transmittance function of phase object located in the front focal plane. Using Ar-laser control illumination (wavelength about 0.5145 mcm) the zero order phase of this spectrum is shifted by π/4. Lens 5 realizes an inverse Fourier-transform, the phase object image occurs in the output plane 8. Thus, phase inhomogneities transform to intensity inhomogeneities. The photograph of the visualized phase contrast is demonstrated in Fig. 17 (dark stripes as phase inhomogeneities).

The optical processor scheme we considered (Fig. 16) may be easily reconstructed to realize any kind of spatial filtering by means of applying to the DFO a corresponding control illumination. Thus, the Hilbert-transformation may be obtained under the condition that the complex structure transmittance is

$$K(u,v) \sim \text{Sign } u \text{ Sign } v \tag{7}$$

Hence, by switching on and off the control illumination (in definite quadrants of the spectral plane) one can obtain either the initial function or its Hilbert-image in the output plane. Such a scheme allows one to recognize an image by using both its Fourier- and Hilbert images. Due to this it is possible to increase, in some cases, the recognition efficiency /53/. A similar scheme was applied in /54/ for the formation of Vander Lugt matched filter.

It should be noted that the first experimental samples of dynamic filters we have fabricated are not of a very high quality. We could not obtain, for instance, a uniform orientation of the liquid crystal layer in large areas, and provide sufficient degree of orientation (contrast about 15). The sensitivity of the semiconductor layers was only equal to 10^{-3} J/cm^2. The possibility that the NLC layer on the photosemiconductor degrades the structural parameters is not fully excluded. But these are technological difficulties, not fundamental ones, and we have partially succeeded in overcoming them. For instance, a dark resistivity of photoconductive layers (CdS-ZnS) having maximum of spectral sensitivity in the range of Cd- and Yr-laser radiation, has been increased to 10^{12}-10^{13} Ohm. cm. at the photoconduction jump of not less than 10^3. It allows us

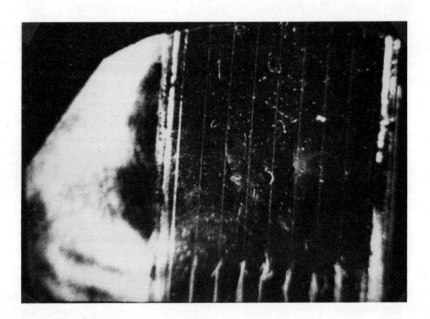

Fig. 17. Picture of the visualized phase object (dark stripes are physe inhomogeneities.)

to use more stable nematic asoxi-mixtures, having resistivity of the order of $10^{10}-10^{11}$ Ohm.cm.

The experiments we have carried out demonstrated the possibility of constructing dynamic filters with uniform NLC orientation on a square of I cm^2, and to realize, with the help of DF, integral transformations of Walsh-Hilbert, phase contrast, Vander Lugt filters, etc.

In addition to liquid crystals, fine grain segnetoceramics can be a successful material for DFO. In combination with a photosensitive layer, they may be used as DFO including the Vander Lugt filter. We have initiated studies of such samples. Another promising material for application in DFO seems to be MOS-structures with electro-optic read-out /54,55/.

References

1. IEEE, Trans on El Dev., 1973, ED-20, N II.
2. Proc. IEEE, 1973, 61, N 7.
3. N.G.Basov et al. Preprint FIAN N 3, 1973.
4. V.N.Selezniov, N.N.Shuikin. Kvantovaya Elektronika,1974, I. N 7, 1485.
5. G.W.Taylor, W.F.Kosonocky. Ferroelectrics, 1972, 3, N 2-4, 81.
6. D.Maydan. Proc. IEEE, 1973, 61, N 7, 254.
7. J.D.Margerum et al. Appl.Phys.Lett., 1971, 19, N 7, 216.
8. T.P.Brody et al. IEEE Trans. on El.Dev., 1973, ED-20, N 11, 995.
9. G.Larunie, J.Roberts, J.Borel. Appl.Opt., 1974, 13, N 6, 1355.
10. I.N.Kompanets, V.V.Nikitin. Preprint FIAN 1973, N 127. Microelectronics 1973, 3, N 5, 441.
11. C.E.Land, P.D.Thacher. Proc. IEEE, 1969, 57, N 5, 751;
12. V.A.Zhabotinska, A.G.Sobolev et al. Avtometriya, 1975, N 3,59.
13. Uchenye zapiski Latvijskogo Gosudarstvennogo Universitita im. P.Stuchki, vol. 230, Riga, 1975.
14. J.R.Maldonado, A.H.Meitzler. Ferroelectrics, 1972, 3, N 2-4, 169.
15. H.M.O'Bryan,Jr. & A.H.Meitzler. Ceramic Bulletin, 1972, 51 N 5, 179.
16. G.H.Haertling, C.E.Land. Ferroelectrics, 1972, 3, 269; IEEE Trans.Sonics and Ultrasonics, 1972, SU-19, 269.
17. C.E.Land, Ferroelectrics, 1974, 7, 45.
18. K.Okazaki et al. Ferroelectrics, 1974, 7, N 153.
19. J.R.Maldonado, A.H.Maitzler. Proc. IEEE, 1971, 59, N 3, 368.
20. A.H.Meitzler, J.R.Maldonado, D.B.Fraser. Bell.Syst.Techn.J., 1970, 49, N 6, 953.
21. M.D.Drake, Appl.Opt., 1974, 13, N 2, 347.
22. L.A.Orlov, Yu.M.Popov. Avtometry 1972, N 6, 8; Kvantovaya Elektronika, 1974, I, N I, 27.
23. L.A.Orlov, Yu.M.Popov. Avtometry, 1972, N 6, 14.
24. L.M.Blinov. Uspekhi Fiz.Nauk, 1974, 114, N I, 67.

25. A.Rozenfeld. "Raspoznavanie i obrabotka izobrazheniy a pomoshjyu vychislitel'nykh mashin", izd. "Mir", Moscow, 1972.
26. E.I.Krutitsky, G.Kh.Fridman. In: Opticheskiye metody obrabotki informatsii, Leningrad, 1974, 78.
27. J.T.LaMacchia, D.L.White. Appl.Opt., 1968, 7, N I, 91.
28. B.E.Khaikin. In: Opticheskiye metody obrabotki informatsii, Leningrad, 1974, 33.
29. A.Vander Lugt. IEEE Tr. of Inform. Theory, 1964, IT. 10, 139.
30. R.F.Edgar. Opt. Technology 1969, I, N I, 183.
31. V.P.Koronkevich, Yu.E.Nestrikhin, P.E.Tverdokhleb. Avtometry, 1972, N 6, 3.
32. J.Goodman. "Introduction to Fourier-Optics", "Mir", Moscow, 1970.
33. L.M.Soroko. Osnovy Golografii Optiki, Izd. "Nauka", Moscow, 1971.
34. A.Papulis. Teoriya sistem i preobrazovany v optike, izd. "Mir", Moscow, 1971
35. J.Berry, Optica Acta, 1967, 14, 269.
36. I.S.Gibin, E.S.Nezhevenko, O.I.Potaturkin, P.E.Tverdokhleb. Avtometry, 1972, N 5, 3.
37. W.K.Pratt, J.Kane, H.C.Andrews. Proc. IEEE, 1969, 57, N I, 58.
38. J.Poncin CNET, Ann. des Tel., 1971, 26, N 7-8, 235.
39. S.Watanabe. Trans. of the 4th Prague Conf. on Inform. Theory, Prague, 1967, p. 635.
40. E.S.Nezhevenko, O.I.Potaturkin, P.E.Tverdokhleb. Avtometry, 1972, N 6, 88.
41. D.Gabor, IBM J.Res- and Devel., 1969, March, 156.
42. M.Sakaguchi, N.Nishida, T.Nemoto, IEEE Trans. on Comp., 1969, C-19, N 12, 1174.
43. I.S.Gibin, M.A.Gofman, E.F.Pen, P.E.Tverdokhleb. Avtometry, 1973, N 5, 12.
44. A.A.Vasiliev, I.N.Kompanets, V.V.Nikitin, In: "Opticheskiye metody obrabotki informatsii", Leningrad, 1974, III.
45. A.A.Vasiliev, I.N.Kompanets, V.V.Nikitin, In "Kvantovaya Elektronika", 1972, N I(13), 120.
46. O.I.Potaturkin, P.E.Tverdokhleb, Yu.V.Chugui. Avtometry, 1973 N 5, 36.
47. B.E.Krivenkov, P.E.Tverdokhleb. Yu.V.Chugui. Avtometry, 1974, N 6, 32.
48. S.Inokuchi, Y.Morita, Y.Sakurai. Appl.Opt., 1972, II, N 10, 2223.
49. A.A.Vasiliev, P.V.Vashurin, I.N.Kompanets, V.V.Nikitin. Report at the All-Union Conference on liquid crystals, Ivanovo, 1974.
50. M.J.Golay, IEEE Trans.Com., 1969, C-18, N 8, 733.
51. K.Preston (Jr.), P.E.Norgen, Electronics, 1972, N 22, 89.
52. N.G.Basov et al. JETP, 1973, 64, N 2, 599.
53. L.M.Soroko. Materials of the Vth All-Union School on Holography (Novosibirsk, 1973), Leningrad, 1973, 40.
54. P.Nisenson, S.Iwasa. Appl. Opt., 1972, II, N 12, 2760.
55. N.G.Basov et al. In: Short Comm. in Physics, 1972, N 6, 34.

KINOFORM OPTICAL ELEMENTS

V.P. Koronkevich, G.A. Lenkova,
I.A. Mikhal'tsova, V.G. Remesnik,
V.A. Fateev, V.G. Tsukerman

Institute of Automation and Electrometry
Novsibirsk, USSR

ABSTRACT

Optical methods have been considered for producing the simplest kinoform elements in thin films of chalcogenide vitreous semiconductors. An elementary fragment of the kinoform is formed by projecting a special mask with the subsequent image displacement relative to the film. When manufacturing lenses, a Fabry-Perot interferometer was used operating in a reflected light. This device allows us to obtain asummetrical distribution of the intensity in rings that simulates a phase distribution in a kinoform lens. The interference pattern by using a contact printing is transferred from a photoplate to a chalcogenide film.

In 1969 Lesem, Hirsh and Jordan suggested a method for the computed-aided manufacturing of a new optical structure and designated it as a kinoform [1]. Kinoform is a phase element of variable optical thickness transforming a wave front in a predetermined manner with minimal losses in light energy. Increments of the kinoform optical thickness do not exceed a light beam wavelength.

Of all the known optical elements most similar to kinoforms are Fresnel lens, Schmidt plates, Michelson echelon and phase diffraction gratings. The historical experience shows that the development of an optical diffraction structure with a required phase distribution presents great technological difficulties. The attempts to replace lenses by Sore and Wood gratings, optical retouching and other techniques have not been put into practice. Only at present due to the advent of computer-controlled multi-level plotters and new promising photomaterials the ways for this problem solution have been outlined.

Several methods for producing kinoform optical elements are known [2-5]. As a rule, a required phase profile is obtained on a photoresistor via contact printing from a negative. A necessary negative transmittance (proportional to the phase) is achieved by photographing special transparencies produced by a plotter or by scanning a multibeam interference pattern. In all the cases known the phase coding method is discrete. Phase quantization provides noises being added to a reconstructed image.

The present paper describes a new optical technique for producing kinoform elements in thin films of chalcogenide vitreous semiconductors (CVS) free of noises caused by phase quantization.

A choice of the material is determined by the following considerations:

1. In thin CVS films under light influence the refractivity is changed by the value of the order of $\Delta n = 0.1$ [6].

2. A great value of Δn allows us to obtain phase shifts up to 4π and high resolution (5000 line/mm) makes it possible to reproduce a thin structure of the phase profile.

3. A kinoform elements can be produced in a real-time operation mode.

The CVS-films do not need development and fixation since their structural changes occur at the moment of exposure.

The CVS-films of 4-10 μm in thickness were prepared by electron-beam evaporation of the initial material in vacuum, the evaporation rate was maintained to be equal to 150 Å/sec. As a substrate oxide glass plates of 1 mm in thickness were used. To provide an efficient recording of the phase profile over the total recording medium volume, it is desirable to fulfil the conditions of uniform absorption of the recording light beam through the thickness. When operating with a kinoform, to reduce losses it is necessary for the light absorption in the medium to be small. The fulfilment of these conditions is realized by choosing a point of operation in relation to the area of the medium eigen--absorption edge. By varying the initial film material composition, the eigen-absorption edge can be displaced over the spectrum from 0.4 to 1 μ m. Figure 1 shows typical spectral characteristics for the CVS-film transmittance. Provided that in recording a kinoform, the film is illuminated by an argon laser (λ = 0.51 μ m), then for this wavelength strong light absorption will occur, since for $As_2 S_3$ the eigen-absorption edge lies in the field of 0.5 m. For a red line (λ = 0.63 μ m) quite the opposite situation is observed. The transmittance will be very high and the wave front due to variations in n will be modulated only in phase.

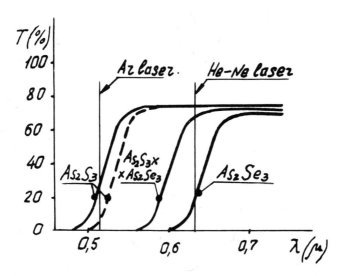

Figure 1. Spectral characteristics of the CVS film transmittance. A dotted line shows the film transmittance after the exposure. The transmittance value is limited to 76% due to a high reflectivity.

For a 2π-phase shift it will be sufficient to evaporate a film of 5-10 m in thickness. A refraction index of the film will lie in the range of 2.5 - 2.8 μm [7]. The exposure results in a change of the refractivity according to the typical curve shown in Figure 2.

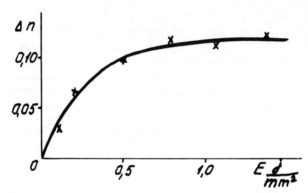

Figure 2. The dependence of the refractivity variation (Δn) of $As_2 S_3$ film on the exposure (E). The exposure was performed by an Ar-laser beam (λ = 0.51 μm).

A direct proportionality between the light power and exposure time is being maintained up to heating of the film under illumination. Heating to the temperature of about 200°C results in erasing of information recorded, in this case the repeated recording made after this erasing changes the refractivity by the value Δn being an order below [6]. It shows that chalcogenide materials can be used for a reversive recording of the phase structure.

A type of the kinoform recording is intended for a certain light wavelength. Therefore for optical elements synthesized on CVS films of fundamental importance is the dependence of Δn on λ (see Figure 3) determining a phase shift value in the region exposured for various wavelengths. From this figure it is seen that the value Δn is significant over the total measured area of the spectrum, and hence the CVS-films can be used for the synthesis of kinoforms and other diffraction optical elements calculated for any wavelength of the above range.

KINOFORM OPTICAL ELEMENTS

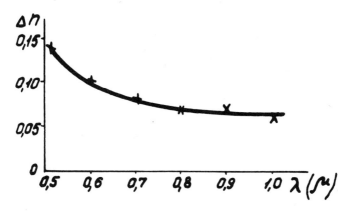

Figure 3. The spectral dependence of the refractivity variation (Δn) of $As_2 S_3$ film.

In an elementary cell of the kinoform element a 2π-phase delay of the incident light is provided. Within the element its phase linearly increases from 0 to 2π. An elementary cell can be simulated by scanning an image of the rectangular triangle on the film under investigation. When scanning the triangle along one of the legs, an exposure time in the element is linearly changes from 0 to the predetermined value. The experimental scheme is clear from Figure 4. Here to the film under investigation an image of the triangular mask is projected by a microobjective. Illumination of the mask is performed by a collimated beam of the argon laser (λ = 0.51 μm). In the production of the elementary cell an $As_2 S_3$ - film of 8 μm in thickness was displaced parallel to one of the triangle legs at a rate of 150 m/sec. As a result a phase profile was obtained according to which the non-linearity of the recording medium characteristics can be determined.

To control a phase profile shape a conventional "Karl Zeiss" shearing-microinterferometer was used. In this device fringe shearing relative to the environmental field corresponds to the course of variation in the phase of the transparent object being investigated. A microphotograph of the film region with a strip of the kinoform elementary cell of 50 m in width is given in Figure 5. The interference fringe profile follows the curve plotted for the dependence Δn on exposure.

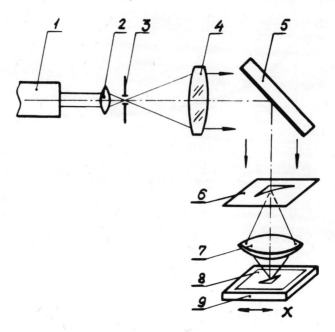

Figure 4. The recording system for a kinoform elementary cell. 1 - Ar-laser; 2,4,7 - lenses; 3 - point diaphragm; 5 - mirror; 6 - mask; 8 - CVS film; 9 - substrate.

Since in the shearing-interferometer the interference is observed of the waves from the object images sheared in respect of each other, in Figure 5 can be seen two phase profiles with different signs. In recording the elementary cell, provided that the exposure has properly been chosen, shearing should correspond to one fringe. In Figure 5 this condition has not been fulfiled and a phase step on the cell edge is equal to 1.3 λ .

When a linear profile of the kinoform element is obtained the mask being projected to the material should be changed. The hypotenuse of the triangle should be replaced by a curve performed according to the interference fringe profile of Figure 5. In this case the non-linearity of the medium parameters will be taken into account, and when scanning the film by such a triangular light spot, the process of phase variation will closely correspond to a linear law. The phase step per λ can be obtained by selecting the illumination and scanning rate; microphotograph in Figure 6 shows a profile of the kinoform elementary cell produced with the help of a figured mask, and on the right an edge of the simplest kinoform element composed of such triangles is shown. A transmission grating with a one-dimensional saw-shaped phase structure is the simplest example for a method of the kinoform synthesis playing the role of a conventional prism. From Figure 6 it is seen that for the prism the condition of 2π-phase step has been met so that the lines of the wave front inclunation of different elementary cells form a continuous line

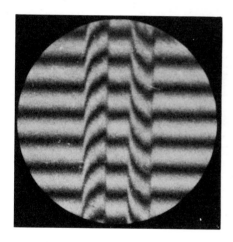

Figure 5. A microphotograph (100x) of an As_2S_3 film fragment with a recorded strip of the kinoform elementary cell illustrating a non-linear response of the medium. The microphotograph has been received with a shearing-interferometer.

Figure 6. A microphotograph (100x) of an As_2S_3 film fragment: on the left - a corrected triangular phase profile of the kinoform elementary cell; on the right - a kinoform - prism combined of triangular profiles.

The efficiency of prism operation depends on the accuracy of performing the phase step and the size of the break regions. A great break and inconsistency of the phase step with 2π result in the fact that these kinoform regions give false diffraction orders. In our case the efficiency of the laser beam ($\lambda = 0.63 \mu$ m) deflection by a kinoform prism was equal to 96% at an angle of deflection of about 1° (the efficiency was measured as the relation between the intensity of the light deflected by the prism and that of the light transmitted through the film without any prism). Figure 7 represents a photograph illustrating the prism operation. The top point is a light beam transmitted through the film with no prism, the bottom ones represent the beam deflected by the prism with low diffraction orders.

Figure 7. The He-Ne laser ($\lambda = 0.63 \mu$ m) beam diflection by a prism. The top point is a light beam transmitted through the film with no prism; the bottom points show a light beam deflected by the prism and weak diffraction orders.

To produce a more complex kinoform element, i.e. a cylindrical lens, a mask was prepared consisting of 24 elementary cells modelling over the Fresnel zone disposition. By scanning the mask image pattern an array of 5 cylindrical lenses was made. An aperture of the lens was equal to 0.8 mm, a focal distance was equal to 15 mm. Figure 8 shows a phase profile of the celindrical lenses, and Figure 9 represents multiplication of the slit image pattern with the help of a lens array. The light energy losses in this lens array were equal to several per cent for λ = 0.63μm.

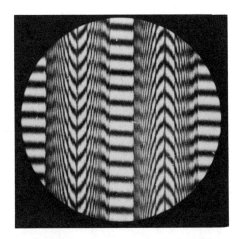

Figure 8. A phase profile of the kinofor cylindrical lens of 24 zones recorded on As_2S_3 film by scanning with a mask image pattern.

Figure 9. Multiplication of the slit by a lens array consisting of 5 cylindrical lenses.

The above method of manufacturing the simplest kinoform elements is free of phase quantization within the boundaries of the elementary cell; by predetermining the mask shape the non-linearity of the light-sensitive media characteristics may be taken into account, and the requirements for the image scanning system precision can reasonably be decreased since scanning is perform by a fragment of the kinoform element.

When producing spherical lenses we were trying to eliminate scanning entirely. For this purpose a special Fabry-Perot interferometer was developed with asymmetric intensity distribution in the interference pattern

[8]. The photographing of this interferometer interference field followed by the transform of an amplitude distribution into a phase one immediately gives the distribution required for a kinoform lens.

An optical scheme of the system is given in Figure 10. Here: 1 is a laser, 2 is a microobjective, 3 is a rotating diffusion scatterer, 4 is an objective projecting the diffusion scatterer to the center of the Fabry-Perot interferometer; 5 is a semi-transparent plate, 6-7 is a Fabry-Perot interferometer, 8 is an output objective, 9 is a photoplate.

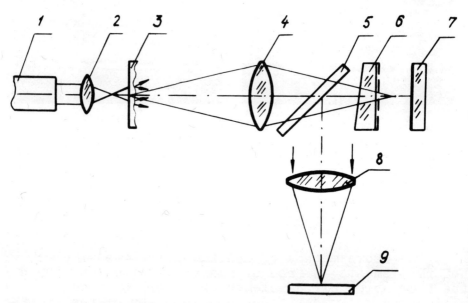

Figure 10. The optical system utilizing a Fabry-Perot interferometer.

An image pattern of the rings in a reflected light was projected to a high-resolution photoplate placed in a focal plane of objective 8. Mirror 6 of the interferometer was covered with an aluminium coating with the transmittance of 0.18 and reflectance of 0.5; mirror 7 was covered with a multilayer dielectric coating with the transmittance of 0.01 and reflectance of about 0.99. The mirror spacing was varied within the limits of a wave-length by utilizing a pieso-ceramical element, so that at the center of the ring-image pattern a minimal value of the intensity was observed. Time of exposure was 20-60 sec. Later on the negative obtained was used as a mask in producing kinoform lenses. By utilizing a contact printing the negative transmission was transformed to a phase profile on a chalcogenide film. To eliminate speckles in the image the illumination was made by a high-pressure mercury lamp.

When developing a kinoform lens we proceeded from the fact that a theoretical phase profile beginning from zone 3-4 is well approximated by a rectangular triangle. With respect to the nature of the CVS and photoplate response non-lineriality, the parameters were chosen for a coating of the front mirror in the Fabry-Perot interferometer according to the data given in [9]. Figure 11 shows: a) the dependence of the reflected light intensity on the interferometer mirror spacing, b) the transmittance of one negative first three zones in respect of the intensity and c) the phase distribution in the kinoform lens zones. The negative transmittance was recorded by a microphotometer. The photograph in Figure 11 "c" was obtained on a Mach-Zehnder interferometer where a kinoform lens was placed into one of its arms.

As a result of the first experiments kinoform lenses were produced with the following parameters: 16 mm in diameter, about 1830 mm in focal distance. The lens focus f_e was pre-calculated according to the formula

$$f_e = \frac{f_{o\beta}^2}{2t} \qquad (1)$$

derived from the radius equalities between the phase lens zones and the rings in the Fabry-Perot interferometer

$$z_K = f_{o\beta}\sqrt{\frac{K\lambda}{t}} = \sqrt{2f_e K\lambda} \qquad (2)$$

Here $f_{o\beta}$ is a focus of the projecting objective, t is a mirror spacing, K is the number of rings (zones). In our system $f_{o\beta}$ = 300 mm, t = 25 mm, f_e = 1800 mm.

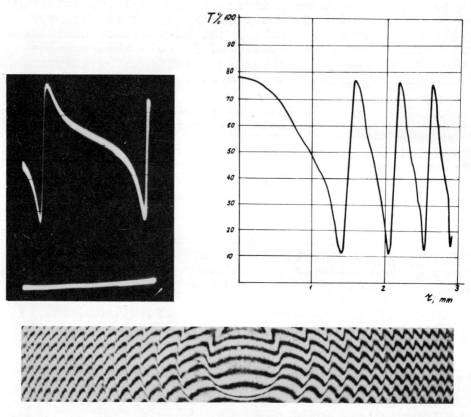

Figure 11. A profile of the ring-zones; a) the dependence of the reflected light intensity on the interferometer length [9]; b) the transmittance of the negative first three zones in respect of the intensity; c) the phase distribution in the kinoform lens zones, a phase step being equal to 2π (10-fold magnification).

KINOFORM OPTICAL ELEMENTS   165

The calculated value had a deviation of 1.6% from the
experimental one.

   Kinoform lens resolution was determined according
to test-miras. For a usual thin lens of ⌀ 16 mm the
theoretical resolution is known to be equal to 9",
while the experimental value obtained for a kinoform
lens was equal to 10", the latter corresponding to visual observation of lines in element 25 of the test-
-mira.

   In Figure 12 mira photographs are given obtained
with a kinoform (f_e = 1830 mm) and a conventional lens
(f = 2000 mm). The line resolution in element 25 for
the kinoform can be seen with the help of an enlarging
lens.

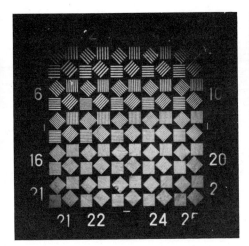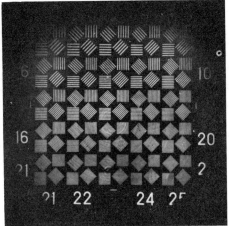

Figure 12. The mira-image patterns a) with a kinoform
          lens, b) with a conventional lens.

   Determine the limiting parameters for kinoform
lenses produced by an optical technique. Figure 13
shows a fragment of the optical scheme of Figure 10,
here 1,2,3 designate beams coming out of the interferometer at an angle φ to the optical axis. From
Figure 13 it is seen that the number of interfering
beams N is limited by the mirror diameter and spacing.
It can be represented to an accuracy of an integer as

$$N = \frac{D}{4t\varphi} \tag{3}$$

Having substituted the value φ from (3) into the formula expressing the dependence between the ring number K, interferometer length and incidence angle φ, we obtain

$$K = \frac{\varphi^2 \cdot t}{\lambda} = \frac{D^2}{16 N^2 \cdot t \cdot \lambda} \tag{4}$$

From the phase lens formula it follows that the number of the last zone is

$$K = \frac{d_e^2}{8 \lambda f_e} \tag{5}$$

where d_e is a diameter of the lens. By equating the right parts of two last expressions with the account of (1) we obtain

$$\frac{d_e}{f_e} = \frac{D}{f_{o\delta} \cdot N} \tag{6}$$

For a scheme in a transmitted light the projecting objective diameter should be not less than that of the interferometer mirrors, i.e. $D_{o\delta} \geq D$. For a scheme in a reflected light (see Figure 13) $D_{o\delta} \geq D(1 + \frac{S}{2tN})$ where S is the distance between the interferometer and the objective. Formula (6) for these two cases can be rewritten in the form

$$\frac{d_e}{f_e} = \frac{D_{o\delta}}{f_{o\delta} \cdot N} \tag{7}$$

$$\frac{d_e}{f_e} = \frac{D_{o\delta}}{f_{o\delta} (N + \frac{S}{2t})} \tag{8}$$

From (6) and (7) it is seen that an angular aperture of a kinoform lens is always less than that of the projecting objective, in this case the proportionality coefficient is equal to $1/N$ or $1/(N + \frac{S}{2t})$.

KINOFORM OPTICAL ELEMENTS

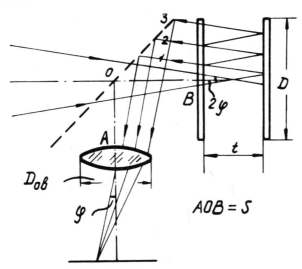

Figure 13. The limitation of the interfering beam number by the diameters of the Fabry-Perot interferometer mirrors and the projecting objective.

In order that the profile of the remote lens zones be slightly different from the that of the central ones, the definite number of the interfering beams N should be predetermined. For a scheme in a transmitted light N should be not less than N_{eff} being determined according to the known formula [10]

$$N_{eff} = \frac{2,98\sqrt{R}}{1-R}$$

where R is a reflectance of the interferometer mirrors. For a reflecting interferometer with $R_1 = 0.5$, $R_2 = 0.99$ and $R = \sqrt{R_1 \cdot R_2} = 0.7$ it may be assumed that $N \simeq 10$. Under the conditions of our experiment $S = 190$ mm, $t = 25$ mm, $\frac{D_{o6}}{f_{o6}} = 1/4.5$, then in accordance with (8) a kinoform lens can be obtained with the angular aperture

$$\frac{de}{fe} = 1/62$$

The experimental value of de/fe = 1/112. A fall in the intensity of the interference fringes as they are receeding from the optical axis (see Figure 14) was the basic reason for the discrepancy between the experimental and calculated values. It resulted in a decrease of the difference between the maximal and minimal negative transmittance by 10-15% in order 10, by 20-25% in order 20 and by 35-40% in order 30. At contact printing from a negative to the chalcogenide film the zones beginning with 40 are not reproduced, since an increase of the negative transmittance in the remote zones results in operation on a more flat slope of the chalcogenide characteristic curve. In this case

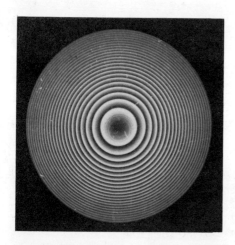 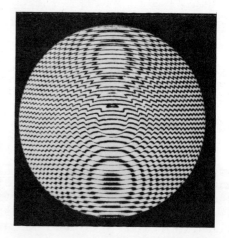

Figure 14. The interference field of the reflecting Fabry-Perot interferometer with an asymmetric intensity distribution in the rings. A fall in the intensity is seen in the direction of the remote rings.

Figure 15. A photograph of the kinoform lens in a Mach-Zehnder interferometer. A phase step is below 2π (4.8 - fold magnification).

the profile asymmetry, i.e. the relation between the
maximal-minimal spacing on a steep slope and the distance between the maximum peaks is kept approximately
constant and equals to 0.27 to the extent of 40 rings.
But in the contact printing of the pattern to chalcogenide films a phase profile (Figure 15) is maintained
only up to 20-25 rings. In our experiments the kinoform
lens diameter was chosen to be 16 mm that corresponds
to observation of about 27 zones.

Conclusion. Our basic effort was directed to the
development of kinoform lenses by a purely-optical technique. As it is seen from the paper, by utilizing this
technique it is difficult to produce high-aperture lenses due to the vignetting of interference beams in the
remote zones of the kinoform. However, for the optical
memory components, e.g. for a lens array, the lenses
with a 2-3 mm aperture and a 15 mm - focus can be obtained by diminishing the initial transparency received
via photographing of the interference image pattern .
These lenses will have a good practicable image quality.

In conclusion the authors wish to acknowledge with
gratitude the valuable collaboration of prof. G.W.Stroke
and prof. Yu.E. Nesterikhin in stimulation and sponsoring of this work.

REFERENCES

1. L.B. Lisem, P.M. Hirsch, J.A. Jordan Jr., IBM J.Res.Dev., v. 13, p. 150-155, 1969.

2. J.A. Jordan Jr., Z.B. Lesem, P.M. Hirsch, D.V. Van Rooy, Applied Optics, v. 9, N 8, p. 1883-1887, 1970.

3. J.-J. Clair, Optics Communications, v. 6, N 2, p. 135-137, 1972.

4. J.-J. Clair, Synthese optique de filtres d'amplitude et de Pase dits "kinoform". These de doctorat, 1972.

5. W.J. Dallas, Optics Communications, v. 3, N 4, p. 340-344, 1973.

6. V.V. Korsakov, V.I. Nalivaiko, V.G. Remesnik, V.G. Tsukerman, Avtometria, N 6, str. 24-31, 1974.

7. W.S. Rodney, I.H. Malitson, T.A. King, J.OPt. Sos. Am., v. 48, N 9, p. 633-636, 1958.

8. Yu.V. Troitsky, Pisma to JETF, v. 11, N 6, str.281--284, 1970.

9. N.D. Goldina, Yu.V. Troitsky in situ.

10. K.W. Meissner, Journ. Opt. Soc. Am., 31, p. 405, 1941; 32, p. 185, 1942.

QUESTIONS ON THE REPORT BY DR. V. KORONKEVICH

Dr. J. Goodman. How many zones has a kinoform lens?

Dr. V. Koronkevich. The lens has 30 zones. The material enables us to obtain resolution up to 5000 lines/min. The limitation of the lens diameter and the number of zones depend on vignetting of beams forming fringes on the lens edges.

Dr. A. Korpel. Due to which process does the absorption band shift occur in chalcogenide materials under illumination with an argon laser?

Dr. V. Koronkevich. The shift of the absorption band edges and the refractivity variation are determined by changes in photostructure of chalcogenide materials. In this case the material does not crystallize and remains amorphous.

Dr. W. Kock. I should like to note that an analogous lens for microwaves was developed by us and described in the Proceedings of the American-Japan Seminar.

ACOUSTOOPTIC SIGNAL PROCESSING

A. Korpel

Zenith Radio Corporation, Research Department

6001 W. Dickens Avenue, Chicago, Illinois 60639

Acoustooptic signal processing may be divided into three areas of application:

I. An acoustic delay line is used as a storage medium to be addressed by light.
II. An acoustooptic device is used as a beam deflector or frequency shifter in a processing operation.
III. Acoustooptic interaction is used to create a coherent light image of a synthetic two-dimensional sound pattern (signal) which may then further be processed.

The first area of application is historically the oldest and I will later show some examples dating back to well before the laser era. First, however, let us take a look at the basic principles involved. Figure 1 shows two examples of what is commonly called Bragg interaction. A transparent sound cell (glass, water etc.) is activated by an acoustic transducer which receives the signal:

$$e'(t)\ \exp(j\Omega_c t) \tag{1}$$

The actual signal to be processed consists of $e'(t)$ which modulates a carrier at frequency Ω_c. The term $e'(t)$ is a complex quantity in the most general case, indicating both phase and amplitude modulation of the carrier. Thus, in the case of simple amplitude modulation at Ω_m, $e'(t)$ may have the form

$$e'(t) \propto 1 + a \cos \Omega_m t \tag{2}$$

Similarly phase modulation may be represented by

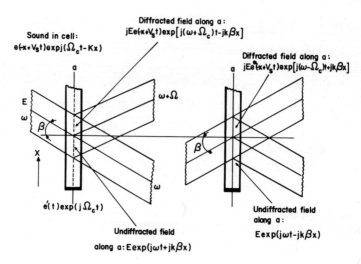

Fig. 1. Upshifted and downshifted Bragg diffraction

$$e'(t) \propto \exp(j\, a\, \cos \Omega_m t) \qquad (3)$$

A simple shift in frequency is given by

$$e'(t) \propto \exp(j\Omega_m t) \qquad (4)$$

etc.

Of particular use for purposes of analyzing the system are the following test signals:

an impulse:
$$e'(t) \propto \delta(t) \qquad (5)$$

a chirp:
$$e'(t) \propto \exp(j\Omega_m t^2/t_o) \qquad (6)$$

As shown in Fig. 1, the signal in the sound cell may be represented by

$$e(-x+V_s t)\, \exp j(\Omega t - Kx) \qquad (7)$$

where V_s is the sound velocity and

$$e(-x+V_s t) \propto e'(t - x/V_s) \qquad (8)$$

I have used the notation $e(-x+V_s t)$ simply for convenience. After these preliminary definitions let us consider the basic sound light interactions. In the ideal case (i.e. $e'(t)=1$ and all beams of sound and light are considered to be plane waves) the incident light must make an angle $\pm\beta$ with the sound wave fronts for interaction to occur at all:

$$\beta \simeq \frac{1}{2}\frac{\lambda}{\Lambda} = \frac{1}{4\pi}\frac{\lambda}{V_s}\Omega_c \qquad (9)$$

The angle β is called the Bragg angle and equation (9) simply states the condition for constructive interference, completely analogous to X-ray diffraction in crystals. The light that is diffracted is either upshifted or downshifted in frequency, depending on whether (9) is satisfied with the sound moving toward the incident light or away from it. The amplitude of the diffracted light is (for the case of weak interaction, assumed here) proportional to the local sound pressure. In our particular case it is thus proportional to $E\, e(-x+V_s t)$ i.e., for upshifted interaction, the diffracted light field along the center a of the sound cell may be written as

$$jEe(-x+V_s t)\, \exp[j(\omega+\Omega_c)\tau - jk\beta x] \qquad (10)$$

Here, ω is the frequency of the incident light. A similar equation holds for the downshifted interaction, as shown in Fig. 1, with the difference that e must be replaced by its conjugate e^* and β by $-\beta$. An objection may be raised to the above model of local sound pressure. It is the following: if indeed (9) has to apply rigorously how is it possible that interaction can take place with frequencies other than Ω_c? Yet that must have occurred in the model described by (10) because $e(t)$ itself exhibits a spread in frequencies so that the signal in the sound cell is likewise characterized by a frequency spectrum. Thus, if $e(t)$ is bandwidth-limited to a highest frequency B_m then the spectrum in the sound cell runs from $\frac{1}{2\pi}\Omega_c - B_m$ to $\frac{1}{2\pi}\Omega_c + B_m$. The fact that this entire spectrum interacts with the incident light in spite of (9), depends on the fact that the sound beam is not really a plane wave as was originally assumed. If the width of the sound column is ℓ, then the beam may be thought of as an angular spectrum of plane waves with a spread approximately equal to Λ_c/ℓ. Due to this fact there exists a tolerance $\Delta\beta$ on the Bragg angle:

$$\Delta\beta = \Delta(\frac{1}{4\pi}\frac{\lambda}{V_s}\Omega_c) \simeq \frac{\Lambda_c}{\ell} \qquad (11)$$

Writing $\Delta(\frac{1}{2\pi}\Omega_c) = 2B_m$ we find

$$B_m \simeq \frac{\Lambda_c V_s}{\lambda \ell} \qquad (12)$$

or

$$B_m f_c = \frac{V_s^2}{\lambda \ell} \qquad (13)$$

Equation 12 or 13 states the maximum allowable bandwidth of the signal to be processed with a sound cell of given length and sound velocity. For a glass sound cell ($V_s \simeq 4000$ m/sec) with an interaction length $\ell = 4$ cm and $\lambda = 0.4$ μm we find from (13):

$$B_m f_c = 10^{15} \qquad (14)$$

Thus, if the carrier frequency f_c equals 100 MHz, then a 10 MHz signal can be processed. By special techniques of beam steering it is possible to further increase the allowable bandwidth. The interested reader is referred to the existing literature.[1-3]

After this discussion of practical limitations, let us return to the diffracted light field as expressed by (10). The term $k\beta x$ (where $k = 2\pi/\lambda$) signifies that the light propagates at an angle β with respect to the sound wave fronts. As seen in Fig.1, the situation is quite symmetrical with respect to the sound-wavefronts. Considering (10) once more, we see that, in essence, this equation represents a moving phase and amplitude image of $e'(t)$ at a light frequency $\omega + \Omega_c$, the "rays" being directed at an angle β upward.

The situation shown in Fig.1, where basically only one diffracted order is being generated, is called operation in the Bragg regime. Is it possible to generate both orders at once while using only one incident beam? Yes, it is, provided that conditions are such that the tolerance on the Bragg angle allows both $+\beta$ and $-\beta$ interaction i.e.

$$\Delta\beta > 2\beta \qquad (15)$$

or, with 11,

$$\frac{\Lambda_c}{\ell} > \frac{\lambda}{\Lambda_c} \qquad (16)$$

which may be written

$$\ell < \frac{\Lambda_c^2}{\lambda} = \frac{V_s^2}{f_c^2 \lambda} \qquad (17)$$

Such a mode of operation (Fig.2) is called "operation in the Raman-Nath or Debye-Sears regime". It is characterized by small interaction lengths and relatively low sound frequencies and was used a great deal in the early history of acoustooptic processing. In this discussion, however, I shall restrict myself to Bragg diffraction operation. This is conceptually the simplest and any Raman-Nath type operation can usually be analyzed by combining upshifted and downshifted Bragg diffraction. Thus, in Fig.2, it is easily seen that the total light field along a is given by the sum of the unscattered and scattered light fields. (It is always assumed that the scattering is so weak that the incident field is not attenuated.) Hence, if

$$e(-x+V_s t) = |e(-x+V_s t)| \exp j\phi(-x+V_s t) \qquad (18)$$

we find, for the total light field:

$$E \exp(j\omega t) [1 + j \frac{\alpha}{2} |e(-x+V_s t)| \cos(\Omega_c t - Kx + \phi)] \qquad (19)$$

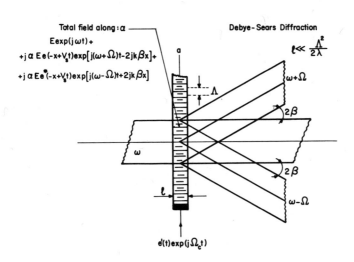

Fig. 2. Debye-Sears or Raman-Nath diffraction

where α is a scattering coefficient. For weak scattering, i.e. α << 1, (19) may be written

$$E \exp\{j[\omega t + \frac{\alpha}{2} e(-x+V_s t)\cos(\Omega_c t - Kx + \phi)]\} \quad (20)$$

where
$$K = \frac{2\pi}{\Lambda_c} = 2k\beta \quad (21)$$

The interpretation of (20) is that the total light emerging from the sound cell carries a moving phase image of the signal $e'(t)$, modulating the carrier Ω_c. Such a phase image may be rendered visible by a variety of techniques (e.g. Fresnel filtering) after which the now visible image may be further processed. To analyze such processing one usually has to decompose the light into its spatial frequency components which, in this case, means going back to the original diffracted orders. Thus, for purposes of analysis it is usually simpler to stay at the level of simultaneous up- and downshifted Bragg diffraction. For explanatory purposes, however, it is sometimes convenient to use the model of a phase image such as provided by (20). Most older experiments have been explained in terms of such models exclusively.

Let us now look at a typical Bragg diffraction processor as illustrated in Fig.3. Here the moving "upshifted" image is focused onto the stationary mask in plane "c" by means of the telescope system $L_1 L_2$. This system also serves to stop the undiffracted light. The light emerging from the transparency or mask $g(x)$ is focused by lens L_3 onto focal plane "d". A pinhole is

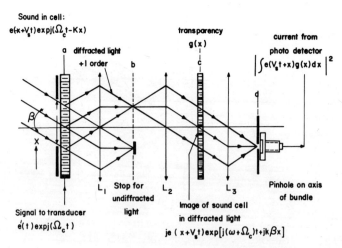

Fig. 3. Image plane processing

ACOUSTOOPTIC SIGNAL PROCESSING

provided at the focus of the diffracted order. It is easily seen that the D.C. current out of a photodetector behind the pinhole is given by

$$i \propto \left| \int_{-D/2}^{+D/2} e(x+V_s t) g(x) dx \right|^2 \quad (22)$$

Thus "i" provides a measure for the finite cross correlation of the signal "e" and the mask "g". Note that the pinhole is an essential element in this scheme. Basically it measures the on-axis intensity in the far field of the incident beam after perturbation by the mask g. Without pinhole the photodetector would have integrated all the light, as shown in Fig.4, and the output would have been

$$i \propto \int_{-D/2}^{+D/2} |e(x+V_s t) g(x)|^2 dx \quad (23)$$

For negative going e's this output does not properly represent a correlation measure. This is because in (23) we have basically integrated power and in (22) amplitude.

When the pinhole is off-set by an angle γ, as shown in Fig.4, it samples the off-axis intensity and the output is given by

$$i \propto \left| \int_{-D/2}^{+D/2} e(x+V_s t) g(x) \exp(-jk\gamma x) dx \right|^2 \quad (24)$$

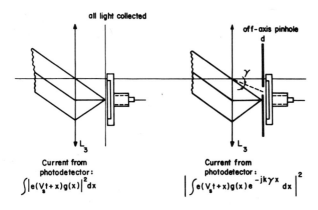

Fig. 4. Processing without pinhole and with off-set pinhole

In this way it is possible to represent more complicated mask functions than simple amplitudes g(x) i.e. the effective mask function in (24) is given by

$$g_{eff} = g(x)\exp(-jk\gamma x) \qquad (25)$$

Going one step further, we can place a "generalized pinhole" $h(\gamma)$ in plane d. The effective mask function would then be given approximately by

$$g_{eff}(x) = g(x)\int h(\gamma)\exp(-jk\gamma x)d\gamma = g(x)s(x) \qquad (26)$$

Expression (26) is only approximate because we ignored the spread of the light in plane "d" due to the finite beamwidth D. At any rate, these examples indicate what can be done with simple single or double aperture masks. Equation (26) provides a wider variety of possible mask functions but, because $h(\gamma)$ is real, they are still limited by the constraint

$$s(x) = s^*(-x) \qquad (27)$$

To achieve a more general phase- and amplitude mask function we have to use the technique of Fig.5. Here the undiffracted light is not stopped but rather modified by an "optical filter" to appear in plane c (the former mask plane), as reference field

$$E_r(x)\exp(j\omega t) \qquad (28)$$

For the sake of simplicity a simple symbolic light path has been shown in Fig.5 for this reference field. It goes without saying

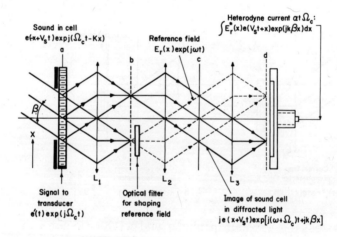

Fig. 5. Processing by heterodyning with reference field

ACOUSTOOPTIC SIGNAL PROCESSING

that in actuality the path could be quite involved and the "optical filter", indicated in Fig.5, quite complex. If we now observe the current at Ω_c, delivered by a photodiode (without pinhole) in plane "d", we find a complex amplitude $I(\Omega_c)$ caused by the heterodyning of the reference field and the Bragg-diffracted field:

$$I(\Omega_c) \propto \int_{-D/2}^{+D/2} E_r^*(x) e(x+V_s t) \exp(jk\beta x) dx \qquad (29)$$

It is clear that we now have a wide choice for the effective mask function:

$$g_{eff}(x) = E_r^*(x) \exp(jk\beta x) \qquad (30)$$

Moreover, the output is not squared (as in 23) but directly proportional to the finite correlation of e and g_{eff}.

One last example will suffice to indicate the versatility of Bragg diffraction processing. In Fig.6 a second sound cell, carrying a signal f is placed in mask plane c. Moreover the undiffracted light from the first sound cell is allowed to pass through unhindered. If we observe the current at $2\Omega_c$ delivered by the photodiode of Fig.5 and resulting from the heterodyning of the twice diffracted light and the undiffracted light we find:

$$I(2\Omega_c) \propto \int_{-D/2}^{+D/2} f(V_s't-x) e(V_s t+x) dx \qquad (31)$$

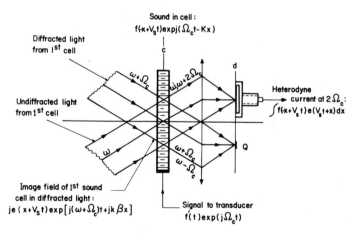

Fig. 6. Processing with second sound cell

To appreciate the operation symbolized by (31), let $D \to \infty$. Then, it may be easily seen that

$$I \propto \int_{-\infty}^{+\infty} f[(V_s' + V_s)t - x] e(x) dx \qquad (32)$$

It is clear that (32) represents the convolution of the *speeded-up* signal f by e. In fact, when $e(V_s t + x) = \delta(V_s t + x)$ we find that

$$I(2\Omega_c) \propto f'[(1 + \frac{V_s}{V_s'})t] \qquad (33)$$

This result is physically plausible because it is easily seen that a traveling impulse in sound cell 1 (i.e. "e") will scan out the signal in cell 2 (i.e. "f"), thereby speeding this signal up or slowing it down depending on the relative sign and magnitude of V_s and V_s'. Hence the arrangement of Fig.6 is a time expander or compressor. Because of the finite width D of the beam it can, in practice only work over a finite time or on a sampled basis.

Note that, although the magnification of the telescope system $L_1 L_2$ is taken to be unity, this is only done for convenience. Any other magnification would result in a different effective sound velocity V_s in the time compressor of Fig.5.

Let us return for a moment to (29). It is clear that $I(\Omega_c)$ can be considered as the convolution of the functions $E_r^*(-x)$ and $e(x)\exp(jk\beta x)$. This strongly suggests that, by considering a Fourier transform plane of "c" such as "b" or "d", for instance, we could arrive at an analysis in terms of a multiplication of frequency spectra rather than in terms of convolutions. Now, rather than performing the mathematical transformations from one plane to another, let us take a look at Bragg diffraction sound cells in terms of frequencies i.e. let us consider the cell to be basically a frequency analyzer rather than a delay line addressed by light. The configuration is sketched in Fig.7. We recognize the same cell as in Fig.1 with the same signal $e'(t)\exp(j\Omega_c t)$ applied. This time, however, we are going to try a frequency analysis, so we write $e'(t)$ in terms of its frequency spectrum $S(\Omega_m)$:

$$e'(t) \propto \int_{-\infty}^{+\infty} S(\Omega_m) \exp(j\Omega_m t) d\Omega_m \qquad (34)$$

A lens L_1 focuses both the diffracted and the undiffracted light in plane b. In particular the light diffracted by the carrier frequency Ω_c is focused at P, corresponding to a deflection angle

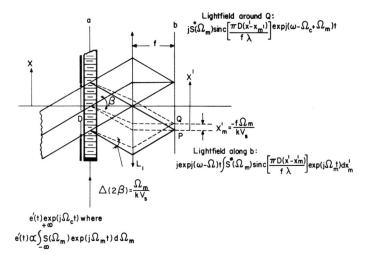

Fig. 7. Fourier plane processing (upshifted)

of β or, what is more relevant, 2β with respect to the incident light. The light at P therefore has a frequency $\omega+\Omega_c$ and amplitude proportional to the carrier i.e. proportional to $S(o)$. Now a frequency $+\Omega_m$ of $e(t)$ will correspond to a frequency $\Omega_c + \Omega_m$ of the total signal in the sound cell. This frequency will diffract light at a somewhat larger angle; *with respect to the incident light*:

$$\Delta 2\beta = \frac{\Omega_m}{kV_s} \tag{35}$$

The light thus diffracted will come to a focus in Q such that the distance $\overline{PQ} = x'_m$ is given by

$$\overline{PQ} = x'_m = \frac{f\Omega_m}{kV_s} \tag{36}$$

where f is the focal length of lens L_1. Thus in Q there exists a light amplitude proportional to $S(\Omega_m)$ with a frequency $\omega+\Omega_c+\Omega_m$.

Continuing this reasoning we find that the entire spectrum of $e(t)$ is displayed around point P. Note, however, that not only are the amplitudes of $S(\Omega_m)$ correctly represented by light amplitudes *but there exists also a one-to-one frequency correspondence*. What is displayed around P is the actual physical spectrum of $e'(t)\exp(j\Omega_c t)$, each frequency being superimposed upon the light frequency ω. We may write for the light field along "b", near P:

$$j \exp j(\omega+\Omega_c) t \int S(\Omega_m) \delta(x'-x'_m) \exp(j\Omega_m) dx'_m \qquad (37)$$

where the relation between Ω_m and x'_m is as in (36). Expression (37) is approximate because no account has been taken of the light spreading due to the finite beam size D. If this is done the δ function in (37) has to be replaced by a sinc function:

$$\delta(x'-x'_m) \to \text{sinc}\left[\frac{\pi D(x'-x'_m)}{f\lambda}\right] \qquad (38)$$

Now assume that a reference field $E_b(x')$ at frequency ω is also incident at plane "b" and that all the light is collected by a photodiode as indicated in Fig.8. The frequency component at Ω_c of the resulting heterodyne current is then given by

$$I(\Omega_c) \propto \exp(j\Omega_c t) \int S(\Omega_m) E_b^*(x') \delta(x'-x'_m) \exp(j\Omega_m t) dx' dx'_m \qquad (39)$$

which may be written as

$$I(\Omega_c) \propto \exp(j\Omega_c t) \int S(\Omega_m) E_b^*(\Omega_m) \exp(j\Omega_m t) d\Omega_m \qquad (40)$$

where
$$E_b^*(\Omega_m) = E_b^*(x'=x'_m=f\Omega_m/kV_s) \qquad (41)$$

Thus, if $S'(\Omega_m)$ represents the frequency spectrum of $I(\Omega_c)$ i.e.

$$I(\Omega_c) \propto \int S'(\Omega_m) \exp(j\Omega_m t) d\Omega_m \qquad (42)$$

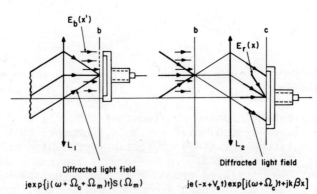

Fig. 8. Relation between image-plane and Fourier-plane processing

Then
$$S'(\Omega_m) \propto S(\Omega_m) E_b^*(\Omega_m) \qquad (43)$$

Equation (43) sums up the essential features of this "Fourier-plane" processing: The reference field $E_b^*(\Omega_m)$ acts as an arbitrary phase and amplitude filter for the signal spectrum $S(\Omega_m)$.

If now, in Fig.8, another lens L_2 is placed between plane b and the photodetector plane c, it is clear that the current from the detector must remain unchanged. (This current basically represents a time variation of the photon flux and this cannot be changed by placing a non- or weakly dispersive element in the path.) A convenient arrangement for calculation is to let "b" and "c" be the front- and back focal planes of L_2. Then the fields in "b" and "c" are related by a simple Fourier transform and, in fact, the fields along "b" in Fig.8 transform to the fields along "c" in Fig.5. The final integral for the current now signifies a convolution, as is to be expected when one Fourier-transforms the product of two functions (43). Thus, there is no basic difference between the "Fourier-plane" processing of Fig.8 and the "image-plane" processing of Fig.5. Any choice is simply a matter of convenience. For sake of completeness the downshifted equivalent of Fig.7 has been shown in Fig.9.

Let us now look at some examples, old and new, of acousto-optic processing. Figure 10 shows a Fourier plane processing

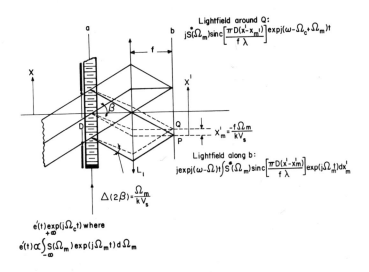

Fig. 9. Fourier plane processing (downshifted)

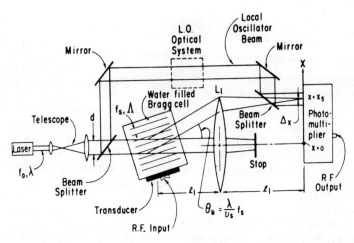

Fig. 10. Fourier plane processing (after Whitman et al[4])

system developed in 1967 in the present author's laboratory.[4] Although the light paths are somewhat complicated, the reader will have no trouble in recognizing the essential elements of Figs. 7 and 8. The reference field E_b was generated by the box labeled "L.O. Optical System" which operated on the undiffracted light in a separate optical path. In a specific application the "L.O. Optical System" consisted of a simple lens as shown in detail in Fig.11. This causes a curved reference field which is equivalent to a simple parabolic phase filter or dispersive delay line. The dispersion can be simply adjusted by varying the position of the lens. Results are shown in Fig.12 in which pulse stretching (a measure of dispersion) is plotted vs radius of curvature of the reference field.

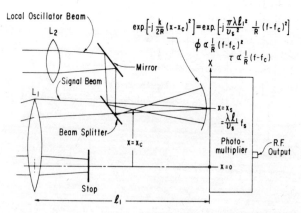

Fig. 11. Details of reference field (after Whitman et al[4])

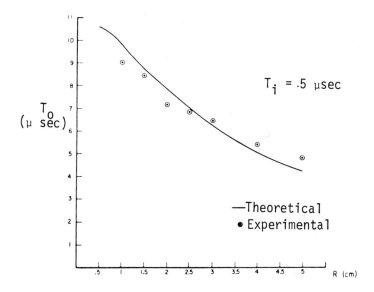

Fig. 12. Pulse broadening vs reference field radius of curvature (after Whitman et al[4])

Figure 13 shows a 1961 correlation processor by Slobodin[5]. An arc lamp was used and the mask (replica) was placed directly on the sound cell rather than in a relayed image plane such as "c" in Fig.3. From the description of the experiment it appears that Raman-Nath operation was involved. However, the knife edge used would seem to make the arrangement somewhat similar to Fig.4 (left) by eliminating undiffracted and downshifted light. In any case, successful correlation was performed between chirp signals and chirp-type gratings.

A slightly different arrangement was used by Rosenthal[6] in 1961 and shown in Fig.14. Again, Raman-Nath diffraction is

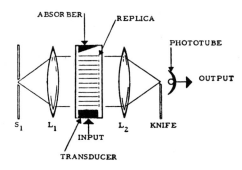

Fig. 13. Early optical correlator (after Slobodin[5])

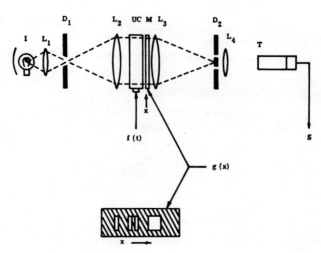

Fig. 14. Early optical correlator (after Rosenthal[6])

involved but this time the "phase-picture" in the sound cell (20) is made visible by using an annular aperture in the detection system. Clearly, anything that will preferentially pass or stop diffracted light will make the phase picture visible and result in some output which is a measure of correlation. Exact analysis of such "ad hoc" systems is, however, not always easy. Thus, Arm et al (Columbia University)[7], in a 1964 discussion of Slobodin's technique, compare it with their own scheme which uses an off-axis pinhole. This is shown in Fig.15; the similarity to Fig.4 (right) will be obvious. They point out that, in their system, a correlation peak is reached when the signal in the sound cell and the replica differ by a constant frequency. This is exactly what

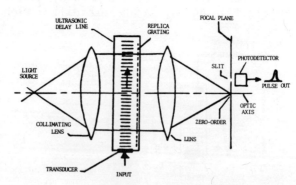

Fig. 15. Correlator with off-set pinhole (after Arm et al[7])

one would expect on the basis of (25) which in fact shows that the effective replica is shifted in frequency from the original one. Other work by the Columbia group may be found in reference 8.

It is hard to say exactly who invented acousto-optic processing but I propose F. Okolicsanyi as the father of the subject. At the end of the thirties Okolicsanyi was connected with the Scophony Television Laboratories and he and his colleagues were investigating ways of acousto-optically modulating and scanning light waves in order to produce a projection television receiver. In the course of his work Okolicsanyi made a thorough investigation of possible sound-cell applications which he published in a beautifully written article in 1937[9]. As an example I show his drawing of what is essentially the time compressor discussed before (Fig.16). His ideas were the basis of the Scophony television projector which was successfully demonstrated but was soon rendered obsolete by the cathode ray tube. A modern version of the Scophony system using laser light and Bragg deflectors was demonstrated in 1966 by the present author and his colleagues[10].

We now have come to the second area of application, i.e. the sound cell used as a beam deflector or frequency shifter in a processing operation. This area has not been quite as thoroughly investigated as the first one. Some applications are rather trivial, such as laser beam scanning of a transparency etc., and we will not discuss them here. The more interesting applications involve heterodyning in some way or other. One proposed system for frequency domain coding[11], for instance, is shown in Fig.17. The sound cell is here activated with a large range of frequencies simultaneously and the diffracted light focused in the Fourier transform plane. As discussed before this plane then displays an actual spread of physical frequencies aligned parallel to the sound cell. If this light distribution is now made to pass through a (one dimensional) transparency g(x) and the emerging light heterodyned with a reference beam at fixed frequency, then the resulting current encodes the transparency in terms of frequencies. That is to say each point of the transparency is

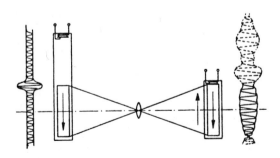

Fig. 16. Time compressor (after Okolicsanyi[9])

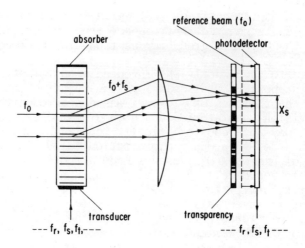

Fig. 17. Frequency domain coding (after Korpel et al[11])

represented by a specific frequency component in the resultant signal. Such a coding is called "frequency domain coding". It is obviously related to Fourier transform holography but the resulting hologram is electronic rather than photographic. There are interesting applications; the reader will find them described in reference (11).

If the frequencies are not fed simultaneously into the sound cell but rather sequentially, then Fig.17 is equivalent to a combination of acoustooptic scanning, followed by heterodyning. Such an experiment was performed in the author's lab[12] in 1969 and is shown in Fig.18. In the experiment we detected essentially the *phase and amplitude* distribution of an arbitrary light field. The light field to be scanned was a simple uniform beam distorted by the "beam perturbing object". Only amplitude was measured and displayed. The particular cross section visualized on the monitor is that in which the scanning beam comes to a focus, *even if that focus is virtual rather than real*. The latter statement follows directly from the fact that the position of the photodiode is immaterial as long as it collects all the light. Thus its output is always that which it would be when placed in the focus of the scanning beam, even though that focus may be virtual, and the above is only possible as a Gedanken experiment. This, at first glance, somewhat startling fact is vividly illustrated in Fig.19 which shows the monitor display of the light field in the plane of a Ronchi ruling. To obtain this picture, a *negative* lens L was used so as to bring the virtual focus in the plane of the ruling. When the scanning beam was collimated (i.e. lens L left out), the monitor displayed the light field at infinity i.e. the far field.

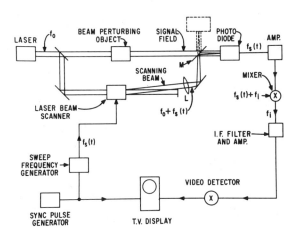

Fig. 18. Laser heterodyne scanning
(after Korpel & Whitman[12])

This is illustrated in Fig.20. It is thus clear that a sound cell may be used to sample amplitude and phase of a light field. The applications to optical processing are not quite evident, but at least such a operation would seem to form a link between linear (coherent) optical processing and linear electrical processing (filtering).

We finally have come to the last area of application: that of Bragg diffraction imaging. Bragg imaging was developed by the

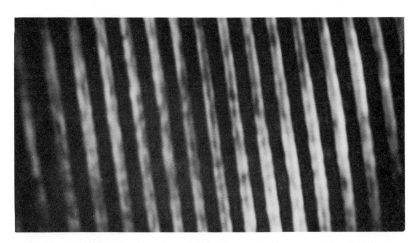

Fig. 19. Laser scanned light field in plane of ruling
(after Korpel & Whitman[12])

Fig. 20. Laser-scanned far field of ruling
(after Korpel & Whitman[12])

present author in 1966 as a method of making sound fields visible[13]. A simple experimental setup is shown in Fig. 21. A convenient beam of laser light traverses a sound cell and results in two spots of diffracted light O' and O". It may be shown that these spots are actually demagnified images of the sound beam cross section. They may be projected onto a screen by suitable optics. An interesting aspect of this imaging is that amplitude and phase of the sound field are preserved in the image. As a consequence of this the imaging is really three dimensional i.e. "volume-to-volume" rather than "plane-to-plane". The immediate application of Bragg imaging was sought in the field of acoustic imaging as applied to medicine and non-destructive testing.

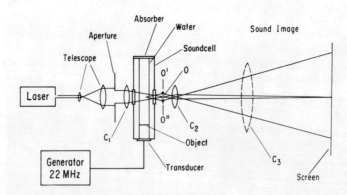

Fig. 21. Experimental setup for Bragg-diffraction imaging
(after Korpel[13])

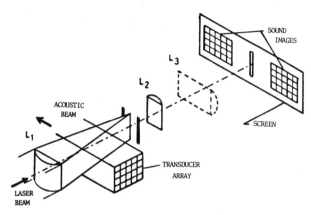

Fig. 22. Acoustooptic page composer (after Szentesi[14])

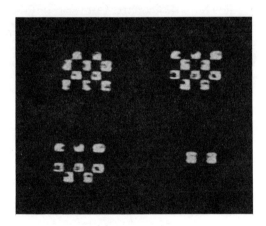

Fig. 23. Images of synthetized acoustic fields (after Szentesi[14])

In 1972 it was shown by Szentesi that Bragg imaging could be used as the basis for a page composer[14]. His arrangement is shown in Fig.22. Each one of the squares in the transducer array can be activated separately and the sound image thus modified. Some experimental results are shown in Fig.23. To my knowledge no further work has been done on this interesting idea; it may be worth further investigation, especially in combination with some of the heterodyning techniques discussed before.

In conclusion then, acoustooptics lends itself to rather unique ways of signal processing. Correlation, convolution and versatile frequency filtering of electrical signals is readily performed. In addition, it provides some new tools for processing optical information. A note of caution should be sounded, however; new integrated electronic techniques have a way of rapidly replacing older schemes based on more complex (and perhaps more interesting) physical phenomena. Acoustooptic signal processing may not be an exception.

REFERENCES

1) R. Adler "Interaction Between Light and Sound", IEEE Spectrum, May 1967, pp. 42-53.

2) E. I. Gordon "A Review of Acousto-Optical Deflection and Modulation Devices", Proc. IEEE, 54, 1931 (1966).

3) A. Korpel "Acousto-Optics" in *Applied Solid State Science*, Vol. 3, R. Wolfe (ed.), Academic Press, New York, 1972, pp. 72-180.

4) R. Whitman, A. Korpel and S. Lotsoff "Application of Acoustic Bragg Diffraction to Optical Processing Techniques", Proc. Symposium of Modern Optics, Polytechnic Press, March 1969.

5) L. Slobodin "Optical Correlation Technique", Proc. IEEE, 51, 1782 (1963).

6) A. H. Rosenthal "Application of Ultrasonic Light Modulation to Signal Recording, Display, Analysis and Communication", IRE Trans. on Ultrasonics Engineering, UE-8, No. 1, 1-5 (1961).

7) M. Arm et al "Optical Correlation Technique for Radar Pulse Compression", Proc. IEEE, 52, 842 (1964).

8) M. King et al "Real-Time Electrooptical Signal Processors with Coherent Detection", Applied Optics, Vol. 6, No. 8, August 1967, pp. 1367-1375.

9) F. Okolicsanyi "The Wave-Slot, An Optical Television System", The Wireless Engineer, October 1937, pp. 527-536.

10) A. Korpel, R. Adler, P. Desmares and W. Watson "A Television Display Using Acoustic Deflection and Modulation of Coherent Light", Joint Issue of IEEE Proc., Vol. 54, and Applied Optics, Vol. 5, October 1966.

11) A. Korpel, S. N. Lotsoff and R. L. Whitman "The Interchange of Time and Frequency in Televison Displays", Proc. IEEE, Vol.57, No. 2, February 1969, pp. 160-170.

12) A. Korpel and R. L. Whitman "Visualization of a Coherent Light Field by Heterodyning with a Scanning Laser Beam", Applied Optics, Vol. 3, August 1969, p. 1577.

13) A. Korpel "Visualization of the Cross Section of a Sound Beam by Bragg Diffraction of Light", Applied Physics Letters, Vol.9, No. 12, 15 December 1966.

14) O. I. Szentesi "An Acoustooptic Page Composer Based on Bragg Imaging", Proc. IEEE, Vol. 60, No. 11, November 1972, pp. 1461-1462.

SOME PECULIARITIES OF PHYSICAL REALIZATION OF OPERATION OPTICAL MEMORY

E.G. Kostsov, V.K. Malinovski,
Yu.E. Nesterikhin, A.N. Potapov

Institute of Automation and Electrometry
Novosibirsk, USSR

ABSTRACT

Some peculiarities of physical realization of the operation optical memory have been considered. It has been shown that most efficiently can operate the medium which operation is based on photoelectric energy transform in recording and electrooptical effects in reading. The application of such system cells consisting of discrete elements allows the third dimension to be efficiently used for their connection. The estimations show that one may achieve the values for the limiting cell packing density of about 10^5 cm^{-2}, the tact frequency of 10^7 cps and the Q-factor of 10^{12} j^{-1}. The characteristic features of operation for an elementary cell model have been investigated experimentally.

1. The recent investigations show that the problem of development of opto-electronic fixed memory devices (FMDs) of large capacity (10^9 - 10^{12} bits) is nearing its technical realization. Not nearly so clear is the problem on using optical means in operation systems. Its cardinal advantages consisting in the possibility of organizing contactless connections with simultaneous disparalleling of data processing channels have not yet been realized as to their particular structural organization.

It is considered that the main problem in developing operation opto-electronic memory devices is the problem of information carriers. In many laboratories practically the whole spectrum of physical effects is investigated potentially applicable for repeated use in the recording-erasing-reading cycles with small time of the cycle ($< 10^{-7}$ sec), minimal energy capacity ($< 10^{-3}$ j/cm^2) and high spatial resolution (> 1000 lines/mm).

In what follows the specificity of functioning of the operation information carrier will be analyzed, a version for an elementary cell of the operation memory will be suggested and the possibility for its realization by utilizing descrete components will be considered.

2. When recording information a light beam affects the medium varying its optical properties (a refraction coefficient, absorption, etc.). In this case energy expenditures are covered either due to the recording beam energy or to a great extent due to the external source (electrical or magnetic field, chemical reactions, etc.). The reading beam responds to variation in the medium properties and reconstructs the information coded.

In fact the problem consists in the control of one light beam (reading) parameters with the help of another (recording) one and the operation medium is an intermediate chain permitting an efficient interaction of two light beams to be realized. It means that the medium should perform the functions of an energy transformer in recording and those of a light modulator in reading. Since it is unpractical to impose all the energy expenditures on the recording beam an external energy source is required intensifying light-initiated changes in the medium up to reasonable values.

Taking into account the above-said, the elementary cell of the operation medium can schematically been shown in the form represented in Figure 1. This representation is conventional (in principle the components may be combined within one layer), but it allows the physical effects forming the basis for the medium functioning to be separated more distinctly.

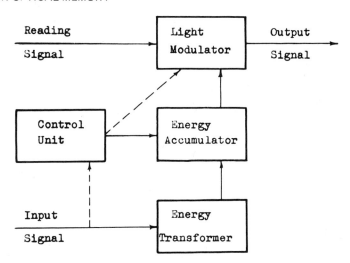

Figure 1. Functional Scheme of Elementary Cell of Operation Medium.

3. Consider the specific features of each element of the cell.

<u>Light Modulator</u>. Out of all the modulation types (amplitude, frequency, phase) preference should be given to a phase one.

The frequency modulation is inefficient due to a low efficiency of the light frequency transform utilizing non-linear effects.

With an amplitude modulation energy losses in the modulator take place that imposes certain limitations on the reading signal level or its duration and results in the problem of heat removal.

In the case of a phase modulation light losses are small, energy evolves outside the analyzer medium. In addition as it is known from the communication theory the phase modulation is the most noise-immunity one [1]. The physical effects permitting the phase modulation (e.g., an electrooptical effect) to be realized, as well as the media having a low half-wave potential (e.g., $Sr_{0.75} Ba_{0.25} Nb_2 O_6$) are known and thoroughly investigated.

Energy Accumulator. Due to a low efficiency of the light wave energy transform into the form required for realizing changes in the media it seems to be more advantageous to use a light beam as an information carrier alone. In this case the light beam functions at the moment of recording reduce to the process of initiating energy accumulation (or evolution). As an illustrative example photomaterials can be given. The intensification realized due to an energy accumulator (chemical development) achieves the values of about $\sim 10^{10}$! For the operation systems it is difficult to choose a reversible chemical reaction with a great quantum yield, therefore "chemical" accumulators are not used.

It seems most advantageous to use the electric field energy. In favour of this statement several arguments may be given: high density of electrostatic energy can readily be achieved, methods are known for a photoelectric transform, fast response of the "capacitive" accumulator can be made great enough ($Rc \leq 10^{-9}$ sec).

The development of the accumulator utilizing the magnetic field energy is low-efficient since the effects of a direct photomagnetic transform are negligible and an achievable density of the energy accumulated in the form of a magnetic field reasonable yeilds to an electrostatic one.

The application of capacitive accumulators and evolution of the accumulated energy initiated by light allows us to realize the intensification with the coefficient $\sim 10^4 \div 10^6$.

Energy Transformer. From the above considerations it follows that the transformer should be photoelectric. A transformer-accumulator combination makes it possible to realize a threshold optical element.

The analysis of operation of cell elements shows that in the best way the system will operate based on the energy photoelectric transform controlled with the help of an electric field in recording and on the electrooptical-type effects in reading.*

4. The above-made consideration of a model of the operation medium elementary cell allows it to be composed of discrete specialized components.

The peculiarities of operation of such a cell utilizing a lithium niobate crystal as a light modulator and an avalanche-type phototransistor as a photoelectric transformer were investigated experimentally. As an energy accumulator the capacity was used formed by the modulator electrodes. Charging the capacity was performed from an external current source via commutator. The basic scheme of the cell is shown in Figure 2, the cycle of its operation consists of the following tacts:

the 1st tact is "erasing" (charging the capacity - variation of the accumulator energetic state).

the 2nd tact is "recording" (feeding of a light signal to the transformer output - initiation of energy evolution).

the 3d tact is "writing" (feeding of a light signal to the modulator output - modulation of a light beam).

The parameters characterizing the operation of the cell: I_o is intensity of the reading light beam; V_{cont} is a value of the control signal (the accumulated energy quantity), Q_m is a state of the memory; t_r is duration of the recording time, t_{st} is time passed from the moment of recording to that of reading (the operation of "information erase" will be considered as one of the control operations). Light intensity on the cell output I_{out} is determined by the memory state Q_m ($I_{out} = f(Q_m)$), while $Q_m = \Psi(V_{cont}, I_o, t_r, t_{st})$ Two counterpositioned cells form a model of a continuous chain of storing elements that allows us to analyze the

---

* Out of the known at present designs the "PROM" - type multilayer structures have already found an application [2]. Their operation is based on utilizing an electrooptical effect in photosensitive crystals of germanate and bismuth silicate, the role of an energy accumulator is played by the electrical field controlled by external devices.

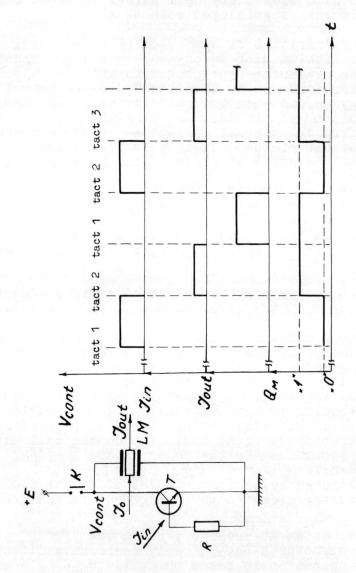

Figure 2. Basic Scheme and Timing Diagram of Cell Operation Cycle.

OPERATION OPTICAL MEMORY

peculiarities of the element-to-element information transfer.

Guided by the same considerations that were used when studying logical quantizers 3 let us write the system of equations characterizing the operation of the above chain of elements

$$Q_m^{out} = \Psi[f(Q_m^{in}), V_{cont}, t_r, t_{st}]$$
$$Q_m^{in} = \Psi[f(Q_m^{out}), V_{cont}, t_r, t_{st}]$$

The unlimited number of cycles of the information cell-to-cell transfer is possible only in case the above system of equations consists of three real roots. The solution of this system allows us to determine the value of the signals "0", "1", and the threshold value Q_m^{th} with which at $Q_m > Q_m^{th}$ in cycling, the process of convergence occurs to the signal level corresponding to "1" or at $Q_m < Q_m^{th}$ to the level corresponding to "0". (It implies that each cell possesses an intensification ability). The solution of the system of equations in combination with the equation characterizing the condition of osculation of the curves $Q_m^{out} = \Psi(Q_m^{in})$ and $Q_m^{out} = \Psi^{-1}(Q_m^{in})$ allows us to determine the minimal values of V_{cont}, I_o, t_r

Note that the value of the minimal parameters V_{cont}, I_o, t_r may serve as a criterion when choosing materials used in memory elements.

For an experimental cell the following values of the basic parameters have been obtained: the minimal control voltage V_{cont} = 100 v, the minimal intensity I_o = 3 · 10⁻⁶ w, $t_r \leq$ 2 · 10⁻⁸ sec, t_{st} = 10⁻³ sec, the maximal coefficient of light intensification ~10³, the Q - factor of the cell 10⁷ j⁻¹.

5. The necessity to use a solid state energy transformer restricts the size and packing density of the memory components to the values characteristic of the modern integral technology. The wave nature of light imposes additional limitations on the above parameters.

Let the light losses due to diffraction phenomena on the edges of the light modulator window be characterized with the help of the wave parameter $\chi = \frac{\sqrt{\lambda h}}{b_1}$, where b_1 is a linear size of the modulator window, h is a distance from the modulator window center to the point of observation. At $X < 1$ the plane light front propagation is described (to the first approximation) within the frames of geometrical optics, and at $X > 1$ the intensity distribution is characterized by a complex directionality diagram.

An expression for the surface packing density

$$N = \frac{X_1^2}{\lambda h (\sqrt{n} X_1 + 1)^2}$$

can be obtained by determining the level of the diffraction losses (it will correspond to the value $X = X_1$) and the minimal distance between the elements b_2 that may be expressed via step of the Fresnel fringes $b_2 = \sqrt{h \lambda n}$ (n is an integer).

From this expression it follows that the extension of optical communication lines should be minimal. The quantitative evaluation given provided that h = 10 mcm (the substate thickness), X = 0.3; n = 9 (according to the Cornu spiral [4] it corresponds to diffraction light losses less than 5%) shows that N will not exceed 10^5 cm^{-2}.

At the sizes of the elements close to limiting the use of an electrooptical effect for light modulation allows us to obtain the Q - factor values of 10^{11} ÷ 10^{12} j^{-1}.

From this it follows that the tact frequency for the functioning of the optical operation memory at the above packing density can be 10^6 - 10^7 cps, i.e. the device comprising 10^8 elements can rprovide the information exchange at a rate of 10^{14} - 10^{15} bit/sec.

REFERENCES

1. A.A. Kharkevitch. Borba s pomekhami. M. Nauka. 1965.

2. P. Nisenson, S. Iwasa. Real Time Optical Processing with $Bi_{12}SiO_{20}$ PROM. Applied Optics. V 11, N 12, 1972.

3. A.W. Lo. Physical Realization of Digital Logic Cireuits Micropower Electronics. Pergamon Press, 1964

4. M. Born, E. Wolf. Osnovy optiki. M. Nauka, 1970.

NON-COHERENT OPTICAL SYSTEM FOR PROCESSING OF IMAGES AND SIGNALS

B.E. Krivenkov, S.V. Mikhlyaev,
P.E. Tverdokhleb, Yu.V. Chugui

Institute of Automation and Electrometry
Novosibirsk, USSR

ABSTRACT

A non coherent optical system has been proposed for performing transforms equivalent to the multiplication of three matrices. The possibility of using such a system has been shown for a spectral analysis of images according to an arbitrary basis with separated variables and for a multichannel processing of signals. For the optical system characteristic parameters the size of the matrices under multiplication is of the order of 100 x 100.

I. INTRODUCTION

The transforms equivalent to the operation of multiplication of several matrices (with the number of rows and columns different from unity) are initial in the solution of some important practical problems. They are used, e.g. in multichannel processing of one dimensional signals and in processing of two-dimensional images (the problems of analysis, filtering, detection and coding).

The problems of simulation of matrix transforms by optical coherent techniques were considered in [1, 2, 3]. Thus, in [1, 2] for the multiplication of two multi-element matrices a system of optical space filtering with a holographic filter was used. The resulting matrix elements were calculated simultaneously and in parallel. Later on for the same purposes a special optical system was proposed more effective in respect of using light energy, calculation accuracy and organization of work in an operational mode [3].

The further investigations [4, 5] carried out in this direction showed that optical systems are able to perform more complex transforms being described by the product of three matrices and that such transforms can be simulated not only by coherent but also by non-coherent techniques. It should be noted, however, that in the newly-developed systems the resulting matrix elements were calculated by the successive [4] and parallel-successive [5] methods involving image interchange of the matrices under multiplication rows (columns).

Below a non-coherent optical system with a parallel calculation of the desired matrix elements is supposed. The transform being realized by the system is described by the matrix equation

$$[D] = [A]\cdot[B]\cdot[C], \qquad (1)$$

wherein $[A]$, $[B]$, $[C]$ are initial matrices of $N \times M$, $M \times P$, and $P \times R$ size with sing-variable elements a_{ij}, b_{jk}, $c_{k\ell}$, and $[D]$ is a resulting matrix of $N \times R$ size with the elements

$$d_{i\ell} = \sum_{j=1}^{M} \sum_{k=1}^{P} a_{ij} \cdot b_{jk} \cdot c_{k\ell}, \quad i=1,2,\ldots,N; \quad \ell=1,2,\ldots,R.$$

The present paper is devoted to the consideration of this system structure and potentialities as well as the results of its experimental study.

II. MODE OF OPERATION

The structural scheme of the system to be discussed is shown in Figure 1, where 1 is a unit of superposition of the first matrix rows, 2 is a unit of integration over the X-coordinate and 3 is an integration-projecting unit. In the planes P_1, P_2 and P_3 the transparancies T_1, T_2 and T_3 with images of the matrices $[A]$, $[B]$ and $[C]$, respectively, are placed.

NON-COHERENT OPTICAL SYSTEM

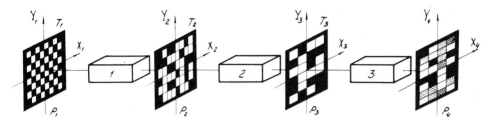

Figure 1. Optical system structure.

If the transparency operation field of the $L \times L$ - size square form is divided into $N \times M$ resolution elements - rectangulars of $\delta x_1 \times \delta y_1$ size, where $\delta x_1 = L/M$, $\delta y_1 = L/N$; then the transmission function with respect to intensity can be represented as

$$T_1(x_1,y_1) = \sum_{i=1}^{N}\sum_{j=1}^{M} a_{ij}\, \text{rect}\left[\frac{x_1 - (j-\frac{1}{2})\delta x_1 + \frac{L}{2}}{\delta x_1}\right]\text{rect}\left[\frac{y_1 - (i-\frac{1}{2})\delta y_1 + \frac{L}{2}}{\delta y_1}\right] \quad (2)$$

In expression (2) a_{ij} is transmission of the (i,j)-th element of the transparency T_1 being proportional to the value of the matrix $[A]$ respective element and

$$\text{rect}[Z] = \begin{cases} 1, & \text{at } |Z| \leq 1/2, \\ 0, & \text{at } |Z| > 1/2. \end{cases}$$

The transmission functions of the transparencies T_2 and T_3 can be described in an analogous way, i.e. in the form of

$$T_2(x_2,y_2) = \sum_{m=1}^{M}\sum_{n=1}^{P} \beta_{mn}\, \text{rect}\left[\frac{x_2 - (n-\frac{1}{2})\delta y_3 + \frac{L}{2}}{\delta y_3}\right]\text{rect}\left[\frac{y_2 - (m-\frac{1}{2})\delta x_1 + \frac{L}{2}}{\delta x_1}\right]$$

$$T_3(x_3,y_3) = \sum_{K=1}^{P}\sum_{\ell=1}^{R} C_{K\ell}\, \text{rect}\left[\frac{x_3 - (\ell-\frac{1}{2})\delta x_3 + \frac{L}{2}}{\delta x_3}\right]\text{rect}\left[\frac{y_3 - (K-\frac{1}{2})\delta y_3 + \frac{L}{2}}{\delta y_3}\right], \quad (3)$$

wherein β_{mn} and $C_{K\ell}$ are transmissions of the (m,n) and (K,ℓ)-th element of these transparencies proportional to the values of the matrices $[B]$ and $[C]$ respective elements.

On the transparency T_1 let us isolate the images of the matrix $[A]$ rows. With the help of unit 1 these images are projected to the plane P_2 over the X-coordinate, are defocused over the Y-coordinate and are superimposed at angles determined by their center coordinates. The one-dimensional light distributions

$$\sum_{j=1}^{M} a_{ij}\, rect\left[\frac{x_2-(j-\frac{1}{2})\delta x_1 + \frac{L}{2}}{\delta x_1}\right] rect\left[\frac{y_2+\frac{L}{2}}{L}\right], \quad i=1,2,\ldots,N$$

obtained in the plane P_2 are modulated by the transparency T_2 transmission function, are integrated over the X-coordinate by unit 2 and are projected to the plane P_3 over the Y-coordinate at different angles. As a result in the plane P_3 is formed a totality of one-dimensional distributions with the intensity

$$R_i(y_3) = \sum_{j=1}^{M} a_{ij} \int_{-L/2}^{L/2} rect\left[\frac{x_2-(j-\frac{1}{2})\delta x_1 + \frac{L}{2}}{\delta x_1}\right] T_2(x_2, y_3)\, dx_2 ,$$

$$i = 1, 2, \ldots, N . \tag{4}$$

Then light distributions (4) are modulated by the transparency T_3 transmission function, are projected over the X-coordinate to the plane P_4 by unit 3 and are integrated over the Y-coordinate with simultaneous space separation of light beams in accordance with the initial row images inclination. In fact, it means that in the system under consideration the plane P_1 is projected over the Y-coordinate to the plane P_4. Then with the account of (3) the intensity in the output plane will be distributed according to the law

$$\mathcal{D}(x_4,y_4) = \sum_{i=1}^{N}\sum_{\ell=1}^{R} rect\left[\frac{y_4-(i-\frac{1}{2})\delta y_1 + \frac{L}{2}}{\delta y_1}\right] rect\left[\frac{x_4-(\ell-\frac{1}{2})\delta x_3 + \frac{L}{2}}{\delta x_3}\right] \sum_{j=1}^{M}\sum_{K=1}^{P} a_{ij}\beta(K,j) C_{K\ell}$$

where $\beta(K,j) = \int_{-L/2}^{L/2}\int_{-L/2}^{L/2} \left\{ \sum_{m=1}^{M}\sum_{n=1}^{P} b_{mn}\, rect\left[\frac{y_3-(m-\frac{1}{2})\delta x_1 + \frac{L}{2}}{\delta x_1}\right] \times \right.$

$$rect\left[\frac{x_2-(j-\frac{1}{2})\delta x_1 + \frac{L}{2}}{\delta x_1}\right] rect\left[\frac{x_2-(n-\frac{1}{2})\delta y_3 + \frac{L}{2}}{\delta y_3}\right] rect\left[\frac{y_3-(K-\frac{1}{2})\delta y_3 + \frac{L}{2}}{\delta y_3}\right] \Bigg\} dx_2\, dy_3$$

$$\tag{5}$$

Intensity distribution (5) is proportional to the resulting matrix $[D]$ if on the transparency T_2 the matrix $[B]$ is represented in a transpositioned form. In this case

$$b(k,j) = \sum_{m=1}^{M}\sum_{n=1}^{M} b_{mn} \cdot \delta_{mj} \cdot \delta_{nk} = b_{jk},$$

where δ_{mj}, δ_{nk} are Kroneker symbols.

According to the initial condition the matrices $[A]$, $[B]$ and $[C]$ are sign-variable, therefore the method for specification and calculation of the matrix negative values should be determined. Since $[A] = [A]^+ - [A]^-$, where $[A]^+$ and $[A]^-$ are matrices with the elements $a_{ij} \geqslant 0$ and $a_{ij} \geqslant 0$, then the matrix $[A]$ representation is achieved by separate recording of the positive matrices $[A]^+$ and $[A]^-$ as images on the transparencies T_1^+ and T_1^-. In a similar way the matrices $[B]$ and $[C]$ are preset on the transparencies T_2^+, T_2^- and T_3^+, T_3^-. Then, if to the planes P_1, P_2 and P_3 of the system transparencies* in the combinations T_1^+, T_2^+, T_3^+ ; T_1^+, T_2^+, T_3^- ; ..., T_1^-, T_2^-, T_3^- are introduced, in its output plane we obtain light intensity distributions proportional to the specific matrices $[D]^{+++}$, $[D]^{++-}$, $[D]^{---}$. The elements of the desired matrix $[D]$ are calculated by algebraic summation of the elements of the specific matrices, i.e.

$$[D] = [D]^{+++} - [D]^{++-} + \ldots - [D]^{---}$$

Thus, the determination of the signs and absolute values of the matrix $[D]$ elements is performed for 8 successive cycles of the optical system operation. Each cycle should involve measurement and reversive accumulation of the intermediate results by electronic techniques. If necessary the above number of cycles can be reduced to two. For this purpose the transparencies T_1^+ and T_1^- should be placed in the plane P_1 along the Y-axis and the transparencies T_3^+ and T_3^- should be positioned in the plane P_3 along the

---

*In the case of using controlled transparencies the operation of forming the required images in the planes P_1, P_2, P_3 is realized by a computer program.

X-axis. If in the plane P_2 of the system there is a transparency T_2^+ (or T_2^-), then in the plane P_4 four unoverlapping light distributions are obtained corresponding to the matrices $[D]^{+++}$, $[D]^{++-}$, $[D]^{-++}$, $[D]^{-+-}$ (or $[D]^{+-+}$, $[D]^{+--}$, $[D]^{--+}$, $[D]^{---}$).

III. OPTICAL SCHEME OF THE SYSTEM

Figure 2 represents in two projections an optical scheme of the system for parallel calculation of the resulting matrix elements in accordance with Ex. (1). The system comprises: an extended light source 1, a unit for superposition of the matrix rows over the Y-coordinate (projecting spheric objectives 3, 5 and a cylindrical objective 4), a unit of integration over the X-coordinate (elements 7, 8, 9), an integro-projecting unit (projecting cylindrical objectives 11, 12 and a condenser cylindrical objective 10), as well as condenser spherical objectives 2 and 6. The course of beams in the scheme is shown for the case when the focal distances of negative cylindrical objectives 4, 8 are equal to two combined focal distances of spherical objectives 3,5 (or 7,9). The projection of images in the optical system is realized to a full scale. The resulting light distribution is reproduced in the output plane as (5) where it is read by a scanning or multi-component photoreceiver 13.

The pulse response of the system under consideration and the possibilities of its use depends on choosing the planes P_1, P_2 or P_3 as an input one. In particular, if one assumes that the transmission functions on the transparencies T_1, T_2 and T_3 in respect to intensity are continuous and that the light source is an ideal diffusor, then the pulse response of the system to illumination applied at the point (x_2, y_2) of the plane P_2 is described to the geometrical optics approximation by the expression

$$h(x_4,y_4; x_2,y_2) = \iint_\Omega \mathcal{J}(\xi,\eta) \cdot T_1(x_1,y_1) \cdot T_3(x_3,y_3) \times$$

$$\times \delta(x_4-x_2-\xi/k)\,\delta(y_4-y_2+\eta/k)\cdot\left[1+\frac{(x_1-\xi)^2}{4F_{c\varphi}^2}+\frac{(y_1-\eta)^2}{4F_{c\varphi}^2}\right]^{-2} d\xi\,d\eta,$$

(6)

wherein Ω is the area occupied by the light source with the intensity distribution $J(\xi, \eta)$, (x_4, y_4) are the coordinates of the point in the output plane and $K = F_ц/2F_{cф}$ ($F_ц$ is a focal distance of the cylindrical objective and $F_{cф}$ is a combined focal distance of two spherical projecting objectives). Since the coordinates of the beam passed through the points $(\xi, \eta), (x_1, y_1)$ and $(x_2, y_2), ... (x_4, y_4)$ in the planes P_0, P_1, P_2, P_3 are connected by the relations

$$\begin{cases} x_1 = x_2, \\ x_3 = x_2 + \xi/K, \\ x_4 = x_3, \end{cases} \qquad \begin{cases} y_1 = y_2 - \eta/K, \\ y_3 = y_2, \\ y_4 = y_1, \end{cases} \qquad (7)$$

then having performed in (6) with the account of (7) integration over the variables ξ, η, we obtain

$$h(x_4, y_4; x_2, y_2) = J(K(x_4 - x_2), K(y_2 - y_4)) \overline{T_1}(x_2, y_4) \overline{T_3}(x_4, y_2) t(x_2, y_2, x_4, y_4),$$

where $t(x_2, y_2, x_4, y_4) = \left\{ 1 + \dfrac{[(1+K)x_2 - Kx_4]^2}{4F_{cф}^2} + \dfrac{[(1+K)y_4 - Ky_2]^2}{4F_{cф}^2} \right\}^{-2}$ —

is a function, describing the deviation of the pulse response shape from an ideal one due to the assumption on the nature of the light source being used and the natural decrease in illumination from the center of the operational field to its edge.

Let us restrict ourselves to small angular dimensions of the source and the system operational field at which $t(\cdot) \approx 1$. Having admitted that $J(\xi, \eta) = const$ for $|K\xi| \leq S/2$ and $|K\eta| \leq S/2$ where $S \times S$ is the source operational area, the pulse response accurate to the constant can be described by the expression

$$h(x_4, y_4; x_2, y_2) = T_1(x_2, y_4) \cdot T_3(x_4, y_2) \qquad (8)$$

From this it follows, that the system response depends on the x_2, y_2 coordinates of "point" illumination acting in (8) as parameters and is expressed as a product of two functions depending separately on the variables x_4 and y_4. As it can easily be seen, these functions are proportional to the "cross-sections" of the transmission functions $T_1(x_1, y_1)$ and $T_3(x_2, y_2)$ at $x_1 = x_2$ and $y_3 = y_2$.

But if the transmission of the transparencies T_1 and T_3 is described by expressions (2) and (3) such "cross-sections" will be represented by images of the rows and columns of the matrices $[A]$ and $[C]$. The optical system with a response of Type (8) obtained under the assumption that its imput plane is represented by the plane P_2 refers to the class of linear space-invariant systems.

IV. EXAMPLES OF APPLICATION

One of the most interesting applications of the above system consists in realization of the spectral analysis of images according to the Walsh basis. Indeed, the process of calculation of the Walsh spectrum is equivalent to the Hadamard transform realization according to the formula

$$[F] = [H] \cdot [f] \cdot [H],$$

wherein $[H]$ is a Hadamard matrix and $[f]$ is a matrix of samples of the image being analyzed [6].

For the Hadamard transform realization in the above optical system it is necessary in the planes P_1 and P_3 to dispose the transparencies with the images of the Hadamard matrix positive and negative values in the combinations T_1^+, T_3^+ ; T_1^+, T_3^- ; T_1^-, T_3^+ ; T_1^-, T_3^- and in the plane P_2 to dispose the transparency T_2 with the image being analyzed.

Likewise to [5] the Hadamard matrix representation on two transparencies is caused by the necessity of determining the sign of the spectral components.

The distributions of intensities proportional to the elements of the specific matrices $[F]^{++}$, $[F]^{+-}$, $[F]^{-+}$, $[F]^{--}$ are determined successively for four cycles of the system operation. Double decrease in the number of cycles as compared with a general case results from the fact that for real images the matrix of samples is positive. The elements of the required matrix $[F]$ are calculated by algebraic summation of the elements of the specific matrices, i.e.

$$[F] = [F]^{++} + [F]^{--} - [F]^{+-} - [F]^{-+}.$$

The realization of expansions over the continious tone functions, e.g. over the Legendre polynomial is performed in an analogous way. For this purpose in each row of the transparency T_1 and in each column of the transparency T_3 the functions of expansion $\varphi_i(x)$ and $\varphi_j(y)$ are written, respectively. One of the system advantages consists in that the above functions can be represented as area-modulated images enabling the accuracy of the analysis to be increased [4].

The system under consideration can also be used for multichannel processing of one-dimensional signals. In this case the input plane of the system is represented by the plane P_1 (or P_3). If, e.g. the elements of the matrix [A] elements are the N - signal samples, [B] is a unit matrix and [C] are arbitrary ones, then the distribution of intensities on the system output is proportional to the result of the linear integral transform of these signals with a kernel determined by the matrix [C]. In case a square-law form used as a criterion in the detection or recognition of signals [7,8] is to be calculated, then [A] as before contains samples of the signals to be processed, [B] is a covariance matrix and $[C] = [A]^T$, where (T) is a symbol of transposition.

V. SIZE OF MATRICES

Since integration realized by objective II (see Figure 2) is based on its focusing properties, when evaluating the system resolution light diffraction should be taken into account that limits the size of the matrices being calculated (mainly over the Y-coordinate).

Let in the plane P_1 the resolution element of the size d over the Y-coordinate be singled out and the changes arizing when projecting this element to the plane P_4 be considered. Since the system between the planes P_1 and P_4 over the Y-coordinate can be replaced by a projecting unit with a characteristic size of the pulse response

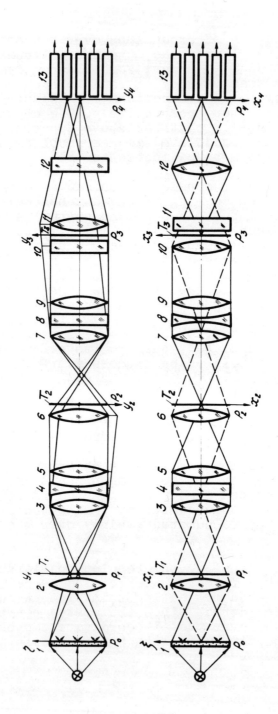

Figure 2. Basic scheme of optical system (in two projections).

$$h_L = 2\lambda f_0/L = 2\lambda S_0,$$

where f_0 is a distance from the system output pupil (the plane P_3) to the plane P_4 and $S_0 = f_0/L$, the size of the resolution element becomes equal to $d + h_L$. In the case $h_L \ll d$ the change in its geometric dimensions can be neglected.

If the transverse size of the row in the plane P_3 equals to δ, the pulse response size is

$$h\delta = 2\lambda f_0/\delta = 2\lambda S_0 L/\delta \qquad (10)$$

Let the rows on the transparency T_1 be separated with opaque intervals of the value Δ. Then the minimum admissible size Δ with which in the output plane the overlapping of images of the adjacent rows does not occur is bound with the pulse response size by the relation

$$\Delta = h\delta \qquad (11)$$

Determine now the admissible sizes of the matrices under multiplication for the system preset parameters L, S_0, Δ and λ (λ is an average wave length of the light being used). From the relations

$$(\Delta + d) \cdot N = L$$
$$P \cdot \delta = L$$

with the account of (10), (11) it follows that

$$P = (L/N - d)/2\lambda S_0, \qquad d < L/N$$

It means that the matrix $[C]$ size to the Y-direction is determined by the matrix $[A]$ size and does not exceed the value $P_{max} = L/2N\lambda S_0$. If the matrices are square their upper limiting size is restricted by the value $N_{np} = 0.7 \cdot \sqrt{L/\lambda S_0}$ that for the system parameters $L = 50$ mm, $S_0 = 2$ and $\lambda = 0.5 \times 10^{-3}$ mm equals to the value of the order of 140.

Thus, inspite of the fact that diffraction effects restrict the size of the matrices being multiplied over one of the coordinates their admissible sizes are large enough. The size of the matrices being multiplied over another coordinate is restricted by the resolution of the projecting spherical objectives.

VI. EXPERIMENTAL RESULTS

To illustrate qualitatively the possibilities of the system proposed the experiments on matrix multiplication were performed as well as the experiments on expansion of the simplest test images into Walsh function and Legendre polynomials.

The results of multiplication of the matrices with the elements taking the values 1 and 0 are given in Figure 3 representing the initial transparencies T_1, T_2, T_3 with images of the matrices $[A]$, $[B]$, $[C]$ and the distributions of light intensities in the system output plane for the cases:

a) Multiplication of a triangular (top left) matrix by two unit ones (with orthogonal disposition of the diagonal elements);

b) Multiplication of a triangular (top left) matrix by unit and triangular (top right) ones;

c) Multiplication of a "chess"-type matrix by unit and triangular (top left) ones.

The cases a), b) correspond to the matrix size 8 x 8, the case c) corresponds to the size 16 x 16.

The results of a spectral analysis of the test images are given in Figure 4. The expansion in the cases a), b) were made by the positive Hadamard matrices and in the case c) by the matrices composed of the biased Legendre polynomials. The figure resperesents the form of the transparencies T_1, T_3 with images of the matrices $[H]^+$, $[H]^+$ and $[L^3]^T$, $[L]^T$. The output continuous tone distributions of light intensities are proportional to the values of the special matrix $[F]^{++}$ elements, in this case the top value (the case a)) corresponds to the Walsh spectrum of the system square field and the medium one (the case b)) correspond to the Walsh spectrum of a "Cross"-image. This figure also represents the Legendre spectrum of the "Cross" image (the case c)).

NON-COHERENT OPTICAL SYSTEM   215

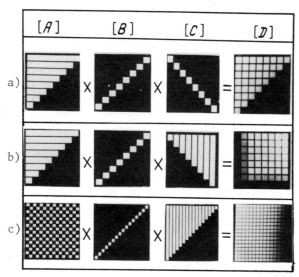

Figure 3. Multiplication of matrices with elements taking values 1 and zero: [A], [B], [C] are initial matrix images, [D] is a type of light distribution in the system output plane.

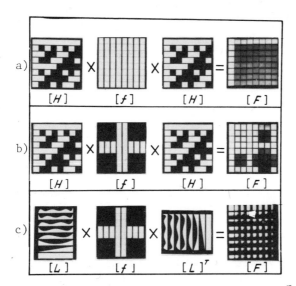

Figure 4. Spectral analysis of test images [f] is a test image, [F] is a matrix of spectral components, [H] is a biased Hadamard matrix image, [L] is a biased Legandre function image.

An error of the matrix transforms performed in the system by an analog method is determined mainly by the quality of the cylindrical optics being used, the accuracy of registration and the size of the initial matrices on the transparencies, the light irregularity of the source and the photoelectric reading accuracy. These factors should be taken into account when developing a special-purpose optical system.

In conclusion note that the system proposed enables the class of integral transforms realized by non-coherent optical techniques to be reasonably extended.

REFERENCES

1. R.A. Heinz, I.O. Artman, S.M. Lee. Matrix Multiplication by Optical Methods. - Applied Optics, 1970, v. 9, N 9.

2. D.P. Jablonowski, R.A. Heinz, I.O. Artman. Matrix Multiplication by Optical Methods. Experimental Verification. - Applied Optics, 1972, v. 11, N 1.

3. E.S. Nezhevenko, P.E. Tverdokhleb. Umnozhenie matrits opticheskim metodom. - Avtometria, N 6, 1972.

4. O.I. Potaturkin, P.E. Tverdokhleb, Yu.V. Chugui. Obobshchennyi spektralnyi analiz izobrazhenii s ispolzovaniem siluetnykh filtrov. - Avtometria, N 5, 1973.

5. B.E. Krivenkov, P.E. Tverdokhleb. Yu.V. Chugui. Opticheskii metod kodirovaniya izobrazhenii pri pomoshchi preobrazovanii Hadamara. - Avtometria, N 6, 1974.

6. Pratt, Kane, Andrews. Hadamard Transform Image Coding. Proceedings of the IEEE, 1969, No. 1.

7. Teoriya svyazi. Perevod s angliiskogo pod redaktskiei B.P. Levina. M., "Svyaz", 1972.

8. N. Nilson. Learning Machines. Toronto London Sydney, McGraw-Hill Book Company, 1965.

QUESTIONS ON THE REPORT BY DR. P. TVERDOKHLEB

1. Dr. G. Stroke. What is the opinion of the specialists on your system application for digital methods of image processing?

Dr. P. Tverdokhleb. The optical system described is under discussion actually for the first time. We will have to know these specialists' opinion. I believe this system will be applied if we improve the accuracy of calculations. For example, this accuracy should be sufficient for good-quality reconstruction of images subjected to spectral analysis in the optical system.

2. Dr. G. Stroke. What advantages are offered by your system?

Dr. P. Tverdokhleb. This optical system is an analog one. It enables us to perform multiplication of three matrices of the order of 100x100 elements in size. The values of the resulting matrix are calculated in parallel and independently. Hence, the main advantages of this system consist in its speed of operation.

3. Dr. G. Goodman. We are also developing space-noninvariant systems, however not for the analysis of images but for the elimination of their geometrical distortions. Have you performed any investigations in this direction?

Dr. P. Tverdokhleb. The experiments illustrating the possibilities of performing geometrical projective transforms were performed at our laboratory 2.5-3 years ago. For these purposes we used a coherent system with a holographic matrix.

4. Dr. G. Goodman. Have you published these results?

Dr. P. Tverdokhleb. Unfortunately, they have not been published. At that time we considered our experiments to be too simple and abstracted from practical applications. Their results were reported at the Conference on Holography held in Tbilisi in 1972.

OPTICAL LOGIC AND OPTICALLY ACCESSED DIGITAL STORAGE

Rolf Landauer

IBM Thomas J. Watson Research Center
Yorktown Heights, New York 10598

I. INTRODUCTION

A computer system can be divided according to its three principal functions:

1. The Processor Unit in which information streams interact. This is where the logic is implemented. Modern computer systems often utilize a number of processors [1].

2. The Storage and/or Memory which acts as a repository for information. Typically this consists of a hierarchy [2] of subunits. The members of the hierarchy correspond to different trade-offs for access speed vs. capacity. Thus small amounts of information can be stored with a very rapid access capability. (There is unfortunately no universally accepted distinction between "Storage" and "Memory." Most frequently, however, Memory refers to the faster portions of the hierarchy, where access can be highly selective, and where information is typically stored in physically distinguishable devices.)

3. The Input-Output Devices which act as transducers to the external world.

There has been tremendous progress in the first two parts of the system, throughout the modern history of the electronic computer. This progress has been permitted by the fact that in this part of the computer system we are free to choose our scale of time, space, and energy, since information only interacts with information. In I/O (abbreviation for input-output) progress has also been impressive, but it has not been as spectacular as in the internal parts of the system. We cannot make printers cheaper and faster by miniaturizing print.

Our basic thesis concerning the potential use of lasers in particular, and of optics, light and holography in general, follows.

As transducers in I/O, lasers have already achieved a place, and we can expect it to grow. I/O inevitably involves contact with optically formed information. In the input we have to scan patterns of various kinds, in the output we have to generate them. Thus, for example, the IBM 3666 Supermarket Scanner [3] uses a laser beam to read a bar code imprint (or label) on grocery items. The IBM 3800 is a high speed printer, with speeds up to 13,360 lines per minute, using a laser to write an image on an electrophotographic surface. It can mix character styles and sizes, and also can flash a copy of a form to superimpose on the variable content written by the laser. A schematic of the 3800 is shown in Figure 1. Xerox's recently announced Telecopier 200 Facsimile System also uses a laser for scanning document input, as well as for writing on a xerographic surface.

As transducers to reach into storage arrays for reading or writing, lasers also offer serious possibilities. But as will be discussed subsequently in detail, storage technology is a very

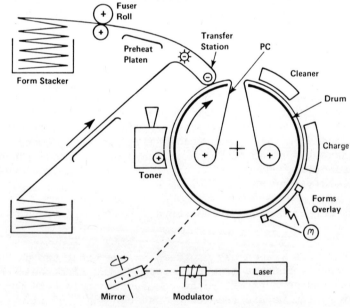

Figure 1. IBM 3800

competitive field with potential for progress in many directions. It is hard to make an overpowering case for the particular advantages of the optical proposals.

Finally the possibility exists for utilizing the interaction between streams of information in radiative form, to do logic. I can only see difficulties here, and no advantages. I am puzzled by continued proposals for laser logic.

In I/O, of course, scanning for reading purposes need not be done by a laser. It has often been done in the past by using the light from a CRT spot. It can be done as in a television camera by interrogating an array of photosensitive elements. Increasingly the semiconductor technology is providing the means to provide these photosensitive elements, and the method of reading them out. Lasers, however, permit us to utilize large f-numbers, which permit a good depth of field, without sacrificing light collecting efficiency [3]. As semiconductor scanners and the associated circuitry improve, however, we are likely to become less concerned with light utilization and signal/noise problems. As a result we can increasingly expect the semiconductor scanning arrays to become attractive since they can eliminate the complexities of scanning a laser beam.

Much of the following material is intended to set a background, by explaining to the optical community what the competition can do. It isn't really fair to compare the most sophisticated optical proposals to what can be turned out in a factory today. For a general introduction to digital devices the reader is referred to a recent elementary text [4], and to a short review [5].

II. LOGIC

Serious proposals for completely radiative logic, using only the interaction between radiative degrees of freedom, without continual conversion back and forth to baseband signals are not new, and date back (at least) to a patent of von Neumann [6], filed in 1954. The same concept was independently discovered by Goto and his collaborators [7]. These early proposals visualized microwave radiation rather than light as the information carrier. The use of light, in incoherent form, was explored in the late fifties in schemes involving photoconductors and neon bulbs or photoconductors and electroluminescent devices [8]. These latter proposals were early attempts to make integrated circuitry, at a time when the full promise of the semiconductor technology was still indistinct. In 1962 the injection laser was discovered, and this gave rise to a series of logic proposals. A recent summary of this field has been

given by Basov in a joint paper with Culver and Shah [9]. Early
proposals and investigations in this area stem from Fowler [10-12],
Lasher and Fowler [13,14], Lasher [15,16], Kosonocky [17,18], and
Nathan, et.al. [19]. Interest in this area stemmed from the fact
that it is possible, in principle, to have radiative systems which
avoid the RC time constants and carrier delay times typical of
ordinary semiconductor device logic.

While injection laser logic seems to be the only widely examined
candidate, there are other possibilities, e.g. Culver and Mehran
have proposed a logic scheme [20] based on the kinetics of self-
induced transparency. A very recent proposal [21] uses dye lasers.

The relative appeal of different approaches to the execution
of logic depends greatly on the energy consumption of the schemes.
Device speed, per se, is not very useful if it comes at the expense
of a great deal of heat dissipation, and thus requires a low
density of components. This in turn means long propagation delays,
which can cost us more in performance than the gain in intrinsic
device speed. This is basically why optical interaction schemes look
unappealing. Semiconductor devices can produce the required signal
interactions at low fields. Typically a voltage a few times kT/e
has to be applied to a p-n junction, typically 10^{-5} cm wide, thus
giving electric fields of the order of 10^5 volt/cm. Nonlinear
optical effects, say parametric effects, give strong nonlinearities
at fields comparable to atomic fields, 10^8 volts/cm, and these
fields must extend over optical wavelengths. Admittedly, in that
case, the signal interaction could take place in a time of the
order of an optical cycle, compared to the much longer time
constants of semiconductor devices. But the high power dissipation
of such fast optical devices would render this high internal speed
useless.

An attempt to be more formal about the above sort of consideration has been given by Keyes and Armstrong [22,23]. They conclude
that for several optical schemes:

$$P > \alpha\, h^2 \lambda^2 c / q^2 a_o^2 t_D^2$$

q is the electronic charge. a_o is a typical atomic dimension, or
else the length associated with the dipole moment of a transition.
t_D is the time required for signal interaction, i.e. the delay of
one stage of logic. α is a numerical multiplier and has different
values for photochromic logic, parametric interactions, or self
induced transparency schemes. It is not clear, however, how
generally valid the Armstrong-Keyes considerations are. In particular they do not really include the case of stimulated emission
interactions, as invoked in injection laser logic. By contrast to
the above equation a given physical transistor structure is generally
presumed to require a power consumption P which varies as $1/t_D$,

as the circuit and power supply are varied. Thus transistor power
rises more slowly with switching speed than in the optical schemes.
(For a deeper analysis, however, of transistor limitations see a
recent discussion by Keyes [24].)

The thermal advantages of transistor, however, seem obvious
without detailed analysis. Only with great difficulty were CW room
temperature injection lasers developed, whereas there has never been
any difficulty with transistors!

Let us try here another way of making the same point. Semiconductor devices utilize electron-electron interactions. These
direct Coulomb interactions can give rise to nonlinear signal interactions at lower energy levels than the radiative interactions
utilized in optical schemes. While we do not know how to state
this fact in a simple and universal analytical way, it seems intui-
tively obvious. More specifically in comparing injection lasers
with transistors we run into the same judgemental problems that
always beset such comparisons: Which hypothetical transistor
structure is fairly compared to which injection laser extrapolation?
While Keyes [22-24] has provided estimates of minimal energy dissipations for optical schemes, these all come with inequality signs
which are associated with factors of great uncertainty. Part of
the great uncertainty stems from the fact that in logic (unlike
memory) it is <u>not</u> easy to compensate for small error probabilities
by a little redundancy. Logic must be relatively error immune.
Without knowledge of manufacturing variability, temperature changes,
and a much much more detailed understanding of all the kinetics than
we have for most of the optical schemes, we can make little progress
in distributing the allowed error budget, and thus in estimating the
numerical factors in the inequalities.

III. COMPETING LOGIC TECHNOLOGIES

The semiconductor technology has, of course, been immensely
successful. Transistors with a raw speed (i.e. not allowing for
fan-out or interconnection delay) of about 150 pico-seconds, have
existed for some years [25,26]. They have not yet seen commercial
utilization in computers, for reasons which can only be stated as a
matter of personal speculation. (It is clear, however, that the
capability to fabricate subnanosecond logic circuits, at reasonable
powers is now generally available [27].) Most likely there were
enough gains to be made by achieving greater densities in integrated
circuitry, to preoccupy the available technical talent. Furthermore
the more delicate higher speed structures would make it harder to
achieve the device yield needed for a high level of integration
(many devices on one chip). The higher device speeds would be wasted
unless a comparably more compact computer structure can be devised,
to allow shorter interconnection distances. Furthermore the

earliest ultra-high speed switching transistors were too high powered, to readily permit dense packing (see Sec. IV) and short interconnection delays. But it is clear that progress in logic speed is not limited - at least at this time - by the internal kinetics of the transistor structure. The ultimate capabilities of transistors are still somewhat uncertain. With due allowance for cooling the whole circuit array, and the interaction of transistors with transmission lines, Keyes [24] predicts a lower limit not much under 100 p.s. for room temperature operation. Thermal consideration for heat flow out of a single device by itself, yield a more liberal 20 p.s. Still more optimistic figures have been presented by Swanson and Meindl [28]. The fact that these latter calculations come within a small factor of kT per switching event, thus implying the motion of very few electrons in a switching event, indicates the very optimistic nature of that particular projection.

The clearest alternative to silicon devices, at this time, comes from Josephson junctions [29-31]. These devices have been observed to switch in less than 38 picoseconds, as far as the internal device kinetics is concerned. (38 picosecond is an upper-bound, due to measurement difficulties.) Actual switching speeds in spread out circuits will, of course, be appreciably slower. These devices are characterized by a product of power dissipation and device delay of as little as 3×10^{-18} joules. That is about six orders of magnitude smaller than present transistors. Josephson junctions have been developed in which stored information changes are represented by a change of a single flux quantum [32,33]. Even in these devices transition speeds of 200-250 picoseconds have been measured [32]. (Limited by measuring equipment, but believed to be faster.) Voltages in these circuits are scaled to the very small superconducting energy gap, the elementary event involves millielectron volts! Naturally further work is needed to make this a reproducible, controllable, high reliability technology. The same, however, can be said of lasers, particularly of injection lasers, with their well known degradation problems. Josephson junctions, with their low dissipation, provide an ideal technology for high density packaging, in accordance with the considerations of Section I.

Finally we must emphasize that optical devices must have dimensions of the order of an optical wavelength. (There is of course nothing to prevent us from utilizing guided TEM waves in an optical frequency range. But is that optics?) It is of course, in principle, possible to do somewhat better than an optical wavelength, as shown by the optical exposure of photoresist materials where dimensions as small as 0.1μ have been obtained [34]. On the other hand Josephson junctions, transistor structures, etc. are <u>not</u> limited to optical dimensions, electron beams can be used to control their geometry [35].

A number of years ago, one could still advocate the use of esoteric semiconductors which exhibit ultrahigh mobilities, e.g. in Schottky barrier transistor structures [36]. For discrete devices, e.g. in microwave use or instrumentation, that may in fact still be an appealing possibility. The continued further advances for integration in silicon have made this a most dubious possibility for logic; after all, it is the total number of logic decisions per second which counts more than their intrinsic speed. The same reasoning, of course, applies to many other novel devices for executing logic.

IV. HEAT DISSIPATION IN LOGIC

Must energy be dissipated in the logic process? It has been argued on the basis of phase-space considerations that throwing away information in a logic stage requires the expenditure of energy [37]. It has also been shown that, in principle, computers can be built which need not throw away any information [38]. But this possibility of constructing reversible computers is not a recipe for practical computer structures. In practical computers we will undoubtedly continue to have a need to discard intermediate results. The basic point about computational reversibility is made in Figure 2 illustrating the phase space of the computer. If the computer can start at two separate states, A and B, and arrive at a single

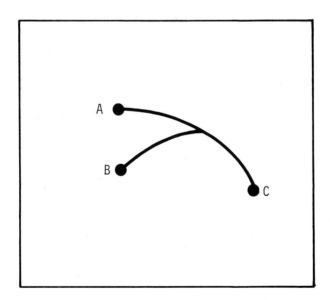

Figure 2. Phase space of computer, showing two states A and B leading to one later state, C.

subsequent state C, it has lost information. Let us assume that this can be done in a reversible system, without losses. In that case the equations of motion are reversible. The path starting at C then, however, leads back to both A and B, which cannot be if the equations of motion are deterministic. The conclusion: Paths cannot converge and information cannot be lost in a lossless system. More detailed considerations [37] show that the minimal loss, which depends on the particular logic function involved, is typically of order kT per operation. The small magnitude of this loss relates to the fact that one need not really do logic with a great many interlocked degrees of freedom, as in a macroscopic transistor. The very much larger losses, however, found in real devices have a closely related origin to the minimal energy losses discussed in the cited phase space analysis [37]. Transistor capacitances, for example, have to be discharged in the switching process, and one cannot design a real computing system in a way such that the capacitive energy can simply be carried along to another device, rather than dissipated.

The real size (and therefore the real energy requirements) of logic devices comes from the fact that the long interconnections between devices must be energized. A computer cannot be a periodic structure in which only neighboring devices interact, such a device is much too hard to program since the programmer must anticipate the complex data motion in great detail. The computer must have a more irregular structure, and this requires interconnections between far away devices. The requirement that these interconnecting links (transmission lines in real computers) have low attenuation and little cross talk leads us into the need for macroscopic interconnections with many degrees of freedom. The logic devices have to provide the energy to drive these long interconnections, e.g. to charge the transmission line capacitances involved.

The phase space considerations which give us a minimal energy dissipation, scaling with the temperature, suggest that we can gain by going to low temperatures, and this is indeed confirmed by Josephson junctions, as already discussed in Section III. Conversely these same considerations suggest that the use of radiative schemes, in which $h\nu$ is likely to replace kT as a phase space measure, is unprofitable. Unfortunately a serious quantum mechanical version of these considerations does not yet exist, and this is only suggestive. (In this connection see also a discussion based on Shannon's channel capacity [39].) Low temperatures are not only favored by power dissipation questions but also by questions of device deterioration. As we continue to miniaturize devices and their circuits they become more susceptible to degradation. At the same time we are likely to have more devices per system, as the devices become smaller, thus aggravating the deterioration problems. An obvious way out of this dilemma: Lower temperatures where thermally activated reaction rates and atomic mobilities are depressed.

Logic devices, whether injection lasers, transistors, or invoking some other signal interaction, generally permit operation over a range of signal levels, S, and the required signal interaction time for production of the logic output will be labelled τ. Typically $\tau \sim S^{-m}$, where m is 1 or 2, or of comparable magnitude. In either case power dissipations are of order S^2. If we assume that a surface can dissipate Q watts/cm^2, the number of logic stages per cm^2 can vary as Q/S^2. The linear separation between stages will then vary as $S/\sqrt{Q}$. The total delay, composed additively of the delay within the stage and the interconnection delay is then of the form

$$\alpha/S^m + \beta S/\sqrt{Q}$$

For modest m, this sum has a minimum when the two terms are comparable. Indeed in practical high speed computing systems the two terms are comparable. Nothing is gained by operating the devices at much higher powers, to gain a short delay time within the device! (Keyes [24] has provided considerations of this sort, in much more detail. The author is also indebted to R. A. Henle for emphasizing this viewpoint in a number of unpublished discussions.)

V. OTHER LASER LOGIC DRAWBACKS

Injection laser logic proposals have specific drawbacks, aside from the power level questions. Not all of these difficulties are characteristic of all possible optical proposals. Nevertheless a discussion of these will help clarify what is needed in a good logic device.

An injection laser is a positive feedback device. Thus variability in the physical structure of the component is amplified, as exhibited for example, by filamentary action. Furthermore the emission wavelength is sensitive to temperature and to carrier injection level. The sensitivity of the wavelength to temperature implies a sensitivity to the exact sequence of previous signals to which the device has been exposed. Furthermore the intensity level of radiation put out by a stage is not very directly controlled by the (presumably easily controlled) B+ voltage of the system. By contrast transistor logic puts out a signal which swings from a level close to some ground voltage to a level close to B+, and the signal swing is thus "restandardized" at every stage. In other words, the close relationship of the energy supply parameters to the signals used makes it easier to establish the necessary signal level standardization. Basov's injection laser logic proposals <u>seem</u> to invoke pulses [9]. Logic schemes which require a delicate control of timing in the arriving signals have been made to work [7,40], but the timing problem has clearly been a major drawback contributing to the demise of such schemes. A logic system where the inputs to a given stage can arrive in any order, and where clock signal distribution is relatively coarse-grained, and not involved at every stage, is much more practical. In particular, such systems permit the "mixing" of levels, i.e. interacting signals can have gone through a different number of logic levels.

Any logic system requires an inversion or negation capability, In laser logic this typically means that an incoming signal has to quench some output mode, by robbing the amplification capability of the device. If the incoming signal is present before the device is turned on, that presents no problem. If, however, the mode to be turned off is already established, this requires a relatively strong input. This leaves little design flexibility in relating the strength of input signals, output signals, and internal cavity signals [13], if we want to be able to use the output of one laser to quench one or more lasers with comparable outputs.

One of the great advantages of transistor circuits, as coupled by transmission lines, is the flexibility they allow in the number of subsequent stages to be driven by the output of a given circuit. The transmission line can either pass by the circuits to be driven, in sequence, or else we can split the line, and divide its current. In either case, after a sufficient delay all the other stages will see the full applied voltage swing. The design of a particular computer, of course, requires that the transmission line delay involved, with due regard for reflections, not exceed some specified amount. This results in a complex body of rules, restricting the wiring patterns and the total number of device capacitances which can be driven from a given circuit. But in a typical computer, for example, the same stage can drive 1, 2, 3, or 4 subsequent stages.

The radiative proposals, typically, do not give us the same flexibility. As discussed above, for example, there isn't much flexibility in the power required by the input to an inversion stage. Other radiative logic proposals [6,7,20,21] seem to suffer from similar drawbacks; they require well defined power inputs. Thus we cannot take the output from a given stage, and as an afterthought use it to drive 3 instead of 2 stages. Logic schemes of this sort then require extra power standardization components, or else a tailor-made design for a stage, depending on the number of subsequent stages. Furthermore splitting an optical signal requires precision surfaces, or accurate waveguide geometries, not just metallic joints. This, it must be remembered, comes in addition to the geometrical requirements in the active device. While a transistor does, of course, require a controlled geometry, these requirements are still far more flexible than those imposed on the resonant cavity of a laser. While mirrors and cleaved laser surfaces may not be needed, as is made clear by distributed feedback injection lasers, this does not avoid the need for geometrical precision.

One recent laser logic proposal [21] typifies the difficulties that can arise in radiative logic. The proposal uses dye lasers in which birefringent elements are used to define two preferred output polarizations. The logical inputs serve as pumps, which orient the dye molecules. There is no separate energy source to provide energy level standardization. The output is not at the same energy level or wavelength as the input. The system does, of course, provide preferred polarization directions and these cannot wander away toward arbitrary polarization. But no means are proposed for standardizing signal energy levels or signal timing, both of which are critical to the proposed logic schemes.

VI. STORAGE

The use of <u>radiative</u> access to storage is also not a new concept and dates back (at least) to the proposals to use spin echoes for information storage [41]. More modern proposals use light largely as a transducer which on the one hand can supply an easily directed beam, and on the other hand permits holographic and other interference schemes. The use of completely radiative storage, using light which is continually recirculated, as in the acoustic delay lines used in some of the very first large scale electronic computers, is also a possibility. But this author is not aware of any publication suggesting advantages for that.

We shall once again, as in our discussion of logic, stress the disadvantages of the proposed, optically accessed memory schemes. As in other advanced technology discussions, most of the existing publications stem from the proponents and stress the advantages. In contrast to the case of logic, there seems to be a better chance in storage, though it is by no means clear that the optical schemes have an assured place.

A few very general points: We have already, in the logic discussion, stressed that electron beam fabrication will permit achievement of submicron dimensions. The dimension of stored bits is therefore not limited by the wavelength of light, if the storage medium is not optically accessed.

As we shall see in the subsequent discussion of magnetic bubbles, there is a new concept in this storage area: The Bubble Lattice Array. Magnetic bubbles, through their interaction, can organize themselves into periodic arrays. Thus information can be stored in dense arrays which do not absolutely require correspondingly miniaturized parts for the information preservation process. None of the light schemes seem to offer comparable possibilities, the information pattern is always determined directly by external imposition. Here once again we run into a property very reminiscent of that involved in the logic discussions: strong interactions are desirable. That is, of course, not typically what one has asked for, in the past, in storage arrays. In the past we wanted non-interacting bits in our storage arrays. It must, furthermore, be admitted that the bubble lattice file is a new concept, and its full significance, as well as the generalizability and extendability of the notion are still hard to assess.

General Storage Requirements

A low cost per stored bit, generally achieved by high bit density is, of course, desirable, but it certainly is not the only parameter that counts. One needs access to a memory, and in over-simplified discussion access time is frequently stressed as the relevant criterion. But most modern memory systems build up qeues of requests at each level of the memory hierarchy. These are qeues waiting to be served and it is generally more important to be able to economically service a high rate of requests, than it is to service them in the shortest possible waiting time. After all, the calculation needing information from a disk has already been interrupted and displaced from the processor, and unless it is a very demanding real time calculation, it isn't going to matter whether the processor gets that information in 100µ seconds or in 10 milli-seconds. Naturally if the computer is tracking enemy missiles, that may be a different matter. We have achieved continually lower costs in disks by achieving higher density, thus spreading the cost of the disk, disk drive, and head over more bits. As a result we have reduced the ratio of access channels to bits in the memory. Thus for two disk storage systems with the same total capacity but using different density technologies, the older system will permit more accesses/second. Unfortunately optical memory designs are subject to much the same pressures. The reason why the motion toward higher density has been possible, and not been as self-defeating as the above argument indicates, is that the continually decreasing cost of random access memories (i.e. the faster integrated circuit memories) has permitted a continually increasing percentage of storage requests to be satisfied without the need to go on to the disks.

It is typical of technologies which involve fabricated devices for each stored bit - magnetic bubbles, integrated circuits, or Josephson junctions, that it is easy to add extra access channels. They come at a relatively low cost. Thus these technologies do not have to be entirely competitive in cost/bit with continuous media, such as disks or optically addressed media, to be appealing.

Storage media can come with varying degrees of a full read, write, and erase capability. Read-only memories (ROM) are unalterable, once generated. These have use in computers, but

OPTICAL LOGIC AND OPTICALLY ACCESSED DIGITAL STORAGE

most often in small memory arrays, giving the computer an internal structure for its operation, rather than as a store for data to be used. Data to be used is generally subject to revision and updating. Even systems code, often suggested as a suitable candidate for read only memories, is subject to frequent correction and revision. A memory storing the shape of Kanji characters (Japanese ideograms) is probably one of the few really valid illustrations of a read only store for data residence.

While the ability to add or modify information is considered very important, this doesn't necessarily require erasability. An old fashioned (and honest) bookkeeper never erased, he just added new information. Memories utilizing lasers to burn holes in metallic films [42,43] are an example of systems which permit updating; additional holes can be put in. When we write a record in such a technology we must, of course, be careful to leave further space. At the very least we should be able to come back later and add a mark meaning: "This record is now obsolete, go to the index to look for the modified version." Once we recognize that magnetic tape is, in general, rewritten in large chunks, and not a few bits at a time, the "postable" storage medium does not seem all that limited. It is, in fact, very much a function of materials costs: Is it cheaper to erase obsolete tape, and reuse it, than to use new "postable" material, to carry the revised records? If the postable material permits _sufficiently_ impressive density advantages, then it could be cheaper, even though it cannot be reused. Of course one doesn't really need to mark a record as obsolete on the record itself, one can do this in the index which leads to the record. Thus nonpostable records are, in fact, useful if the indices are kept on an erasable medium.

The writing process also comes with varying degrees of complexity, and in some of the optical proposals requires a development step. This means that:

a) The material is unavailable for checking, immediately after writing, and unavailable for interrogation, during development. The whole memory hierarchy concept [2] is based on the established fact that successive request to memory are highly correlated. Thus the material most likely to be needed is being developed. One can, of course, keep a duplicate copy elsewhere, until development is complete, and clearly the faster the development process is, the less of a burden this becomes.

b) A developed record is not "postable," it cannot be revised. In some proposals, of course (e.g. thermoplastics), the record can be erased completely, but that only saves raw material cost through recycling, it does not add to the system's flexibility.

c) Development is a batch process and lumps information together.
We may thus have to deal with larger information groups than we would otherwise like. Or - to take the example of development by heating for thermoplastics - we need a great many independent heating elements and the circuitry required to provide access to them, and selection between them.

VII. COMPETING TECHNOLOGIES

a) Magnetic Surface Recording

Large storage arrays in current systems reside on magnetic surfaces. Magnetic surfaces permit, first of all, a great variety o format. They can come in the form of magnetized rigid disks, read and written by a transducer, moving in and out radially, much as a phonograph pickup. Magnetic storage can also come in a great variet of more flexible media. Tapes are well known and provide a very compact way of packaging information three-dimensionally. Tapes hav the very useful property that they do not need to be written and rea by the same piece of equipment, at the same temperature or even by equipment made by the same manufacturer. Consider the specification required to permit that same flexibility for holographic storage! Optically sensitive media can, of course, also be deposited on flexible substrates, and the metallic film hole burning technology [42-44], for example, does that. But that is a lot easier with a system relying on the reading of directly recorded digital information, than in holography, where radius of curvature and dynamic stretching of the medium under motion would have to be carefully controlled. Mylar, for example, exhibits long term dimensional instability, and its static dimension can exhibit changes of the order of 1%. Even a 1D hologram, with the information transverse to the direction of motion, will be affected by the forces related to t motion of the medium. We have learned, over decades, how to store information on magnetic surfaces with ever greater density, by simultaneous miniaturization of the reading and writing apparatus, its distance from the magnetic surface, and the thickness of the recording layer [45].

Aside from ordinary magnetic tape on reels, flexible magnetic media permit many other configurations. Some of these have been achieved, others are in the proposal stage. Pohm and Zingg [46] hav suggested stacking thin flexible mylar disks on a spindle, with about 40 disks per cm along the spindle axis. Sharp edged blades would reach into the pack, to separate the disks, and thus allow room for entry for the reading and writing head. Other configurations [44] include, for example, the Ampex terabit memory, with a capacity of 3×10^{12} bits. This device uses up to 64 standard videc tape reels. Each reel is 5 cm wide, is read and written transversel at 2000 cm/second. The tape is searched (for the proper record) at a speed of 2500 cm/second, along the length of the tape. There are other massive automated tape-like libraries listed in Reference [44]. A part of IBM's 3850 is pictured in Figure 3. In this storage system automated picker arms select one out of (up to) 4720 cartridges. Each cartridge has about 2000 cm of tape, about 7 cm wide, with 4×10^8 bits of data per tape. Each of 2×10^{12} bits can be reached in about ten seconds. In these high density systems, using flexible media, densities of the order of 10^5 bits/cm^2 are utilized. In Sony's color video cassette system about 10^6 flux reversals per/cm^2 are recorded. Note, of course, that a television recording system can tolerate a much lower signal/noise ratio.

Modern disks (e.g. IBM's 3340) can have as many as 120 tracks/cm, and about 2200 bits/cm along these tracks. The data rate is

Figure 3. Data cartridges housed in the honeycomb storage compartments of the IBM 3851 mass storage facility. A control locates a needed cartridge and transfers it to a recording device where the information can be sent to magnetic disk drives.

7 Mbit/sec. The average time for arm motion to another track is 25 msec, and the disks rotate at a speed of 3000 r.p.m., and thus require about 10 msec to get to a desired location half-way around the disk.

Proposals for the television equivalent of phonograph records [47-50] invoke information densities equivalent to about 10^8 bits/cm^2. In these schemes the medium is not reversible, and this allows a wider choice of technologies but that, by itself, is not the only reason for the much greater density. The higher density comes from:

(1) These "records" are read sequentially, there is no need for <u>quickly</u> accessing a randomly selected track.

(2) The signal/noise ratio can be much poorer in a pictorial representation, than in your bank balance.

It is widely accepted that progress in magnetic recording beyond the current products is possible. One prediction, dating back quite a few years [51], anticipates densities of 10^7 bits/cm^2 on rigid disks. Mee has more recently arrived at a similar prediction for a 1980 product [52]. Densities along a more slowly moving tape track as high as 16,000 cycles/cm were achieved in audio recording well over a decade ago [53], and more recently in high density data recording [54].

Digital magnetic recording uses saturated magnetization, and relatively simple coding schemes. It is, in principle, possible to use the analog capability of magnetic media, and thus go to multi-level recording. It is in principle also possible to apply alternating fields to the surface of a magnetic layer in such a fashion as to give magnetic penetration to varying depths, and thus to record "three-dimensionally." These concepts and various variation on this theme, are not necessarily dead ends. The fact that digital recording has never pushed seriously in these directions is instead symptomatic of the fact that there have been more promising avenues of improvement, elsewhere. We raise these points here to point out that advantages of the sort claimed for holography, are not necessarily that unique.

b) Magnetic Bubbles

In recent years a new storage technology has emerged, using cylindrical domains of reversed magnetization, in a uniaxial magnetic material, with an externally applied field, which opposes the domains of reversed magnetization. A detailed discussion can be found in a new book [55]. In the most common version of the bubble scheme a rotating field, in the plane of the material, magnetizes permalloy

patterns deposited on the uniaxial material, and these time
dependent magnetization patterns cause the domains to be moved
along. Information is denoted by the presence or absence of a
domain. This is not a "random-access" memory, we cannot reach in
electronically to read a particular bit position, but must move the
bits along the permalloy pattern, until the desired information
comes to the reading apparatus, e.g. a magnetoresistive element.
Since the information in these schemes typically moves at mega-
cycle rates, it will take 1 millisecond, for example, to get at a
bit 1000 bit position away from the read apparatus. Figure 4
characterizes these schemes, in the lines labelled (1) and (3).
Data closely related to that of Line 1 is also presented in Figure 5.
Line (2) of Figure 4 characterizes a relatively new proposal: The
bubble lattice array. Here the interaction between bubbles is used
to form a hexagonal bubble "crystal", which is then moved along in
its entirety, as shown in Figure 6. The bubbles form a periodic
structure, thus the information is not contained in the presence or
absence of a bubble, but rather in its domain wall structure. We
see here, in this proposal, the handling of bits, in a form in which
the bit pattern is a metastable structure. No delicate circuitry or
manufactured pattern is needed for the preservation of each bit; we
can store in a continuous medium. This proposal, while in an early
state of development, and therefore speculative in its feasibility
and promise, demonstrates the possibility of storing information in
patterns whose density is not, in any simple way, limited by our
ability to manufacture small devices. Such self-organizing schemes
will, of course, require strong interaction between bits, and this
seems an unlikely effect for photons, except perhaps at very high
energies, and therefore at very high energy dissipations.

Figure 4 shows the achievability of information densities of
$10^9/cm^2$, and this is not any real limitation, but reflects only the
more plausible near term electron beam capabilities. Thus we can
expect to go well beyond optical wavelength limitations, in a
technology which in view of its high storage densities can become a
competitor to our magnetic surface recording in cost, but has far
more flexibility in providing access to the information. This
technology provides a possible road to achieve densities beyond
lithographic capabilities for circuit formation. By the same token,
however, the densities do relate to material properties; in contrast
to transistors we cannot reduce dimensions simply by improvements in
fabrication technique, but must look to new materials to achieve
smaller domains.

c) Integrated Circuit Memories

Information can, of course, be stored in circuits, in which
there is a conductive path to each bit position. In fact, if we
are to build a random access memory, which can very quickly reach

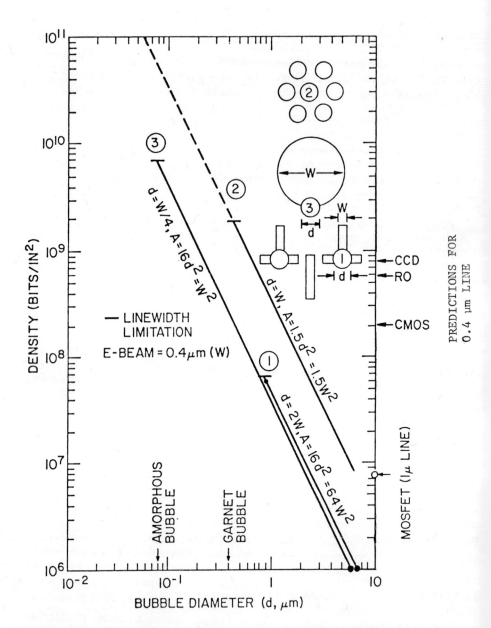

Figure 4. Arrows along the bottom denote the smallest domains observed. Line (1), (2), and (3) characterize the different methods of bubble control illustrated in the upper right hand corner. Line (1) is the "T" and "I" bar pattern, (3) the centiguous disk pattern which requires far less demanding metallurgy. (2) is the bubble lattice array. The arrows along the right edge describe semiconductor schemes. From Reference 55.

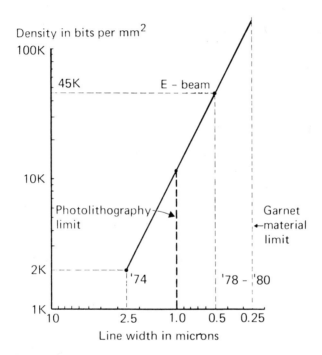

Figure 5. An independent, but closely related estimate of T-I bar densities shown in Figure (4), Line 1. From Reference 52.

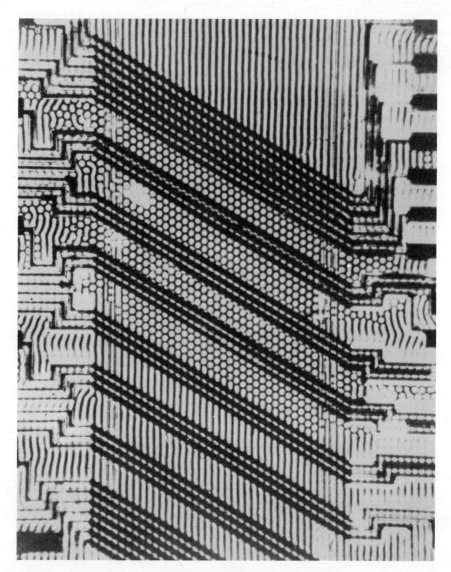

Figure 6. Photomicrograph of an experimental bubble lattice device. The small circles are bubbles (actual size ~ 7 μm). This device, used for investigating the dynamic behavior of bubble lattices, produces the regular array of bubbles and translates it along the vertical direction. It does not perform the other functions of a complete memory.

into any desired bit position, it is hard to see how to avoid this. The circuits involved can be silicon integrated circuits, or can be Josephson junctions [32,33,56]. In the latter case no standby power is required to preserve information, energy is needed only to switch information, or to read it. (Some circuits do use power, even when not switching.) Even in the silicon schemes, however, continually lower power consumption is being achieved, and in both technologies there is progress toward higher densities.

In the case of silicon, Figure 7 illustrates how circuit and device ingenuity has reduced the storage area required for one bit, without demanding narrower device dimensions. It looks, however, like we have almost exhausted that opportunity. Indeed one of the technologies shown in Figure 7, Charge Coupled Devices (CCD), is not a random access technology, but passes charges along a silicon surface, much as cylindrical domains are moved in the magnetic bubble technology. CCD technology can potentially go beyond binary storage, to storage of a wider range of charges. In integrated circuit chips, 1/2 cm or less on a side, it is now common practice to provide between 4000 and 9000, and soon 16000 bits of information [57-59]. Information stored on silicon chips is, except in the case of the CCD schemes, available in hundreds of nanoseconds, or less.

These densities are achieved typically with lithographic techniques whose minimum dimensions are 5µ. Clearly more advanced lithography, which has existed in the laboratory for many years already, can achieve major further gains. Advanced laboratory devices with 1µ dimensions, made by optical techniques, have been in existence for years [36], and electron beam techniques can go much further than that [35].

VIII. OPTICALLY ACCESSED MEMORIES

Optically accessed memories have been discussed in great detail in a set of conference papers [60,61], and in other reviews [44,62-66]. We will not repeat these discussions. The proposals fall into two major groups: Those read and written a bit at a time, and the holographic schemes. Those accessed a bit at a time essentially replace the mechanically movable magnetic head by a deflectable laser beam. Most typically they use the laser beam simply for its localized heating action. Even if a suitable physically reversible effect can be found this requires, for complete erasure:

a) A broader or more intense beam in erasing, or
b) extreme accuracy in positioning for erasing.

Furthermore, the temperature cycling must, of course, be large compared to unintentional fluctuations. This raises questions about the stability of the typically thin storage film, after prolonged

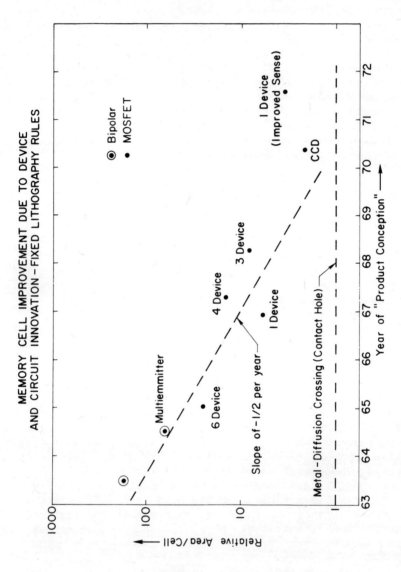

Figure 7. Improvement in area per bit, due to device and circuit ingenuity. The horizontal dashed line is the minimum needed for contact between separate electrodes running in one direction and semiconductor diffusions in the other, and is about $9W^2$, where W is the minimum metallic linewidth. (Supplied to the author by D. Critchlow).

cycling. After all, to get efficient and fast response to the laser beam we must utilize an effect in which we only have time to heat up a thin active region, and not a great deal of its environment. We must also heat that region by an amount large compared to thermal fluctuations in the ambient. That sounds like a prescription for thermally induced stress cycling.

Nevertheless, in contrast to developmental holographic storage, practical bit accessed, optical stores do exist. An early version (actually written with an electron beam and read optically) was IBM's Photo Digital Mass Storage System [67]. This memory used silver halide emulsions and involved the drawbacks of a development process discussed in Sec. VI, and later again in this section.

Precision Instrument's Unicon already cited in connection with metallic film hole burning technology in Secs. VI and VII, is postable and thus more versatile. It is a clear demonstration of the high densities that are available through optical techniques. In that case 5μ holes are burned in a thin rhodium film, and a density of about 4×10^6 bits per cm^2 is obtained. Indeed this may be the limit of what can easily be done in a system (at practical cost), requiring fast motion and random access, without running into excessive depth of field problems. The clear advantage of optically accessed memories: It is much easier to servo a light beam quickly and reliably onto a relatively high density of tracks than a massive magnetic head. On the other hand, it's likely, in the long run, that magnetic bit densities along the track can outdo optical schemes [51,52]. Unicon demonstrates an obvious advantage of the "spot at a time" schemes. With the high available light intensity at the spot one can utilize highly nonlinear phenomena, with a clear threshold. This eliminates the need for a developing and/or fixing process; there is no need to protect the information in subsequent reading, or in exposure to stray light.

A holographic memory typically has four components:

1. A latrix, or page composer, i.e. an electrically controllable digital image, for subsequent holographic transformation.

2. A deflector to permit random access to the desired hologram.

3. A reversible holographic memory.

4. A detecting array to hold and sense the output.

While none of these really exist today, the momentum and versatility of the silicon technology would undoubtedly allow us to develop item 4. Deflectors certainly exist, and will improve. Their status is best indicated by the fact that practical apparatus, of the sort discussed in the latest I/O devices of Section I, requiring scanning over many spots, but without random access, still use mechanical scanners. There is, in fact, the possibility of further progress in mechanical scanners, e.g. replacing moving mirrors by rotating holograms [68]. We do not believe that the page composer problem has been solved. Liquid crystals are too slow. Ferroelectrics suffer from problems which have always caused difficulty with ferroelectric memories. These include fatigue [69], (loss of reversible polarization with repeated switching), and difficulty in getting really high speed switching compared to other modern memory technologies. Finally there is the "disturb" problem, related to the fact that ferroelectrics don't really have a very sharply defined coercive force, and small unintentional signals, arising from voltages applied intentionally to neighboring elements, can cause some switching. This latter effect would not be too serious in a page composer application, where there is a very limited amount of switching in the neighborhood of a given element, before the element is intentionally reset to receive the next image. Nevertheless ferroelectrics are probably the most serious candidate for this purpose, though many crystalline ferroelectrics, as currently grown, do not have adequate optical quality. Media which are deformable under charge deposited by an electron beam represent another possible approach [70]. The reversible medium, (3), also seems to elude us. Thermoplastic materials have a limited cyclability and require development. Materials which use light to induce charge redistribution and electric fields in electrooptic materials suffer from

a. Questions about complete reversibility. Do the electrons in the long run always find their way back to their original trapping sites?

b. If the information is simply stored in electronic displacements the memory is volatile.

c. If the memory is developed by ionic motion, then a slow development step is required. Furthermore once the ionic motion neutralizes the electronic charge, the electrooptic field will depend on the exact amount of subsequent exposure to reading light.

Proposals exist [71] for using switching in ferroelectrics as a development method after illumination has produced an electronic charge distribution. This can, in principle, be fast development. But the use of domain behavior will limit the resolution and raises questions about the long term stability of the resulting domain walls. After all, it is well known that ferroelectric domain walls do move slowly in fields less than the 60 c.p.s. coercive force. Additionally, it would seem that subsequent illumination could free the trapped carriers and permit simultaneous motion of these carriers

and the associated domain walls. Indeed, it was already observed
two decades ago [72], that illumination of $BaTiO_3$ with light undid
the effects of fatigue [69]. Nevertheless the high sensitivity of
some of the "optical damage" schemes in ferroelectrics seems to
warrant further investigation. In KTN for example [73] photo
excited carriers can be made to move over 1μ, giving this material a
sensitivity comparable to silver halide emulsions. Ref. [73] also
demonstrates that there is still room for major conceptual symmetry
questions concerning photoexcited bulk currents.

The delicacy of holographic schemes is signified by the fact
that in demountable development equipment they invariably require
heavy optical benches. Their celebrated immunity to localized
defects seems offset by their sensitivity to thermal expansion,
directional orientation, and other sources of dimensional changes in
the storage medium. As already mentioned, magnetic tape schemes
are insensitive to the stretching arising from the machinery moving
it, temperature changes or the typical 1% long term aging changes in
mylar dimensions.

The general need, however, for a fixing process seems to be
the most fundamental drawback, and not only because of the time,
complication, and expense of the fixing process. It means that data
is unalterable, except by completely and simultaneously erasing all
the information in the hologram. Data also cannot be written over
a part of the hologram, it must be written at one time, as one
record. These disadvantages of the fixing process become most
pronounced in three dimensional Bragg type holograms, where a
particularly large amount of information is involved.

Many of the optical schemes require expensive single crystal
materials, and even in the case of ceramic materials (e.g. PLZT),
will be demanding in terms of material perfection. These materials
are more likely to be comparable in cost to silicon, or amorphouss
bubble materials, than to magnetic tape. In addition we must
address the optical materials with complex beam generation and
addressing equipment. By contrast in a silicon or bubble memory we
can use such apparatus in the manufacturing process, and don't ship
it along with the chip!

IX. STANDARDIZATION IN MEMORY

Standardization was already discussed in Sec. V, on Logic.
There standardization referred to the history of a signal as it went
through successive logic stages. A bit goes into storage and out of
it; in the process of detecting the stored state standardization is
an almost inevitable ingredient. In view of the general complexity
of the read-out process, it is not much of a complication. There

are, of course, memory schemes (e.g. Sec. VIIc, Charge Coupled
Devices) where information is not held in bistable devices and thus
deteriorates with time. In these the bit must be restandardized
periodically, as part of the memory operation.

Standardizing the information on its way out of the memory is
typically not a formidable chore, but the requirement that the
memory not change after many cycles of storage, may be more restrictive. Thus, for example, ferroelectricity was the original candidate
for an integrated, electrically accessible, memory. One of the difficulties of this technology was the "fatigue" phenomenon [69] that
we have cited. After repeated switching, less of the crystal
became reversible.

Standardization of the memory cell is, of course, aided if we
are switching back and forth between two states, each of which is
favored against small disturbances. Even if only one of the two
states is so favored, it can still prevent a long range slow drift
arising from, say, unequal switching forces.

The bistable state is illustrated in Fig. 8(a). In many
practical systems, e.g. ferromagnetic recording, the energy terms
favoring the "0" and "1" state are slight, and we are really dealing
with a situation more like Fig. 8(b), where we rely on strong switching to drive the material toward a well defined "saturated" state.
Indeed even if only one of the two states show this saturation, as
in Fig. 8(c), and we repeatedly drive hard against the barrier shown
that is adequate to insure standardization. At least it is if the
material doesn't suffer structural or chemical changes in other
ways, as an unintentional byproduct of the switching.

The almost ideal memory component is therefore bistable, and is
easily coupled to an electrical circuit through its bistable degree
of freedom. Furthermore as it is switched, it should not suffer
major structural or chemical changes, and switching should involve
modest energies. To aid heat dissipation, and to be accessible
through macroscopic machinery, the degree of freedom should couple
the behavior of many atoms or crystalline cells. All this is essentially a prescription for the typical cooperative phenomena such
as ferroelectricity, ferromagnetism, and superconductivity, which
have been utilized in storage technology development. The strong
similarity between these effects and various dissipative phenomena
has been stressed elsewhere [74] and will not be elaborated here.

Unfortunately the optical storage schemes, by and large, do
not offer these advantages. Indeed a holographic memory, almost by
definition, must record a continuous interference pattern and cannot be easily bistable. Holographic proposals have, in fact,
invoked bistable materials but at the expense of higher order dif-

fraction effects, and otherwise tightening the available design trade-offs. A binary holographic medium would not, for example, permit simultaneous recording of independent Bragg patterns in three dimensions. Admittedly one of the difficulties with the typical electrically accessible solid state phenomena is the small available effect on the refractive index. This makes it hard, in planar holograms, to get a good resolution vs. efficiency compromise.

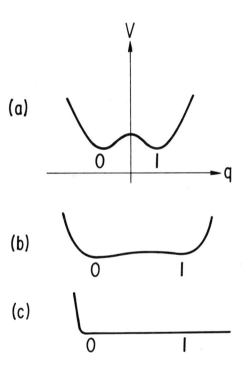

Figure 8. a) Bistable potential. b) Potential with steep walls, providing well defined limits to switching. c) Potential with one steep wall, providing one well defined limit against which system can be driven.

More generally let us assume that in holographic <u>recording</u> in a bistable medium the reference beam is strong compared to the image beam, and that the interference pattern takes us over the full binary range. Then we have a system which is very sensitive to reference beam variations. Conversely if the information beam is comparable to the reference beam, then there will be interference effects between different parts of the image.

Standardization, in the case of holographic schemes involving electron transport - just to give one example - requires that the erasing process return all the carriers exactly to their original position. We cannot, for example, allow the slow motion of carriers over large distances, due to a nonrandom sequence of illumination patterns.

Schemes for FEPC (Ferroelectric-photoconductor storage) are one example of a digital, photosensitive medium. Storage here is ferroelectric, the photoconductor is simply a switch or gate, controlling [66,75] the switching of an ordinary electrically reversible ferroelectric. We can think of the light as supplying a catalytic effect, i.e. as a reduction of the field that has to be applied to the series combination of photoconductor and ferroelectric to permit switching.

The FEPC schemes, however, have certain difficulties. Not the least of these is the dependence on two classes of materials: Ferroelectrics and Photoconductors, neither of which has been shown to be entirely free of "fatigue" problems. Nevertheless the use of light as a gating tool, with gain, and the reliance on an intrinsically bistable storage medium, seems to be a step in the right direction. The photoconducting-thermoplastic schemes are similar, but rely on a set of materials whose ability to cycle back and forth is even more suspect.

One recent proposal [76] performed holographic erasure by recording the negative of the original hologram. Quite aside from the fact that such a technique is incompatible with the existence of a "fixing" process, it clearly requires a very delicate relationship between the original recording and the erasing signals, and is devoid of any "standardization" feature. Repeated cycles of slightly incorrect erasure will just cause a build-up of noise.

Another proposal which deserves mention here is the use of color centers in alkali halides, where the localized arrangement of defects can be reoriented through illumination by polarized light [77]. In this approach we cycle between equivalent ground states of the defect in a manner reminiscent of Figure 8a, and the optical field acts as a switching force. The sensitivities, however, while appreciably better than for thermal schemes, are not comparable to those for effects which provide real quantum gain. Development and/or fixing is avoided by using reading light at a wavelength at which the refractive index changes are very small. As a result volume holography is needed.

CONCLUSION

While lasers seem to have an established place in the computer, their clearest role is in the I/O area. In our discussion of logic and storage we have gone out of our way to emphasize the difficulties and also the competition faced by optical proposals. Storage is, in a sense, the most interesting area, since the inventions, materials, and concepts which give the laser a serious place are probably not yet in existence. There are no general arguments, however, to rule them out, as in the case of logic. Our somewhat negative evaluation reminds one of the fate of other negative technological predictions. A. L. Schawlow, some years ago [78] quoted from a 1909 assessment of the future of the airplane, and I recommend this charming passage to the reader's attention. Unfortunately, however, pessimistic technological assessments are usually right.

REFERENCES

1. K. Ganzhorn, Elektronische Rechenanlagen 15, 263 (1973).

2. R. L. Mattson, Applied Optics 13, 755 (1974).

3. L. D. Dickson and R. L. Soderstrom, to be published in Applied Optics with Proceedings of Conf. on Laser Engineering and Appl., Washington D.C., May 28, 1975. A closely related manuscript has also been submitted to the IBM J. Res. Develop.

4. W. V. Smith, Electronic Information Processing (Artech House, Dedham, Mass., 1974).

5. S. Triebwasser, Proceedings of the 5th Conf. (1973 International) on Solid State Devices, Tokyo, 1973, Supplement to the Journal of the Japan Society of Applied Physics, Vol. 43 (1974).

6. J. von Neumann, U. S. Patent 2,815,488 (1954).

7. E. Goto, Proc. I.R.E. 47, 1304 (1959).

8. G. Diemer and J. G. Van Santen, Philips Res. Repts. 15, 368 (1960).

9. N. G. Basov, W. H. Culver, and B. Shah in Laser Handbook, F. T. Arecchi and E. O. Schulz-DuBois eds. (North Holland Pub. Co., Amsterdam, 1972).

10. A. B. Fowler, U. S. Patent 3,303,431.

11. A. B. Fowler, Appl. Phys. Letters $\underline{3}$, 1 (1963).

12. A. B. Fowler, Journal of Appl. Phys. $\underline{35}$, 2275 (1964).

13. G. J. Lasher and A. B. Fowler, IBM J. Res. Develop. $\underline{8}$, 471 (1964).

14. A. B. Fowler and G. J. Lasher, U. S. Patent 3,509,834.

15. G. J. Lasher, Solid-St. Electron. $\underline{7}$, 707 (1964).

16. G. J. Lasher, U. S. Patent 3,427,563.

17. W. F. Kosonocky, U. S. Patent 3,431,437.

18. W. F. Kosonocky, IEEE Spectrum 2, 183, March 1965.

19. M. I. Nathan, J. C. Marinace, R. F. Rutz, A. E. Michel, and G. J. Lasher, Journal of Appl. Phys. $\underline{36}$, 473 (1965).

20. W. H. Culver and F. Mehran, U. S. Patent 3,643,116.

21. E. J. Johnson, L. A. Riseberg, A. Lempicki, and H. Samelson, Appl. Phys. Lett. $\underline{26}$, 444 (1975).

22. R. W. Keyes and J. A. Armstrong, Applied Optics 8, 2549 (1969).

23. R. W. Keyes, Science, 168, 796 (1970).

24. R. W. Keyes, Proc. IEEE, $\underline{63}$, 740 (1975).

25. J. J. DeCillo, Proc. IEEE, $\underline{56}$, 1608 (1968).

26. H. N. Ghosh, K. G. Ashar, A. S. Oberai, D. DeWitt, IBM J. Res. Develop. $\underline{15}$, 436 (1971).

27. See, for example, papers 10.3, 11.1, 11.2, 1975 IEEE Internation. Solid-State Circuits Conference, Digest of Technical Papers (Lew Winner, N. Y. 1975).

28. R. M. Swanson and J. Meindl, Paper 10.5 in 1975 IEEE Internation. Solid-State Circuits Conference, Digest of Technical Papers (Lew Winner, N. Y. 1975).

29. W. Anacker, Paper 14.1 in 1975 IEEE International Solid-State Circuits Conference, Digest of Technical Papers (Lewis Winner, N. Y. 1975).

30. W. Baechtold, Paper 14.2 in 1975 IEEE International Solid-State Circuits Conference, Digest of Technical Papers (Lewis Winner, N. Y. 1975).

31. W. H. Henkels, IEEE Trans. Mag., Mag-10, 860 (1974).

32. H. Zappe, Appl. Phys. Lett. 25, 424 (1974).

33. P. Gueret, Appl. Phys. Lett. 25, 426 (1974).

34. F. H. Dill, J. A. Tuttle, A. R. Neureuther, paper in 1975 IEEE International Solid-State Circuits Conference, Digest of Technical Papers (Lewis Winner, N. Y. 1975).

35. A. N. Broers and M. Hatzakis, Scientific American 227, 34 (May 1975).

36. R. Landauer, Phys. Today 23, 22 (July 1970), translated in Uspekhi Fizicheskikh Nauk 106, 125 (1972).

37. R. Landauer in Physik 1971, p. 286 (B. G. Teubner, Stuttgart 1971).

38. C. H. Bennett, IBM J. Res. Develop. 17, 525 (1973).

39. R. Landauer and J. W. F. Woo in Synergetics, H. Haken ed. (B. G. Teubner, Stuttgart, 1973), p. 105.

40. E. Goto, K. Murata, K. Nakazawa, K. Nakagawa, T. Moto-oka, Y. Matsuoka, Y. Ishibashi, H. Ishida, T. Soma, and E. Wada, IRE Trans. on Electronic Computers 9, 25 (1960).

41. A. G. Anderson and E. L. Hahn, U. S. Patent 2,714,714.

42. W. B. Riley, ed., Electronics (Feb. 14, 1972) p. 91-2.

43. H. R. Dell, Computer Design, 10, (Aug. 1971) pp. 49-54.

44. G. Feth, IEEE Spectrum 11, 28 (Nov. 1973).

45. J. M. Harker and M. Chang in Proc. 1972 Spring Joint Computer Conference, p. 945 (AFIPS Press, Montvale, N. J., 1972).

46. A. V. Pohm and R. J. Zingg, IEEE Trans. on Mag., Mag-8, 574 (1972).

47. J. C. Gilbert, Wireless World 76, 377 (1970).

48. A. J. Boyle and J. McNichol, The Electronic Engineer 30, 38 (February 1971).

49. D. N. Kaye, Electronic Design 3, 35 (February 1973).

50. K. Compaan, P. Kramer, unpublished paper, 1973 Intermag Conf.

51. E. Hopner, AGEN, Nr. 10, p. 48, (September 1969).

52. D. Mee, "A Comparison of Bubble and Disk Storage Technologies," 1974 Intermag Conference, Canada, submitted to IEEE Trans. on Mag.

53. J. J. Brophy, IRE Trans. on Audio, AU-58 (1960).

54. I. P. Breikss, IEEE Spectrum 12, 58 (May 1975).

55. H. Chang, Magnetic Bubble Technology (IEEE Press, N. Y. 1975).

56. H. H. Zappe, IEEE Journal of Solid State Circuits, SC10, 12 (1975).

57. Computer Decisions 7, No. 2 (Feb. 1975), p. 16.

58. W. K. Hoffman and H. L. Kalter, IEEE Journal of Solid-State Circuits, SC8, 298 (1973).

59. R. Foss and R. Harland, Paper 10.1, 1975 IEEE International Solid-State Circuits Conference, Digest of Technical Papers (Lewis Winner, N. Y. 1975).

60. A. Kozma and E. S. Barrekette, Applied Optics, 13, 747 (1974), and succeeding papers.

61. A. L. Hammond, Science 180, 287 (April 20, 1973).

62. O. N. Tufte, D. Chen, IEEE Spectrum 10, No. 2 (Feb. 1973), pp. 26-32; IEEE Spectrum 10, No. 3 (March 1973), pp. 48-53.

63. Fundamentals of Amorphous Semiconductors (National Academy of Sciences, Washington, D. C. 1972), Chapter VII.

64. A. H. Eschenfelder in Electronic Materials, N. B. Hannay and U. Colombo, eds. (Plenum Press, N. Y. 1973), pp. 603-638.

65. R. E. Matick, Proc. IEEE 60, 266 (1972).

66. L. K. Anderson, Ferroelectrics 3, 69 (1972).

67. J. D. Kuehler and H. R. Kerby in Proceedings Fall Joint Computer Conference (IEEE Computer Society, N. Y. 1966) p. 735.

68. R. V. Pole and H. P. Wollenmann, Applied Optics 14, 976 (1975).

69. J. C. Burfoot, Ferroelectrics (D. Van Nostrand Co., Ltd., Princeton, N. J. 1967), pp. 201, 202, 216.

70. B. Kazan, and M. Knoll, Electronic Image Storage (Academic Press, N. Y. 1968), p. 261.

71. F. Micheron, C. Mayeux, and J. C. Trotier, Applied Optics $\underline{13}$, 784 (1974).

72. C. Karan, private communication.

73. A. M. Glass, D. von der Linde, D. H. Auston, T. J. Negran in a paper presented at the Conference on Defect-Property Relationships in Solids, Princeton University, March 24, 25, 1975. To be published with the proceedings of the conference in the Journal of Electronic Materials.

74. H. Haken, Revs. Modern Phys., $\underline{47}$, No. 1, (1975), p. 67.

75. D. W. Chapman and R. Mehta, Ferroelectrics $\underline{3}$, 101 (1972).

76. J. P. Huignard, J. P. Herran, and F. Micheron, Appl. Phys. Lett. $\underline{26}$, 250 (1975).

77. W. E. Collins, I. Schneider, M. E. Gingerich, and H. P. Klein, "Information Storage in Photodichroic Crystals," Naval Research Laboratory Memorandum Report 2858, Naval Research Laboratory, Washington D. C., August 1974.

78. A. L. Schawlow, American Scientist $\underline{55}$, 197 (Sept. 1967), see p. 218.

Supplement to "OPTICAL LOGIC AND OPTICALLY ACCESSED DIGITAL STORAGE"
by Rolf Landauer

A few additional comments concerning optical storage, particularly holographic storage, may be in order. The reader should note some of the other papers in this volume. A thorough and realistic review of optical modulators and page composers is provided by D. Casasent. A. Vander Lugt stresses the use of small one dimensional holograms, and thus points out a route which avoids many of the difficulties we have stressed. The papers by S. H. Lee and I. Kompanets also are closely related to our subject matter.

One of the shortcomings of ferroelectric light modulating systems: Almost all insulators subject to a D.C. component, in a time dependent field, will exhibit instabilities. These arise from the motion of electrons in deep traps and from ionic motion. The presence of an additional small ordinary electronic conductivity, without blocking effects at the electrodes can, in principle, be used to maintain a uniform field despite the displacements of ions and trapped charges.

In connection with the ferroelectric-photoconducting sandwiches there is a problem explained in the literature, but apparently not widely appreciated. To achieve a low coercive force for the ferroelectric, and a stable polarization after switching, it is necessary to bring the compensating charge in the electrodes exceedingly close to the ends of the polarization pattern in the ferroelectric. [M. Drougard and R. Landauer, J. Appl. Phys. $\underline{30}$, 1663 (1959)]. Gaps of even atomic dimension between the ends of the switchable polarization and the electrode charge can produce serious depolarization effects. This fact was rediscovered for optically addressed memories and is described by R. R. Mehta, B. D. Silverman and J. T. Jacobs, J. Appl. Phys. $\underline{44}$, 3379 (1973). Photoconductors with their limited carrier density will not permit the close approach of the compensating charge that is permitted by a metal with its screening length of order 1 Å. Discontinuous metallic films at the interface between a photoconductor and a ferroelectric can help provide the necessary density of states for accumulation of the compensating charge, and still permit lateral isolation between adjacent bits.

Recent work by D. von der Linde and A. Glass, presented at the 1975 Symposium on Applications of Ferroelectricity (to be published in "Ferroelectrics"), points to a new method of controlling image formation, with emphasis on refractive index control in holograms formed by drifted, photoexcited carriers. But clearly their scheme is of wider applicability. A hologram is formed at a frequency ω_1, and this frequency has inadequate energy to generate free carriers. An additional uniform illumination, at a second frequency, then allows absorbtion and carrier generation. This takes place either through a nonlinear two-photon process, or through sequential one-photon processes, via an intermediate level inaccessible through irradiation by ω_1 alone. Reading, at ω_1 will then be non-destructive, or at least not conspicuously destructive.

Finally, in case it is not clear from our text, we wish to stress that a volume hologram is rather like a library in which replacement of one book necessitates replacement of the whole library.

QUESTIONS AND COMMENTS

<u>Dr. P. Tverdokhleb</u> (question to Dr. R. Landauer)

a) I should like to know your opinion on the possibility of practical application of holographic memories of 10^{10} bits in capacity and a module memory organization.

<u>Dr. R. Landauer.</u> To my opinion, such types of memories are efficient when the capacity of the memory module with high-speed access is of the order of 10^6 bits. Since in the case under consideration the number of modules is equal to 10^4 the time of module replacement should not exceed 0.1 sec.

b) Will the architecture of computers be changed with the advent of large-capacity memories?

<u>Dr. R. Landauer.</u> Optical memories serve as external units of the computer memory. Their connection to the computer needs only a special interface. Thus, it is unlikely that the architecture itself will undergo any changes.

NONLINEAR OPTICAL PROCESSING

Sing H. Lee

Department of Applied Physics and Information Science
University of California, San Diego
La Jolla, California 92037

I. INTRODUCTION

It is well known that coherently illuminated optical processing systems are very useful for performing such linear operations as convolution, cross-correlation and Fourier spectral analysis because the Fourier transform of an optical signal exists physically and can, therefore, be measured and modified [1]. With properly synthesized composite grating filters and binary filters, the mathematical operations of addition, subtraction, differentiation, integration and generalized orthogonal transformations can easily be carried out [2-4]. To increase the computing capability and flexibility of optical processors, digital electronic computers are incorporated to form hybrid processors [5-10]. If nonlinear operations can be implemented easily, optical processing will be even more useful. Several schemes are presently available to perform nonlinear processing without employing nonlinear devices. These include the theta modulation technique [11], the half-tone process [12], the holographic approach [13] and the method utilizing active or amplifying elements [14]. This paper discusses, however, only optical processing involving real-time nonlinear devices, with and without feedback [2,15], together with the current research activities at UCSD on nonlinear processing.

II. OPTICAL PROCESSING WITH REAL-TIME NONLINEAR DEVICES

Nonlinear optical materials can be used to form nonlinear devices. Those which are reusable and have a fast response time are of special interest because they are potentially useful in

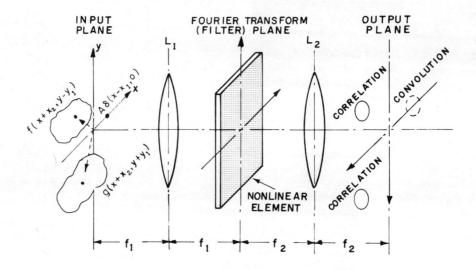

Figure 1. A coherent optical system that performs linear operations with a nonlinear element in the filter plane.

NONLINEAR OPTICAL PROCESSING

real-time processing. Nonlinear materials have been utilized to perform feasibility studies on linear as well as nonlinear optical processing.

II.A The Use of Nonlinear Devices to Perform Linear Processing

To perform linear processing, nonlinear devices may be used in the filtering plane of a coherent processor to avoid the need for the fabrication of complicated spatial filters [16]. For example, the convolution or correlation of two image functions $f(x,y)$ and $g(x,y)$ can be obtained using the scheme shown in Fig. 1. The nonlinear device in the filtering plane will provide the necessary mixing or multiplication of the Fourier spectra from the two inputs, $f(x,y)$ and $g(x,y)$. Nonlinear devices generally introduce some unwanted output also (see Appendix for analysis). To separate the convolution and correlation functions spatially from the unwanted outputs, a reference delta function is introduced in the input plane. Another useful role of the reference delta function is to provide a tilted plane wave illuminating the nonlinear device and biasing it at the proper operating point.

Saturable absorbers such as organic dyes are nonlinear materials. Figure 2 shows the amplitude transmittance vs. input intensity for cryptocyanine dye, which has a response time of several nanoseconds. Note that while the amplitude transmittance is linearly proportional to input intensity, it is nonlinearly related to

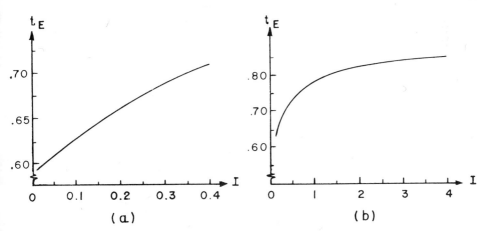

Figure 2. A typical amplitude transmittance characteristic of cryptocyanine dissolved in methanol (a) at low input intensities (b) at a wide range of input intensities.

the input electric field. Figure 3 illustrates some experimental results obtained with cryptocyanine at the ruby laser wavelength for some simple pattern functions. In Part (a) of the figure, both f and g in the input are chosen as delta functions to help determine the output locations of the correlation dots (circled by solid lines) and the convolution dot (circled by dashed lines). In Part (b) both f and g in the input become two dots aligned vertically. The autocorrelation of f and g in the output is three dots. In Part (c) f is two dots aligned vertically and g is two dots aligned horizontally. The cross correlation of f and g is shown as four dots.

The nonlinear filtering element was also used to correlate a spiral with a composite pattern consisting of a spiral, a cross and a tilde as shown in Fig. 4a. The more intense dot, appearing closest to the x-axis in the correlation functions, indicates the detection of the spiral (Fig. 4b). When a cross instead of a spiral is the pattern to be detected, the more intense dot occurs in the center of the correlation pattern corresponding to the position of the cross in the composite pattern (Figs. 4c, d). Recently these experiments were successfully duplicated with a flash-lamp-pumped R6G dye laser (30 pulse/sec.) and with DODCI dye as the nonlinear material [17].

II.B The Use of Nonlinear Devices to Perform Nonlinear Processing

To perform nonlinear processing, nonlinear devices can be employed in the image plane. Figure 5 shows a possible setup for the experiment. A pattern function at plane P_1 is illuminated by a collimated laser beam and imaged onto the nonlinear device by lens L_1. This image passes through the nonlinear device and is then imaged onto plane P_3 by lens L_2. Thus, the image on P_3 is obtained from that on P_1 through the nonlinear device.

When cryptocyanine dye is employed as the nonlinear material, a transfer characteristic curve can be derived from Fig. 2 with input and output intensities plotted on a log-log scale as illustrated in Fig. 6a. The transfer characteristics can be subdivided into three regions. At the low and high intensity ends of the curve, the slope γ is 1. But in the middle region γ is greater than 1. Passing an image of intensity variations in the middle region of the curve will result in the nonlinear operation, $I_{out} \propto (I_{in})^{\gamma}$. The γ in the middle region of the transfer characteristic curve is concentration dependent (Fig. 6b). Higher concentrations generally give greater slopes in the middle region. Operating a nonlinear device with three different concentrations in its middle range yielded the experimental results shown in Fig. 7. Figure 7(a) is the reference image. Figure 7(b) is for low concentrations, and

NONLINEAR OPTICAL PROCESSING 259

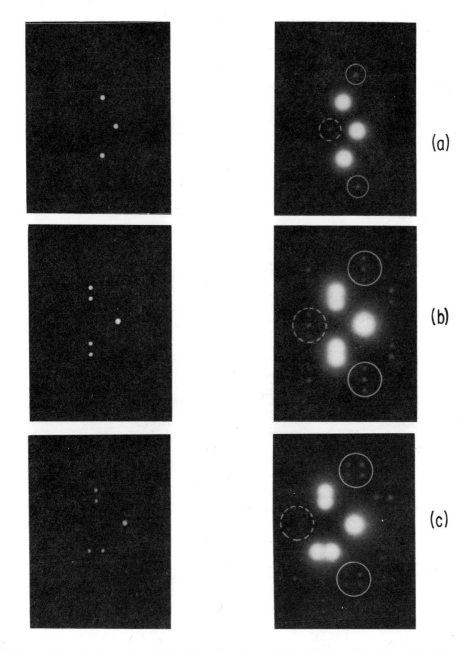

Figure 3. Left column is the input; right column is the output. The correlation functions are exhibited in the regions circled by solid lines, while the convolution functions in the regions circled by dashed lines.

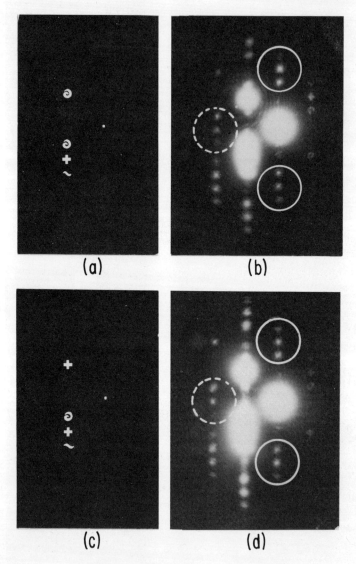

Figure 4. Left column is the input; right column is the output.

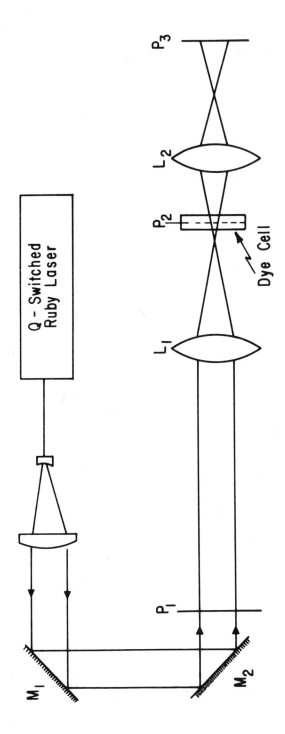

Figure 5. An experimental set up with the nonlinear element in the image plane for contrast enhancement.

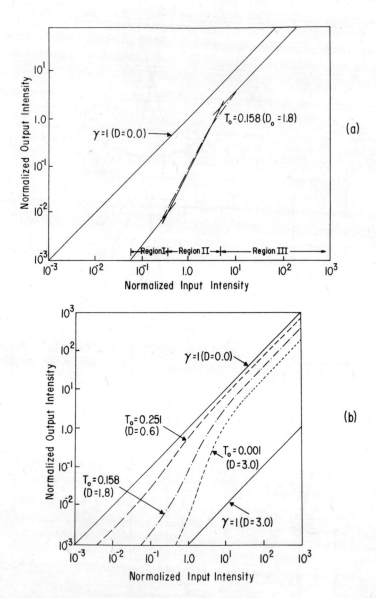

Figure 6. (a) A typical transfer characteristic curve for organic dye solution.
(b) The transfer characteristics for three different dye concentrations.

Figure 7. Experimental results demonstrating the effects of dye concentrations on image contrast. (a) The reference image. (b) For low concentration. (c) For moderate concentration. (d) For high concentration.

Figs. 7(c) and 7(d) for higher concentrations. As expected, the highest concentration solution produces the highest contrast. This effect can be seen clearly by comparing the step tablets located at the bottom of each picture, as high contrast implies that it takes fewer steps to get from one density level to another.

III. CURRENT RESEARCH ACTIVITIES AT UCSD

Several research topics on nonlinear optical processing are under investigation. The characteristics of nonlinear materials and devices are being examined and improved. Also the application of nonlinear device for thresholding and logic operations, which are essential to the optical-digital approach to optical computing, are being investigated.

III.A Nonlinear Optical Material Study

Organic dyes such as cryptocyanine and DODCI have been utilized as the nonlinear materials to perform feasibility studies on linear as well as nonlinear processing, as described above. Both of these dyes have large saturation intensities and in practice are not quite satisfactory for optical processing. Other materials of higher sensitivities are under study. Iodine and sodium are among the list of materials, which promise to have saturation intensity 10^3 times lower than those dyes previously studied (Table I).

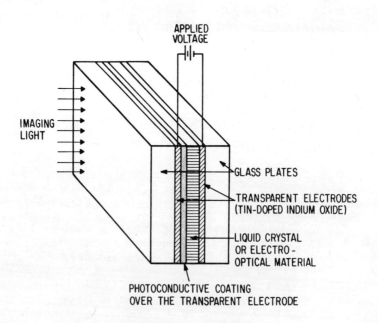

Figure 8. Schematic diagram of a device under investigation for nonlinear as well as linear processing.

TABLE I

NONLINEAR OPTICAL MATERIALS

Material Parameter	Cryptocyanine	DODCI	R6G	Iodine	Iodine	Sodium
	Dye	Dye	Dye	Solution	Vapor	Vapor
Absorption Wavelength (Å)	6943	5800	5300	Solvent dependent	5145	5890
Bandwidth (Å)	500	500	500	300	Pressure dependent	Pressure dependent
Absorption Cross-section (cm^2)	7.2×10^{-16}	8.8×10^{-16}	2×10^{-16}	9×10^{-18}	3×10^{-17}	2×10^{-14}
Life-time (sec)	0.04×10^{-9}	0.3×10^{-9}	5×10^{-9}	3×10^{-6}	3×10^{-6}	5×10^{-9}
Saturation intensity (Watts/cm^2)	10^7	1.3×10^6	3.7×10^5	1.4×10^4	4×10^3	3.3×10^3

III.B Nonlinear Optical Device Study

In addition to the search for sensitive optical materials which respond nonlinearly in real-time, several devices which utilize photoelectric and electro-optic effects for nonlinear processing are under investigation. Figure 8 illustrates a device configuration which utilizes CdS (cadmium sulphide) or PVK (poly-n-vinyl carbazole) as the photoconductor and $LiNbO_3$ (an electro-optic material) or mixed liquid crystal as the modulated material. It is known that photoconductor-activated liquid crystal devices using CdS photoconductor and MBBA (liquid crystal) can be operated *nonlinearly* [18,19] with good resolution (better than 50 lines/mm), high sensitivity (1 to 50 $\mu W/cm^2$ at 510 nm), large contrast ratio (100:1) and low operating voltages (20 to 100 V rms) [20]. But, the known devices have relatively slow response times (10 msec turn-on and 30 msec turn-off). To improve on the response time, mixed liquid crystals (cholesteryl nonanoate, MBBA and a cyano Schiff's base) and PVK photoconductor are being employed. The mixed liquid crystal [21] promises to provide a response time of 60 μsec, and the photoconductor 20 μsec. [22]. $LiNbO_3$ responds even faster, but requires higher operating voltages.

One of the most important operations for nonlinear devices is thresholding. Optical feedback schemes similar to that illustrated in Fig. 9a are being examined to enhance nonlinearity of optical devices for thresholding. With optical feedback, there will be multi-reflections passing through the nonlinear material. These multi-reflections will interfere with one another constructively or destructively, depending on the mirror separation. The transmission characteristic of the device is thereby affected. Figure 9b shows the transmission vs intensity characteristic of a saturable resonator [23]. It is interesting to note that thresholding operations are attainable with mirrors of high reflectances. With threshold devices, optical signals can be converted from analog to binary forms (Fig. 10). Optical signals in binary form will be essential to processing with optical-digital techniques [24]. Calculations based on 2" × 2" devices with 50 lines/mm resolution and 100 μsec cycle time show that analog to binary conversion rates can be greater than 6.25×10^6 bits/frame and 62.5×10^9 bits/sec.

III.C Planar Logic Device Design

Besides investigating optical threshold devices for analog to binary conversion, planar optical logic devices are also under development for parallel processing with optical-digital techniques. Figure 11 illustrates a device configuration designed for a planar logic AND gate. It utilizes birefringent electro-optic materials and appears quite similar to the device configuration of Fig. 8, except that the photoconductive layer is no longer a uniform layer,

NONLINEAR OPTICAL PROCESSING

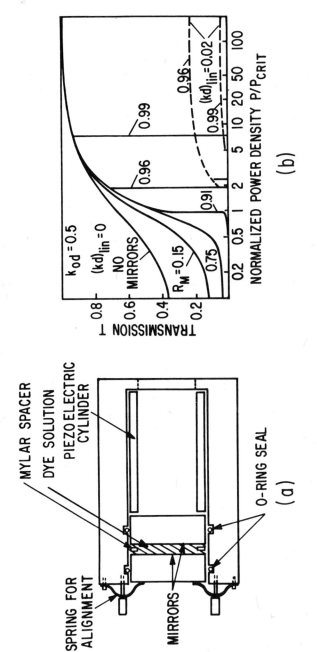

Figure 9. (a) Construction of the saturable resonator.
(b) Transmission of a saturable resonator versus the incident power density for different mirror reflectivities. The full curves are for the case that all losses are saturable; the dashed curves are for additional linear losses $(kd)_{lin} = 0.02$. From Reference 23.

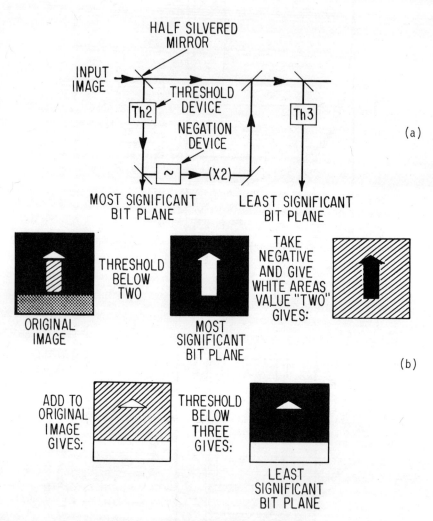

Figure 10. (a) Analog to digital conversion hardware schematic for a four level image.
(b) An example of analog to digital conversion algorithm for a four level image.
Courtesy of D. H. Schaefer and J. P. Strong of NASA Goddard Space Flight Center.

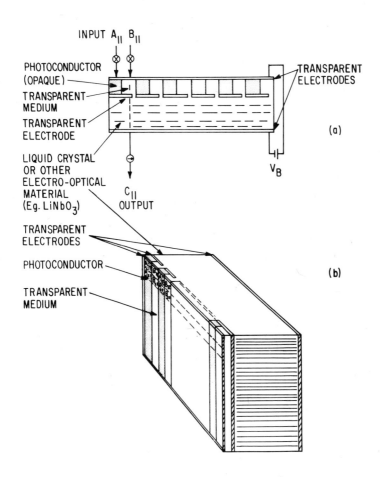

Figure 11. Planar logic AND gate.
(a) Bottom view. (b) Perspective view.

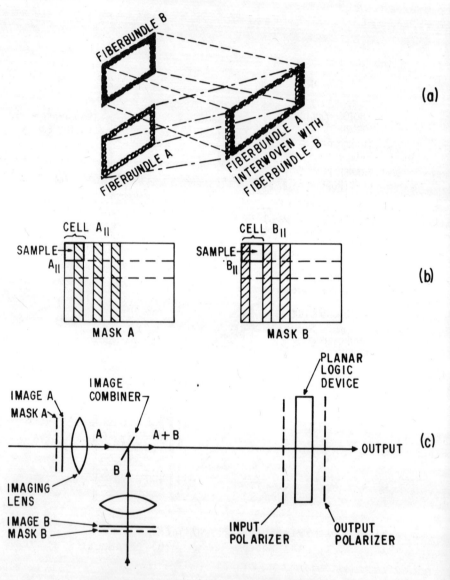

Figure 12. (a) Image combining using optical fiber interweaving;
(b) Masking or sampling of input images;
(c) Planar logic device with associated optical system.

but has a pattern, and that between this patterned photoconductive layer and the electro-optic layer a patterned electrode has been added. To explain how this AND gate works, let us first sample the two binary input images with fiber bundles (Fig. 12a) or photomasks (Fig. 12b). When the sampling rate satisfies the well-known sampling theorem, there will be no loss of image information. The sampled images are then combined by interweaving the fiber ribbons (Fig. 12a) or by an image combiner (Fig. 12c). In either case, the combined image consists of alternating columns from the two original inputs as illustrated in Fig. 11. Now, this combined image passes through a polarizer and is incident upon the planar logic AND gate. The planar logical AND gate performs an AND operation because there will be light coming out from cell B_{ij} with orthogonal polarization relative to the input polarization only when there is light incident in both cells A_{ij} and B_{ij}. The light incident on cell A_{ij} causes the potential across the photoconductive material in the cell to drop to zero. This drop in potential will be transformed into an increase in voltage across the electro-optic material in cell B_{ij}, since the electrode pattern above the electro-optic layer links cells A_{ij} and B_{ij} together. The voltage increase across the electro-optic material rotates the polarization of the light through cell B_{ij} by 90°. In other words, had it not been for light incident on cell A_{ij}, the potential across the photoconductive region in cell A_{ij} would be high and that across the electro-optic material in cell B_{ij} low. Then the light passing through cell B_{ij} will not be rotated in polarization. An output polarizer orthogonal to the input polarizer will easily convert the output polarization information from cell B_{ij} into binary intensity information.

Planar devices for other logical operations such as NEGATION, OR and EXCLUSIVE OR can be similarly designed. Based on the assumptions of 20 cells/mm resolution on 3" devices and 100 μsec switching time, logic operations can be performed at the rate of 2.25×10^6 bits/frame and 22.5 giga-bits/sec. Faster rates can be attained with electro-optical materials of faster switching time, e.g. $LiNbO_3$.

III.D IOC Logic Design

Optical switching and logic operations can also be performed with integrated optical circuits (IOC), which utilize both electro-optic and photo-conductive effects [25,26]. The electro-optic effect is utilized with waveguides made of electro-optic materials as illustrated in Fig. 13. In the absence of an applied electric field, the waveguides are identical. By selecting the proper coupling length and separation between the guides, light propagating in one can be coupled or switched to the other (Fig. 13a). When an electric field is applied, the waveguides no longer have identical indices of refraction and the coupling coefficient is greatly reduced. The

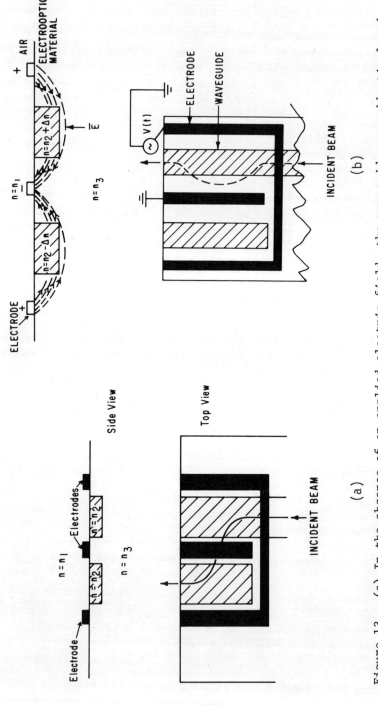

Figure 13. (a) In the absence of an applied electric field, the waveguides are identical and optical switching occurs;
(b) In the presence of an applied electric field, the waveguides are nonsynchronous and light tends to remain in the initially excited waveguide.

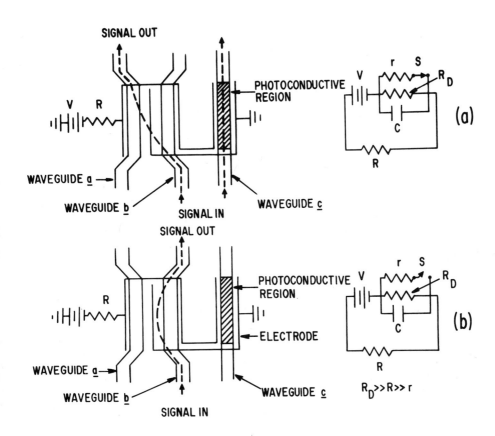

Figure 14. An optical switch. (a) when light is in waveguide c, the resistance of the photoconductive region (r) is very small. The voltage between electrodes is effectively shorted out and the two optical waveguides are almost identical. Coupling or switching occurs. (b) When no light is in waveguide c, the dark resistance of the photoconductive region R_D is very large and most of the applied voltage V appears across it. Then the two waveguides have different indices of refraction. No switching can occur.
Courtesy of Henry Taylor of Naval Electronics Laboratory Center at San Diego.

light remains in the initially excited waveguide (Fig. 13b).

When photoconductive regions are incorporated in the guides, application of the electric field to the guides can be controlled by sending optical signals down the guides. For example, when light is present in guide $\underline{c}$ of Fig. 14a, the applied voltage and field to all waveguides becomes negligibly small because (i) guide $\underline{c}$ has a photoconductive region and (ii) the electrodes across the photoconductive region are connected in parallel with those across the electro-optic switch capacitance (C), which encompasses waveguides $\underline{a}$ and $\underline{b}$. Light propagating in guide $\underline{b}$ will be switched to guide $\underline{a}$ due to strong coupling in the absence of a significant applied field. Now, when no light is propagating in guide $\underline{c}$ (Fig. 14b), the electrical resistance between the electrodes is high. Most of the applied voltage V will appear across the guides. The guides will have different indices of refraction and no coupling or switching will take place. Thus, the optical signal in $\underline{c}$ can control the light in $\underline{b}$ to exit either from $\underline{a}$ or $\underline{b}$.

The IOC switch of Fig. 14 can also be employed as an AND gate, since a signal emerges from waveguide $\underline{a}$ only if signals are incident in both waveguides $\underline{b}$ and $\underline{c}$. When no signal is incident on waveguide $\underline{b}$ or $\underline{c}$, no signal can emerge from $\underline{a}$. Therefore we have the AND operation. IOC configurations for other logic operations such as NEGATION, OR, EXCLUSIVE OR can be similarly designed.

The fabrication techniques for IOC's are quite similar to those of integrated electronic circuits. Therefore, the resolution limit is a few microns for the optical waveguides and about 50µ for each circuit. Assuming the width of the electro-optic substrate is 2 inches, it should be possible to fabricate 10^3 logic circuits on a single device. Since electro-optic materials have switching times that are less than 10^{-7} sec, it should be possible to perform logic operations with IOC's at a rate faster than 10^{10} bits/sec.

ACKNOWLEDGEMENT

The research is supported by the National Science Foundation, NASA at Goddard Space Flight Center and the Naval Electronic Laboratory Center at San Diego.

Bruce Bartholomew and Lew Goldberg are acknowledged for their valuable discussions, and Jack Cederquest for reading this manuscript.

APPENDIX

This appendix analyzes the operation of the nonlinear element in Fig. 1. When biased properly, the nonlinear element has an amplitude transmittance $t_E(f_x, f_y)$ proportional to the incident intensity $I(f_x, f_y)$:

$$t_E(f_x, f_y) \propto I(f_x, f_y) \tag{1}$$

Let the incident field $I(f_x, f_y)$ corresponding to the Fourier transform of the input amplitude be

$$E^-(f_x, f_y) = A \exp[-j2\pi f_x x_1] + F(f_x, f_y) \exp[j2\pi(f_x x_2 - f_y y_1)] + G(f_x, f_y) \exp[j2\pi(f_x x_2 + f_y y_1)]. \tag{2}$$

According to Eq. (1), the amplitude transmittance of the nonlinear device is

$$\begin{aligned} t_E(f_x, f_y) \propto & |A|^2 + |F(f_x, f_y)|^2 + |G(f_x, f_y)|^2 \\ & + AF^* \exp[-j2\pi\{f_x(x_1 + x_2) - f_y y_1\}] + c.c. \\ & + AG^* \exp[-j2\pi\{f_x(x_1 + x_2) + f_y y_1\}] + c.c. \\ & + FG^* \exp[-j2\pi f_y(2y_1)] + c.c. \end{aligned} \tag{3}$$

The field amplitude transmitted by the nonlinear element will therefore be

$$\begin{aligned} E^+(f_x, f_y) &= E^-(f_x, f_y) \, t_E(f_x, f_y) \\ &\propto (2A) \{F^*G \exp[-j2\pi(f_x x_1 - 2f_y y_1)] \\ &\quad + FG^* \exp[-j2\pi(f_x x_1 + 2f_y y_1)] \\ &\quad + FG \exp[j2\pi f_x(x_1 + 2x_2)]\} + 24 \text{ other terms.} \end{aligned} \tag{4}$$

When this field amplitude occurs in the front focal plane of a Fourier transform lens (here the output lens of the coherent processor), we have at the back focal plane

$$\begin{aligned} e_o(x,y) \propto & (2A) \{f*g*\delta(x-x_1, y+2y_1) + f*g*\delta(x-x_1, y-2y_1) \\ & + f \otimes g * \delta(x + x_1 + 2x_2)\} + 24 \text{ other terms,} \end{aligned} \tag{5}$$

where $e_o(x, y)$, $f(x, y)$ and $g(x, y)$ are the Fourier transforms of $E^+(f_x, f_y)$, $F(f_x, f_y)$ and $G(f_x, f_y)$ respectively. Therefore, provided that an appropriately chosen nonlinear element is placed in the filtering plane of a coherent processor and is illuminated by a field distribution described by Eq. (2), the correlation function of $f(x, y)$ and $g(x, y)$ will be obtained in the regions centered at $(x_1, 2y_1)$ and $(x_1, -2y_1)$, and the convolution function will be located at $(-x_1 - 2x_2, 0)$. Since the contributions of light from the other twenty-four terms in Eq. (5) will appear elsewhere in the output plane, these correlation and convolution functions can be readily detected. [It should be pointed out that some of the other twenty-four terms overlap with each other and hence one may not be able to differentiate between all of them in practice.] To provide the field illumination corresponding to Eq. (2), the pattern functions $f(x, y)$ and $g(x, y)$ are displayed along with a reference δ-function of amplitude A in the optical processor input (see Fig. 1).

REFERENCES

1. A. Vander Lugt, "Coherent Optical Processing." Proc. IEEE 62, 1300 (1974).

2. S. H. Lee, "Mathematical Operations by Optical Processing." Opt. Eng. 13, 196 (1974).

3. S. H. Lee, S. K. Yao, and A. G. Milnes, "Optical Image Synthesis in Real Time by a Diffraction Grating Interferometric Method." J. Opt. Soc. Am. 60, 1037 (1970).

4. S. K. Yao and S. H. Lee, "Spatial Differentiation and Integration by Coherent Optical Correlation Method," J. Opt. Soc. Am. 61, 474 (1971).

5. D. S. Sand, M. Faiman, and W. J. Poppelbaum, "A Real-time Electro-optical Fourier Transform System for Video Images." IEEE J. of Quan. Electronics, Q.E.-9, 708 (1973).

6. F. B. Rotz and A. Vander Lugt, "A Hybrid Optical Data Processing System." Report of an NSF Workshop on Optical Computing Systems, held at Carnegie-Mellon University, p.26, (September 1972).

7. D. Casasent, "A Hybrid Image Processor," Opt. Eng. 13, 228 (1974).

8. E. L. Hall, R. P. Kruger, and A. F. Turner, "An Optical-Digital System for Automatic Processing of Chest X-rays." Opt. Eng. 13, 250 (1974).

9. H. Stark "An Optical-Digital Image Processing System," *Opt. Eng.* 13, 243 (1974).

10. R. A. Meyer, D. G. Grant, and J. L. Queen, "A Digital-Optical Hybrid Correlator." Technical Memorandum TG1193A, Appl. Phys. Lab., Johns Hopkins University, (September 1972).

11. J. D. Armitage and A. W. Lohmann, "Theta Modulation in Optics." *Appl. Opt.* 4, 399 (1975).

12. H. Kato and J. W. Goodman, "Nonlinear Transformations and Logarithmic Filtering." *Opt. Comm.* 8, 378 (1973).

13. K. Preston, Jr., *Coherent Optical Computer*, McGraw Hill, Ch. 8 (1972).

14. N. G. Basov, W. H. Culver, and B. Shah, "Applications of Lasers to Computers." *Laser Handbook*, edited by F. T. Arecchi and E. O. Schule, DuBois, North Holland, Amsterdam (1972).

15. S. H. Lee, "Optical Computing with Laser Light." Presented at the Third Conference on the Laser, April 22-25, 1975 in New York. The Conference was sponsored by the New York Academy of Sciences.

16. S. H. Lee and K. T. Stalker, "A real-time Optical Processing System Employing Nonlinear Optics." *J. Opt. Soc. Am.* 62, 1366 (1972).

17. K. T. Stalker and S. H. Lee, "The Use of Nonlinear Optical Elements in Optical Information Processing." *J. Opt. Soc. Am.* 64, 545 (1974).

18. W. H. Leighton, "Electrical Properties of CdS Thin Films in Thickness Direction," *J. Appl. Phys.* 44, 5013 (1973).

19. J. P. Strong, "Two-Dimensional Image Computation for Earth Resource Observation Program." Report of an NSF Workshop on Optical Computing System, p. 28 (September 1972).

20. T. D. Beard, et. al., "AC Liquid Crystal Light Valve," *Appl. Phys. Lett.* 22, 90 (1973).

21. E. Jakeman and E. P. Raynes, "Electro-Optic Response Times in Liquid Crystals," *Phys. Lett.* 39A, 69 (April 10, 1972).

22. P. Regensburger, "Optical Sensitization of Charge Carrier Transport in Poly (N. Vinyl Carbazole)," *Photochem and Photobio.* 8, 432 (1968).

23. E. Spiller, "Saturable Resonator for Visible Light," J. Opt. Soc. Am. 61, 669 (1971).

24. D. H. Schaefer and J. P. Strong, "Tse Computer". NASA Technical Report X-943-75-14 (January 1975).

25. Most of our present ideas on IOC switches and logics are originated from Dr. Henry Taylor and his group at the Naval Electronic Laboratory Center, San Diego.

26. H. F. Taylor, "Optical Switching and Modulation in Parallel Dielectric Waveguides." J. Appl. Phys. 44, 3257 (1973).

DISCUSSION

Stroke   Can you give some examples of nonlinear processing and their application areas?

Lee   Some examples of nonlinear processing are $I_{out} \propto (I_{in})^\gamma$, thresholding and logic operations. Based on thresholding and logic operations, other nonlinear operations such as logarithmic, exponential, square-root and square can be carried out on images or other two-dimensional information.

Thompson   The logarithmic operation would be useful in image restoration problems involving multiplicative noise.

Goodman   Nonlinear processing might be important in solving image processing problems involving space variant optical systems

Landauer   Is the two dimensional processing format a fixed one and how can the standardization of signal levels be carried out?

Lee   The only requirement on the format is that the image or two dimensional information to be processed must be sampled at a rate which satisfies the sampling theorem. To standardize the signal levels, the device configuration of Fig. 11 can be utilized. For example, let (a) A_{ij} ($i,j = 1,2\ldots$) represent the intensity pattern, which contains noise and which is to be standardized, (b) The intensity of all B_{ij} coming into the device be at a desired standardized level. If the photoconductor is operated in its nonlinear region, the outgoing C_{ij} from the device will be the standardization of A_{ij} and the noise contained in A_{ij} eliminated. Another comment I wish to make concerning the device of Fig. 11 is about the power consumption, though this point has not been raised in the question. Preliminary calculations show that the electrical power required to operate a logic element is about a few microwatts, which is much smaller than that required in the radiative logic approach proposed in Ref. 14.

Korpel   What are the major differences in the two approaches utilizing planar logic and IOC logic?

Lee   The degree of parallelism in planar logic is higher than that obtainable in IOC logic. But, the switching speed of IOC logic can be made faster by proper design to minimize the electrode capacitance.

QUESTIONS AND COMMENTS

Comments by Dr. P. Tverdokhleb on the report by Dr. S. Lee.

A possibility of using optical elements for digital information processing is often under discussion at our laboratory. We do not understand quite clear the advantages of optical logical elements over electronic ones. In addition, we believe that besides the technological problems noted by Dr. R. Landauer in the discussion the problems of data transmission (without energy losses) should be solved both in the plane of one "layer" of processing and in one "layer" of another one. In the latter case image intensifiers should be placed between the "layers". However the devices suitable for these purposes have not been developed yet.

OPTICAL FOUNDATIONS OF DIGITAL COMMUNICATION AND INFORMATION PROCESSING*

Prof. Dr. George W. Stroke
and Prof. Dr. Maurice Halioua

State University of New York, Stony Brook, N. Y. 11794

ABSTRACT

Modern communications require the transmission, storage and display of information in pictorial form, both for data (documents) as well as for information in actual "image" form. Digital implementations are likely to be the most practical, in general cases, when certain current limitations are surmounted. Both the transmission as well as the processing of the information represent, currently, digital implementations of fundamental optical principles. Therefore, the solution of the digital communication and processing problems requires the consideration and application of solutions which have reached a very advanced stage in the optical implementations, generally known as "coherent optical processing", "holography" and "optical computing".

This paper provides an introduction to the status of optical information processing, illustrated with the aid of specific examples, for which optical processing methods have reached a degree of practical perfection in real-world cases, as yet unmatched by digital computer methods. Recommendations for further perfection of digital methods are included.

(*) An early version of some parts of this article may be found in G. W. Stroke "Optical Computing", Keynote Address in International Optical Computing Conference, IEEE Cat. No. 75-CH0941-5C, held in Washington, D. C. April 23-25, 1975.

INTRODUCTION

Recent advances in digital communication and in digital computers now promise the possibility of a widespread implementation of processes which are fundamentally optical in nature. The flexibility of digital methods justifies the current attempts to transpose into the digital domain the required solutions of the problems which have now, in fact been solved to a large extent in optics, thanks to the development of the three foundations: the laser, holography and optical computing.

Among the elements to be considered in the digital adaptation one may mention, for illustration the following, among others: image "capture" (i. e. the suitable recording of the imaging information in some electrical form), analogue-to-digital (A/D) conversion, digital data compression (e. g. sampling of picture elements, grey-level compressions, trade-offs, etc.), processing algorithms (computer software, aperture shaping, deconvolution, correlations, etc.), coding and security, storage (magnetic, mass memories, holographic memories, solid state memories, electron beam recorders, etc.) display (CRT, microfilm, fast printers, etc.), computer architecture (parallel processing, minicomputers, etc.).

Unlike optical image processing, for example, which is readily implemented with only three basic components (laser, lenses and photographic film), it is evident that digital image processing requires the use of a much larger number of basic components, and therefore the solution of many more and different problems. The comparison is already somewhat more favorable if the communications links are included.

Much can be learned, in fact, from optics which should be of fundamental interest for the digital implementations which we seek.

THE COMPUTATIONAL CAPABILITIES OF OPTICS

Few instruments in the history of mankind have contributed more to the extension of human capabilities in fundamental and in applied scientific research than optical instruments. The development of atomic and nuclear physics, the spectral determination of the composition of stellar and planetary atmospheres, are just two among many major examples which range from metallurgy to biology and medicine. The search for more effective use of optics has kept up with a steady increase in the need for its capabilities in solving problems through its image processing contributions.

Early recognition of the computational capabilities of optics may be traced back to Joseph von Fraunhofer (Germany, 1823) and to Ernst Abbe (Germany, 1873) among others, who recognized some of the remarkable Fourier transform relations associated with lenses and which are currently taken almost for granted in contemporary applications. However, it was probably the collaboration of Albert A. Michelson (U.S.A. 1905) with Lord Rayleigh (Great Britain) that firmly introduced the Fourier transform formalism and concepts into optics in a form still used in current work. Michelson, who in 1907 was the first American scientist to be awarded a Nobel Prize, explicitly used the two-dimensional Fourier transform relation in an operational notation in a paper published in the Philosophical Magazine in April 1905. Subsequently, he also constructed his famous Fourier synthesizer for use in spectroscopy, thus setting the foundations for the field of Fourier transform spectroscopy which was again the subject of such intense interest between the middle 1950's and the late 1960's. J. B. Fourier (France, 1768-1830) thus made fundamental contributions in mathematics, as now used in modern optics, almost two centuries ago (1). A major contribution to the use of Fourier transforms in optics was made in 1946 by the Frenchman M. Duffieux (1).

COMPARISON OF OPTICAL AND DIGITAL COMPUTATION

The current use of optical computing methods has proven to be especially successful in some problems which in fact were first tackled with the aid of digital electronic computers. Side-looking synthetic aperture radar and optical image deblurring (restoration) are two of the most well known examples (2). They illustrate in fact a logical continuation of computing along the lines already recognized by Lord Rayleigh and Michelson, among others, around the turn of the century (3). However, in spite of more than a century of success in optical image processing since Abbe's microscope diffraction work in 1873, the validity of the description "optical computing" still appears to be challenged. There has been an apparently erroneous impression that optical processing deals only with actual optical images as such. In fact, optical computing is used to synthesize and process images and signals acquired with radiations which range from X-rays and electron beams (as in electron microscopy) to radio and radar waves (as in synthetic aperture coherent radar), and indeed signals and images acquired with ultrasonic radiations as well (4). Also the obvious success of optical computing is often incorrectly placed into doubt by questions regarding such readily surmountable difficulties as photographic dynamic range or signal-to-noise problems or preceision among others.

A number of direct comparisons between methods of optical and digital image processing have been underway for a number of years (2). In all cases, as far as it is known, the optical processing people make extensive use of the most advanced forms of digital electronic computers as well, wherever it appears to be most suitable in their work: in most cases these groups are known to be dealing with "real-world" problems, where they are called upon to solve certain problems by the best means available, and not to deal with some "model" or "simulation" experiments for which digital electronic computer methods are so ideally suited.

One may therefore ask what may be the explanation for the fact that results in holographic image improvement (image deblurring) and in side-looking radar, for example, continue to require optical computing in order to obtain the most perfect attainable form. There are in fact several reasons for this in addition to the computing speeds. These remarkable speeds remain unmatched, of course, in optical computers, because of the enormous superiority of the parallel processing capability of coherent optical processing systems, notably those using holograms as input (the case of side-looking radar) or as complex filters (the case of image deblurring).

In the case of the optical image deblurring computer, for example, as applied to the improvement of high-resolution electron micrographs (5) one finds that the electron micrographs even with the outstanding present-day electron guns may permit the discrimination of only two grey levels at the specimen in a representative case. While the corresponding electron micrograph may be used directly as the input in the optical computer, without additional loss in the S/N ratio, the situation is generally quite a bit less favorable in the processing of scanning out the photographic transparencies and digitizing them in view of digital computing processing.

The reasons for the several degradations which are so commonly found when one scans and digitizes a photograph are quite a bit more complicated than may have been generally considered in the past. There are not only the quantization problems and the convolution (transfer function) loss due to the scanning, but there is also the granularity-noise loss, which is introduced when scanning with the smaller apertures that would otherwise help in diminishing the scanning degradations. Such granularity noises are surmountable, of course: we have found, for example, that copying of the original photographs in white light with a good lens that does not resolve the original grain produces an essentially grain-free image on the high-resolution plate which is then directly used in the optical computer. At the same time such copying is in fact essential in order to produce by transmission through the second plate a field which is proportional to the original recorded intensity; this step is required in order to deal appropriately with the linear convolution integrals, for example, which characterize the photographs in these cases. It is conceivable of course to go through this same copying process on high-resolution fine-grain plates and respecting the gamma conditions before scanning out the photograph in view of digitizing and of digital electronic computer processing. However, once the careful copying procedure has been performed, one has in fact already gone all the way in view of the direct optical processing which yields the final result (e.g. the deblurred image) directly in a simple optical projector like arrangement. On the other hand, digital processing calls for the use of scanning of the photograph (now in an appropriate form), digitizing, digital electronic computation and print-out on some CRT or on some other display device, including now also several types of laser print-out systems: this is evidently much more involved.

The essence of these observations should be to indicate that digital electronic image processing may be expected to produce results at least comparable and hopefully, one day, even superior to those presently achievable in the optical computers. However, it should not be surprising to find that the attainment of such results with the aid of digital processing will require the same care and the same science in the pre-processing as those presently available on optical computing, thanks to the refinements produced in bringing the optical computers to their present stage of perfection.

The general remarks presented here have been deliberately formulated in non-mathematical terms for clarity and simplicity. One should hasten to state, however, that each of the arguments

a.

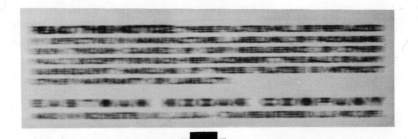

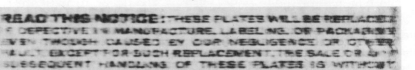

b.

FIGURE 1. Deblurring of Accidentally-Blurred Photographs. [Blind deconvolution using holographic optical computing).

Early test results showing the use of optical computing to holographic improvement of motion blurred photographs using the principles described by G.W.Stroke and M.Halioua in Physics Letters (1972), Vol. 39A, pp. 269-270 and illustrating in a. the results obtainable with a well adapted deblurring filter and in b. the fact that even an imperfectly adapted filter (incorrectly scaled by about 25%) may still produce a reasonably readable image from a totally illegible text. Both examples show improvement in so-called "blind" cases where there exists no previous knowledge about the type and/or degree of blurring: the decoding key (blurring function) is determined experimentally from the actual blurred photograph by use of laser Fourier transformation and comparison of the corresponding spectrum with a previously determined library of typical blurrings. The appropriate library blurring function may then be used easily and equally quickly to construct the required holographic image-deblurring filter. (Further details may be found in 2,5,6,7).

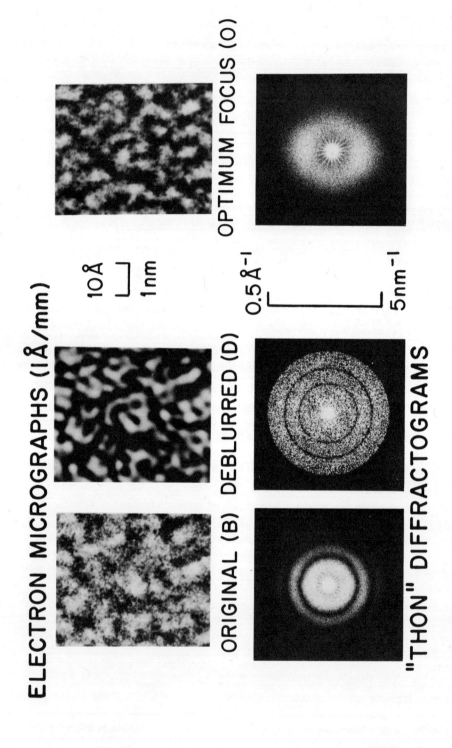

FIGURE 2. Image Improvement in High-Resolution Electron Microscopy Using Holographic Image Deconvolution (Optical Computer Processing) according to G.W.Stroke, M.Halioua, F.Thon and D.Willasch [Optik(1974), Vol. 41, pp. 319-343](5).

Shown are greatly enlarged portions of unprocessed and processed electron micrographs of a thin(50Å) amorphous carbon-foil specimen recorded in a high-resolution SIEMENS electron microscope Elmiskop 102 at 125kV (as a part of a collaborative project between SIEMENS Berlin Electron Microscopy Division and the SUNY Stony Brook Electro-Optical Sciences Laboratory). Micrographs of thin specimens, e.g. also those of biological interest, show essentially no contrast in unstained specimens when recorded in focus. Blurring in such cases must be introduced deliberately (by the method commonly known as "defocusing phase contrast") in order to obtain any adequate image contrast in the first place. Unfortunately contrast is thus introduced in the original micrographs (B) at the expense of severe blurring and strong contrast inversions (2,5) which tend to make electron micrographs uninterpretable below resolutions of some 10Å to 20Å, whereas the intrinsic resolution capabilities of the microscopes are from 2.5Å to 3.5Å in such strongly-defocused cases. Also shown with each micrograph are the so-called "Thon diffractograms" (2,5), which display the amplitude of the CTF (contrast transfer function) in the form extracted directly by Fourier transformation from the actual micrographs (respectively improved image) and which characterize its degradation (respectively improvement). The holographically improved image and its corresponding diffractogram are shown in (D) in the form extracted by optical computing deconvolution from the original blurred electron micrograph (B). Also shown for comparison in (0) is an original unprocessed defocused electron micrograph recorded in defocusing phase-contrast in the so-called "optimum" (Scherzer) focus, where the CTF has only one positive lobe, so that there are no inversions: the conclusions in (5) were that details near the limit of resolution (about 3Å) have been made visible in the holographically improved images that could not have been detected in the original strongly defocused micrograph (with its broad spatial frequency range, as required for the highest resolutions) and that they are superior clearly also the micrograph previously considered as 'optimum' (with its smaller spatial frequency range, and therefore smaller resolution capability). Similar results have been obtained with electron micrographs of viruses (2,4).

ORIGINAL ELECTRON MICROGRAPHS: F.Thon and D.Willasch, SIEMENS, Berlin

HOLOGRAPHIC DEBLURRING : G.W.Stroke and M.Halioua, SUNY Stony Brook

presented and all the steps used in actual case of optical computing need to be, as they are in fact, carefully supported by detailed mathematical analysis in every case (1-6).

SOME RECENT ACHIEVEMENTS IN OPTICAL COMPUTING *

A number of representative examples of some new results recently obtained by optical computing are shown in the two illustrations which follow. Many others, of great interest and variety, form part of this Seminar.

Figure 1. shows some early test results illustrating the use of optical computing to holographically improve images of motion-blurred photographs using the principles described by G. W. Stroke and M. Halioua in Physics Letters (1972), Vol. 39A, pp. 269-270 and illustrating in a. the results obtainable with a well-adapted deblurring filter and in b. the fact that even an imperfectly adapted filter (incorrectly scaled by about 25 per cent) may still produce a reasonably readable image from a totally illegible blurred text. Both examples show improvements in so-called "blind" cases where there exists no previous knowledge about the type and/or degree of blurring: the decoding key (blurring function) is determined experimentally from the actual blurred photograph by use of laser Fourier transformation and comparison of the corresponding spectrum with a previously determined library of typical blurrings. The appropriate library blurring function may then be used easily and equally quickly to construct the required holographic image-deblurring filter. (Further details may be found in 2, 5, 6, 7).

Figure 2 presents image improvement in high-resolution electron microscopy using holographic image deconvolution (Optical Computer Processing) according to G. W. Stroke, M. Halioua F. Thon and D. Willasch (Optik(1974), Vol. 41, pp. 319-345)(5). Shown are greatly enlarged portions of unprocessed and processed electron micrographs of a thin (50Å) amorphous carbon-foil specimen recorded in a high-resolution SIEMENS electron microscope Elmiskop 102 at 125kV (as a part of a collaborative project between SIEMENS Berlin Electron Microscopy Division and the SUNY Stony Brook Electro Optical Sciences Laboratory). Micrographs of thin specimens, e.g. also those of biological interest, show essentially no contrast in unstained specimens when recorded in

---

(*) See illustrations at the end of the article.

focus. Blurring in such cases must be introduced deliberately (by the method commonly known as "defocusing phase contrast") in order to obtain any adequate image contrast in the first place. Unfortunately contrast is thus introduced in the original micrographs (B) at the expense of severe blurring and strong contrast inversions (2, 5) which tend to make electron micrographs uninterpretable below resolutions of some 10 Å to 20 Å, whereas the intrinsic resolution capabilities of the microscopes are from 2.5Å to 3.5Å in such strongly-defocused cases. Also shown with each micrograph are the so-called "Thon diffractograms" (2, 5), which display the amplitude of the CTF (contrast transfer function) in the form extracted directly by Fourier transformation from the actual micrographs (respectively improved image) and which characterize its degradation (respectively improvement). The holographically improved image and its corresponding diffractogram are shown in (D) in the form extracted by optical computing deconvolution from the original blurred electron micrograph (B). Also shown for comparison in (O) is an original called "optimum" (Scherzer focus), where the CTF has only one positive lobe, so that there are no inversions: the conclusions in (5) were that details near the limit of resolution (about 3Å) have been made visible in the holographically improved images that could not have been detected in the original strongly defocused micrograph (with its broad spatial frequency range, as required for the highest resolutions) and that they are superior clearly also the micrograph previously considered as 'optimum' (with its smaller spatial frequency range, and therefore small resolution capability). Similar results have been obtained with electron micrographs of viruses (2, 4).

CONCLUSIONS AND RECOMMENDATIONS

In conclusion, it may be fair to state that the well known and quite remarkable results which have been obtained in image processing in the last ten or twenty years with direct digital computing means as well as with optical computer means tend to be encouraging enough to describe them as of practical interest and of certainly great promise in the many applications already considered, and probably in many others as well. Many new developments are likely to result from contributions such as those presented at this Conference.

Many among us who have been using both digital and optical computing from the outset tend to believe that it would be particularly desirable at this time to work on singling out specific areas of practical interest for which either optical or digital image processing appear to be naturally suitable "par excellence" and may be likely to stay so for some time to come.

It would be equally desirable to try to single out with the aid of theoretical and experimental analysis what developments either in optical or in digital computing can be rapidly improved within a reasonable time and cost to give one or the other of the methods a decisive degree of improvement and capability in view of solving either general or specific problems. One example among several such comparisons which may be cited for illustration is the case of X-ray crystallography for which the digital structure synthesis computation has long been considered as the only tool. New developments in the easy and very rapid generation of complex holographic image deblurring filters (5) now make it a valid question to consider once more if optical computing methods could not again be used to solve the structure synthesis problem directly in more complicated cases along the lines already used for the simple cases of centro-symmetrical crystals by Sir Lawrence Bragg and Martin J. Buerger in the late 1930's (6) and which in fact were later cited by Dennis Gabor in 1948 as having been a point of departure for the work on holography (7) which in due course earned him the Nobel Prize in 1971.

Background on digital image processing and a number of basic references are included in the bibliography of references (2), (5) and (6). Among the most recent reviews, see references (8) and (9).

REFERENCES

1. a. A Maréchal, "Optique Géometrique Générale" in Handbuck der Physik, edited by S. Flugge (Berlin and Heidelberg: Springer Verlag, 1956, Vol. 24)
   b. M. Duffieux, L'Intégrale de Fourier et Ses Applications a l'Optique (Rennes: Société Anonyme des Imprimeries Oberthur, 1946)
   c. G. W. Stroke, "Diffraction Gratings" in Handbuch der Physik, edited by S. Flugge (Berlin and Heidelberg: Springer Verlag, 1967, Vol. 29)
2. For general background see e.g. G. W. Stroke "Optical Computing", IEEE Spectrum, Vol. 9, No. 12 (December 1972) pp. 24-41.

3. a. A. A. Michelson, Light Waves and Their Uses (Chicago: The University of Chicago Press, (1902)(Reprinted: Phoenix Scien ce Series, The University of Chicago Press).
   b. A. A. Michelson, Studies in Optics (Chicago, The University of Chicago Press 1927)(Reprinted: Phoenix Science Series, The University of Chicago Press, 1962).
4. Ultrasonic Imaging and Holography (Medical, Sonar and Optical Applications) edited by G. W. Stroke, W. E. Kock, Y. Kikuchi and J. Tsujiuchi (New York: Plenum Press, 1974).
5. G. W. Stroke, M. Halioua, F. Thon and D. Willasch, "Image Improvement in High Resolution Electron Microscopy Using Holographic Image Deconvolution", Optik $\underline{41}$ (1974) 319-343.
6. G. W. Stroke, An Introduction to Coherent Optics and Holography (New York, Academic Press: 1969, Second Edition).
7. D. Gabor, W. E. Kock and G. W. Stroke, "Holography" Science $\underline{173}$ (1971) 11-23.
8. B. R. Hunt, "Digital Image Processing", IEEE Proc. $\underline{63}$ (1975) 693-708.
9. T. G. Stockham, Jr., T. M. Cannon and R. B. Ingebretsen "Blind Deconvolution Through Digital Signal Processing", IEEE Proc. $\underline{63}$ (1975) 678-692.

ACKNOWLEDGMENTS

This work was carried out with support from the National Science Foundation, the National Aeronautics and Space Administration and, in part, thanks to a collaborative effort under way with SIEMENS AG(Electron Microscopy Division, Berlin). The participation of Dr. F. Thon, Dr. D. Willasch and of Dr. V. Srinivasan is acknowledged with pleasure.

APPENDIX:

I. ESSENTIAL BACKGROUND AND FUNDAMENTALS

INTRODUCTION

It has long been assumed in so-called "Fourier-transform optics" that a perfect thin lens performs a linear transformation known as a spatial Fourier transformation.

This F. T. relation is in general derived heuristically by a simple application of a so-called "Huygens principle" as follows.

Consider a number of coherently radiating points $P_0(x=0, y=0)$, $P_1(x_1, y_1), \ldots P_n(x, y)$ in the x,y plane, in the

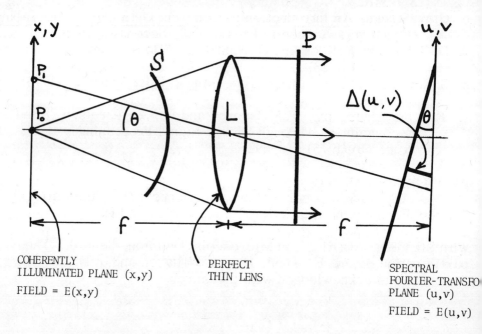

COHERENTLY ILLUMINATED PLANE (x,y)
FIELD = E(x,y)

PERFECT THIN LENS

SPECTRAL FOURIER-TRANSFO PLANE (u,v)
FIELD = E(u,v)

front focal plane of a perfect thin lens L, forming a field $E(x,y)$. Each point P_n radiates a spherical wave S which is transformed into a plane wave P by the lens, such that, according to "Huygens principle" the resulting field $E(u,v)$ is equal to the "superposition" of all the plane waves P, weighted by $E(x,y)$.

One notes that the off-axis plane waves are dephased with respect to the on-axis wave by a phase

$$\phi(u,v) = \frac{2\pi}{\lambda} \Delta(u,v) \tag{I.1}$$

where

$$\Delta(u,v) = \frac{1}{f}(ux + vy) \tag{I.2}$$

Accordingly, one has immediately

$$E(u,v) = \iint_{-\infty}^{+\infty} E(x,y)\, e^{i\frac{2\pi}{\lambda f}(ux+vy)}\, dxdy \tag{I.3}$$

within constants.

(Eq. (I.3) derived as it is, heuristically, is the so-called "fundamental Fourier-transform equation" used quite generally, without question, in all image processing work.

Unfortunately, the "Huygens principle" on which the equation is based is, in fact, only the first-order linear approximation of rigorous electro-magnetic theory, and, more specifically, of the applicable rigorous form of Huygens' principle for electromagnetic waves, namely the Maxwell-Lorentz theory.

The need for use of a more exact formulation may be best shown with reference to the applicable theory of diffraction gratings. Serious errors of the heuristically-derived F. T. equation can result, in particular, in the application to real-time spectrum analysis devices even though it may have been assumed that "compensation" of certain "at rest" transducer wavefront imperfections by itself (e.g. according to holographic principles) would result in a perfectly compensated real-time spectral analysis device.

The essential theoretical background and fundamentals relevant to this paper follow herewith with particular reference to real-time optical spectrum analyzers used for illustration.

II. ELECTROMAGNETIC THEORY OF DIFFRACTION (RIGOROUS)[*]

INTRODUCTION

It has long been noted [**] that an optical spectrum analyzer is comparable to an optical grating spectrometer using monochromatic radiation and an imperfectly spaced diffraction grating.

---

[*] This section is based in part on G.W.Stroke, "Diffraction Gratings" pp. 426-754 in Handbuch der Physik Vol. 29, ed.S. Flugge [Berlin and Heidelberg, 1967:Springer Verlag]

[**] G.W.Stroke, An Introduction to Coherent Optics and Holography, Second Edition (New York, 1969:Academic Press)pp.86 ff.

In an optical spectrum analyzer, the grating 'imperfections' contain the significant information, recorded on photographic film, for instance with the help of a cathode-ray oscilloscope or with electron or laser beams. Seismic vibration recordings are among early examples of signals recorded and analyzed in this manner. The analysis aims at determining the power spectrum of the recorded signal which forms the 'imperfect' diffraction grating.

The principle of spectrum analyzers is based on the fact that the amplitude distribution of the light in the diffraction patterns formed by an optical grating is equal to the spatial Fourier component of the electric field transmitted through the grating (perfect or imperfect) when the grating is illuminated by a plane monochromatic wave [*]. In other words, the intensity distribution in the diffraction patterns formed by the grating is nothing but the spectrum of the imperfections in the wavefront diffracted by the grating. When the 'imperfections' are deliberately introduced into the grating, for instance in the form of an intensity grating, recorded on the face of a storage tube, e.g. (LUMATRON), GE light valve, etc. then the spectrum (formed by optical Fourier transformation) will immediately provide the spectral distribution of the signal-intensity variations as a function of the chosen coordinate on the storage surface (film, light valve, etc.). The preceding remarks make the assumption that a "perfect" diffraction pattern (ideally a delta-function) will be formed by the spectrum-analyzer system described.

The optical error analysis for the spectrum analyzers is thus contained in the analysis of wavefront imperfections and of their effects on the spectral diffraction patterns as carried out for diffraction gratings in the reference [*] noted above.

For the sake of clarity, portions of the relevant section are recalled in part below.

The significant results of this work are directly applicable to the analysis of spectrum analyzers in the terms considered here.

One of the results is that the significant spectral image-forming information is contained in the phase of the wavefronts: the phase of the wavefronts can, in turn, be readily measured with the aid of a Michelson Twyman-Green interferometer.

Another result of the work on gratings is that there exists now a catalogue of wavefront errors which are particularly damaging in terms of the spectral imaging considered here. These include the so-called 'linear' wavefront errors and the various periodic and multiply-periodic errors, as well as the errors of the substrate itself. The spectral effect of such errors can be immediately assessed from simple measurements in the wavefront interferograms, without the need for digital computation: this type of computation will however always permit one to get the detailed amplitude (and intensity) distribution in the spectral image, notably for the general case of arbitrarily distributed errors.

Another important result of this work is the existence of simple relationships between the observable wavefront imperfections, on the one hand, and, on the other the corresponding imperfections in the storage film and the substrate, to the extent that they may be considered separately.

For example, in view of assessing the existing properties, as well as the needed characteristics of a transducer for the attainment of representative operating goals, one needs to make use of the results of such work. Careful use of the need for "apodization" also needs to be made [*]. This has already appeared as necessary in order to reduce the side lobes to tolerable levels. It should be evident that the normal side lobes which characterize diffraction patterns are not acceptable.

For example, for a square aperture, the diffraction pattern may be written in the form of a $sinc^2 x$ function, i.e., for the ampltitude

$$\frac{\sin x}{x}$$

For large values of x, we may examine the values of x for which sin x=1. For example, the value of -60dB in intensity requires an amplitude of 10^{-3}. This will be at the values of x=1,000.

The first minimum of sinc x occurs for $x=\pi$, and the first secondary maximum of $x=1.5\pi$. If we consider the value of $x=\pi$ as the "diffraction unit", arbitrarily, we see that we would have to go out as far as some 300 diffraction units in order to reduce the normal 'secondary' diffraction peaks to a value of -60dB in intensity without apodization. Very similar values apply to circular apertures.

It should be possible, at least in principle, to carry out the required 'apodization' by an adaptation of the 'image deconvolution' principles. Consideration may be given to this possibility and to any possible advantages.

NOTE: Space-variant and Space-invariant optical computers

It is essential to note that the rigorous theory which follows is quite general. It may be applied to space-variant, as well as to space-invariant systems of the more usual type. Attempts

---

[*] P. Jacquinot and B.Roizen-Dossier, "Apodisation" pp. 31-188 in Progress in Optics, Vol.III, ed.E.Wolfe(New York,1964:John Wiley)

proposed for solving the space-variant problems by means of simple (heuristic) Fourier-transform formalism are insufficient, by themselves, of course, as they are for the solution of space-invariant systems, in the terms discussed here and in the rigorous electromagnetic theory of diffraction which follows.

1. <u>Rigorous form of Huygens' principle for electromagnetic waves (Maxwell-Lorentz theory)</u>. In order to show the domain and limits of applicability of the so-called Kirchhoff-Huygens' principle in grating image formation, one needs to recall that the Huygens' principle can be derived from Maxwell's electromagnetic equations, provided that certain approximations are made at appropriate steps in the derivation. The approximations made then clearly describe the domain and applicability of the Fourier-transform formulation of diffraction.

Before proceeding with the derivation of the approximate formulations of the diffraction theory of image formation, one needs to show that it is possible to derive without approximation, a rigorous form of Hyugens' principle (Eq. 1.7) from Maxwell's equations. After that, an approximate form of the rigorous Huygens' principle is derived (Eq. 2.20) and finally the desired Fourier-transform formulation is given in Eq. 3.1 The approximations involved will clearly appear in the course of the derivations.

Many authors have dealt with aspects of this problem.[1] Among them are H. A. Lorentz (1896), Larmor (1903), Kottler (1923), Baker and Copson (1939, 1950), Bouwkamp (1940, 1954), Born and Wolfe (1959), E. Wolf (1951) Croze and Darmois (1949), Maréchal (1956), A. Rubinowicz (1957), Maréchal and Francon (1960)[2], Robieux (1960). Extensive bibliographies are given by Baker and Copson[3], Bouwkamp (1954) and Rubinowicz[4] among other. Here we follow in part the developments of Lorentz, Rubinowicz and Maréchal.

Maxwell's equations, in augmented Gaussian form are

$$\text{curl } H = \frac{4\pi}{c} j_e + \frac{1}{c} \varepsilon_o \frac{\partial E}{\partial t} \quad , \tag{1.1}$$

$$\text{curl } E = - \frac{4\pi}{c} j_m - \frac{1}{c} \mu_o \frac{\partial H}{\partial t} \tag{1.2}$$

---

[1] For early references, see for ex. C.J.Bouwkamp: Rep. Progr. Phys. 17,35-100(1954)
[2] A.Maréchal et M.Francon: Diffraction, edit.Revue d'Optique,Paris 1960
[3] B.B.Baker and E.T.Copson: The Mathematical Theory of Huygens' Principle. Oxford: Clarendon Press 1950.
[4] A.Rubinowicz: Die Beugungswelle in der Kirchhoffschen Theorie der Beugung. Polska Akademia Nauk, Warszawa, 1957.

DIGITAL COMMUNICATION AND INFORMATION PROCESSING

where j_e = electric current density and j_m = magnetic current density. Let

$$H = H \exp(-i\omega t),$$
$$E = E \exp(-i\omega t). \tag{1.3}$$

In vacuo, where $\varepsilon_o = 1$ and $\mu_o = 1$, Maxwell's equations are

$$\text{curl } H = \frac{4\pi}{c} j_e - \frac{i\omega}{c} E,$$
$$\text{curl } E = -\frac{4\pi}{c} j_m + \frac{i\omega}{c} H \tag{1.4}$$

Let E', H' be an electromagnetic field that also satisfies Maxwell's equations, and let j'_e, j'_m be the corresponding electric, respectively magnetic current densities. We have

$$\text{curl } H' = \frac{4\pi}{c} j'_e - \frac{i\omega}{c} E', \tag{1.5}$$

$$\text{curl } E' = -\frac{4\pi}{c} j'_m + \frac{i\omega}{c} H'. \tag{1.6}$$

With the help of the well known vector identity div (ExH')-div(E'xH)= H curl E-E curl H'-H curl E'+E' curl H and Gauss' theorem

$$\int_R \text{div } D \, dv = \int_S D \cdot n \, dS$$

and one obtains finally

$$E_{p'} = \int \left[-\frac{ik}{4\pi} (n \times H) \frac{e^{-ikr}}{r} - \frac{i}{k} \text{grad div } \frac{1}{4\pi} (n \times H) \frac{e^{-ikr}}{r} + \text{curl } \frac{1}{4\pi} (n \times E) \frac{e^{-ikr}}{r} \right] dS \tag{1.7}$$

where $k = 2\pi/\lambda$, $i = \sqrt{-1}$ and n = normal to S. Eq. (1.7) is a rigorous form of Huygens' principle for electromagnetic waves. It involves no approximations. It states that the electric field $E_{p'}$ at a point P' can be obtained in terms of appropriate integration of the contributions at P' of the E and H field distribution on the surface S enclosing the point P'. It is clear that no other sources are distributed within the region R enclosed by S.

2. <u>Approximate form of rigorous Huygens' principle</u>. Eq. (1.7) is a rigorous expression of Huygens' principle obtained from Maxwell's equations without approximations. We next derive an approximate form of Huygens' principle which gives E_o, the complex amplitude of the electric field, in the vicinity 'O' of the center of a quasi-spherical wave front Σ.

The equation to be derived is

$$E_{0'} \simeq \frac{i}{\lambda} \frac{e^{-ikR}}{R} \iint_\Sigma E_0 e^{ik\Delta} dS \qquad (2.1)$$

where Δ is the wave-front aberration of the quasi-spherical wave front Σ from the spherical wave front S, of radius R and centered at O'.

The approximations involved in deriving Eq. (3.1) will be stated in the process of derivation.

First, one assumes that

$$\lambda \ll r \qquad (2.2)$$

that is that kr is large. Let the surface S be a wave front Σ described by

$$L(x,y,z) = \text{constant} \qquad (2.3)$$

One also may admit that S is very close but not necessarily coincident with Σ. We next show, according to Maréchal and Francon (1960) that E and H are tangent to the wave front Σ [i.e. to the phase front $L(x,y,z)=$ constant] and that they form an orthogonal triad with grad which is in the direction of n (in the vacuo grad $L=n$). In the Heaviside-Lorentz system of rational units, Maxwell's equations in vacuo ($u=1$) can be written as

$$\text{curl } H = \frac{\varepsilon}{c} \frac{\partial E}{\partial t}, \qquad (2.4)$$

$$\text{curl } E = -\frac{1}{c} \frac{\partial H}{\partial t}. \qquad (2.5)$$

With the time dependence stated explicitly, one has in complex notation

$$\left. \begin{array}{l} E = E_0(x,y,z) \exp\left[+ i\omega\left(t - \frac{L_E(x,y,z)}{c}\right)\right] \\ \\ H = H_0(x,y,z) \exp\left[+ i\omega\left(t - \frac{L_M(x,y,z)}{c}\right)\right] \end{array} \right\} \qquad (2.6)$$

where L_E and L_M describe the local phase of E and of H respectively so that E_0 and H_0 are real vectors. We shall find that $L_E=L_M(x,y,z)$ is the optical path length. In Eq. (2.6) one only assumes that the x,y,z components of E and of H, respectively, are in phase, which does not restrict the argument.

It should be clear that $L_E=L_M=L$ is the optical path. We finally have the equations

DIGITAL COMMUNICATION AND INFORMATION PROCESSING

$$\varepsilon E_o = - \text{grad } L \times H_o \tag{2.7}$$

$$H_o = \text{grad } L \times E_o \tag{2.8}$$

from which we conclude that E_o, H_o and grad L indeed form a system of orthogonal vectors, and also that E_o and H_o are tangent to the wave front $L(x,y,z)$ = constant under the stated assumptions.

In vacuo, grad L = n, and we now return to Eq. (1.7) where we let $E \to E_o$ and $H \to H_o$. Eq (1.7) can now be written after some simplifications as

$$E_{o'} \simeq \int_S \left\{ + \frac{ik}{4\pi} E_o \frac{e^{-ikr}}{r} + \frac{i}{k} \text{grad div } \frac{1}{4\pi} E_o \frac{e^{-ikr}}{r} + \text{curl } \frac{1}{4\pi} H_o \frac{e^{-ikr}}{r} \right\} dS \tag{2.9}$$

where we may also write

$$\phi(r) = \frac{1}{r} e^{-ikr} . \tag{2.10}$$

With the use of appropriate vector identities and the relation

$$\text{grad } \phi(r) = i_r \frac{\partial \phi(r)}{\partial r} \simeq - i_r ik\phi(r) \tag{2.11}$$

with i_r being the unit vector of the radial direction, and therefore

$$\text{div } \frac{E_o}{4\pi} \phi(r) = - \frac{ik}{4\pi} E_o \cdot \phi(r) \, i_r \tag{2.12}$$

and by noting that curl $H_o = 0$, because H_o = constant, one has

$$E_{o'} \simeq \frac{ik}{4\pi} \int_S [E_o - (E_o \cdot i_r) i_r + H_o x i_r] \phi(r) dS. \tag{2.13}$$

Since

$$E_o = E_{oyz} + (E_o \cdot i_r) i_r \tag{2.14}$$

where E_{oyz} is the projection of E_o onto the yz plane, it further follows that

$$E_{o'} = \frac{ik}{4\pi} \int_S [E_{oyz} + H_o x i_r] \phi(r) dS \tag{2.15}$$

With $i_r \simeq n$, $H_o x i_r \simeq E_o$ and with $k = 2\pi/\lambda$, we can write (2.15) as

$$E_{o'} \simeq \frac{i}{\lambda} \int_S E_o \phi(r) \, dS \tag{2.16}$$

which has almost the desired form of Eq. (2.1)

We need only to note that

$$r = R - \Delta \tag{2.17}$$

where R is the radius of the reference sphere S, and Δ the aberration of the image-forming wave front Σ from the reference sphere S.

With these relations, we have

$$\phi(r) = \frac{e^{-ikr}}{r} = \frac{1}{R-\Delta} e^{-ikR} e^{ik\Delta} \tag{2.18}$$

or approximately

$$\phi(r) \simeq \frac{e^{-ikR}}{R} e^{ik\Delta} , \tag{2.19}$$

We have finally

$$E_{o'} \simeq \frac{i}{\lambda} \frac{e^{-ikR}}{R} \int\!\!\!\int_\Sigma E_o e^{ik\Delta} dS \tag{2.20}$$

which is indeed the approximate form of the rigorous Huygens' principle which was to be derived.

It only remains to express Eq. (2.20) in the Fourier-transform formulation.

3. <u>Fourier-transform formulation of the image formation problem (physical optics approximation)</u>. Eq. (2.20) gives a useful approximate form of the Huygens' principle derived from the rigorous form of Huygens' principle given in Eq. (1.7). A considerable further simplification in the use of the approximate form of Huygens' principle results in expressing that <u>the complex amplitude of the electric field at any image point in the focal plane of an image-forming pupil is equal to the Fourier-transform of the complex amplitude distribution of the electric field distribution within the image forming pupil</u>.

We recall that three important approximations are involved in particular:

1. The formulation applies to the center and the near-vicinity of a quasi-spherical wave front of radius R.
2. The wavelength λ is such that $\lambda \ll R$.
3. (a) The electromagnetic field vectors E and H are assumed tangent to the wave front $L(x,y,z)$=constant.
   (b) E, H and grad L form an orthogonal triad.

The equation to be derived is

$$E_{p'}(k\alpha, k\beta) \simeq \frac{i}{\lambda} \frac{e^{-ikf}}{f} \iint_{\substack{\text{pupil}\\(\text{coordinates}:x,y)}} E_o e^{ik\Delta(x,y)} e^{ik(\alpha x + \beta y)} dx dy \qquad (3.1)$$

where $E_{p'}$ is the complex amplitude of the electric field vector at a point P' in the image, such that the coordinates of P' are

$$x' \simeq \frac{fx}{\gamma}, \quad y' \simeq \frac{f\beta}{\gamma}$$

with

$$\gamma \simeq 1.$$

The complex amplitude distribution in the pupil (coordinates x,y) is given by

$$E_o e^{-ik\Delta(x,y)},$$

and α, β, and γ are direction cosines in the pupil, on the plane-wave side of a perfect lens (or mirror). Here the radius R of the image-forming wave fronts has been identified with the focal length f.

Let P' be a general point in the x'y' image plane, such that P' is near O'. Let S be the sphere centered at P' and let ρ be a vector in the pupil describing a point P(x,y) on the plane-wave side of the pupil. We recall that it is preferable to deal in the image-formation problems with the plane-wave side of the pupil of a "perfect" image forming system. A perfect image forming system is defined as having the property of transforming plane waves in the direction i into spherical waves centered on P' in the focal plane of the optical system. Moreover, by virtue of the theorem of Gouy[1], aberrations Δ from a reference plane on the plane-wave side of the optical system remain equal to Δ on the same "ray" when measured with respect to the corresponding reference sphere on the spherical-wave side of the optical system. With respect to the tilted plane S and referring to the aberration Δ of Σ from S in Eq. (2.20) we thus have to substract from Δ an additional path difference Δ such that

$$\Delta = i_r \cdot \rho \qquad (3.2)$$

where i_r is the unit vector in the direction from O to P'. We have

---

1. cf. A. Marechal, "Optique Geometrique Generale" in Handbuch der Physik, Vol. 24, pp. 44-170, Ed. by S. Flugge (Berlin and Heidelberg, Springer Verlag, 1956).

$$\rho = xi_x + yi_y \tag{3.3}$$

and

$$i_r = i_x\alpha + i_y\beta + i_z\gamma \tag{3.4}$$

where α, β, γ are the direction cosines of i_r. Eqs. (3.2), (3.3) and (3.4) give

$$\Delta_\theta = \alpha x + \beta y \tag{3.5}$$

and we must write

$$\Delta \rightleftharpoons \Delta - \Delta_\theta \tag{3.6}$$

in Eq. (2.20) and henceforth. We also note that

$$dS = dxdy \tag{3.7}$$

in the pupil. With these equations, and with $R = f$ being the focal length of the image forming system, Eq. (2.20) immediately gives

$$E_{p'}(k\alpha, k\beta) \simeq \frac{i}{\lambda} \frac{e^{-ikf}}{f} \iint_{\substack{\text{pupil} \\ (\text{coordinates};x,y)}} E_o e^{ik\Delta(x,y)} e^{-ik(\alpha x + \beta y)} dxdy \tag{3.1}$$

which is the desired Fourier-transform formulation of image formation in the approximate formulation of the rigorous Huygen's principle, as derived from Maxwell's equation with the stated approximations.

Eq. (3.1) has the form of a Fourier transformation and can be read as follows: the complex amplitude of the electric field vector at a point in the image plane is equal to the Fourier transform of the distribution of complex amplitude of the electric field within the image-forming aperture. Clearly here, the field vectors in the aperture and the field vectors in the image plane project along the same direction in the image plane. One Fourier transformation needs to be carried out for each point in the diffraction pattern.

For example, for a perfectly uniform plane wave front within a rectangular aperture of width A along the x-axis, one has $|E_o|=1$, $\Delta=0$, and therefore

$$E_p'(k\alpha) \simeq \frac{i}{\lambda} \frac{e^{-ikf}}{f} \int_{-A/2}^{+A/2} e^{-ik\alpha x} dx \tag{3.8}$$

which immediately gives upon integration

$$E_{p'}(k\alpha) = \left[\frac{i}{\lambda} \frac{e^{-ikf}}{f} \frac{A}{2}\right] \left[\frac{\sin(k\alpha A/2)}{(k\alpha A/2)}\right]. \tag{3.9}$$

The complex amplitude $E_{p'}(k\alpha)$ thus varies according to the well-known $\sin x'/x'$ function along $x' \simeq f\alpha$ in the image plane and has a first minimum for

$$\alpha_o = \frac{\lambda}{A} \text{ radians} \tag{3.10}$$

or for

$$x = f\frac{\lambda}{A} \quad (f = \text{focal length of focussing system}) \tag{3.11}$$

In general only the time-averaged intensity (we omit the time-average sign)

$$I_{p'} = E_{p'} \cdot E_p^* \tag{3.12}$$

is detectable (with the help of photoelectric, photographic or other receivers). $I_{p'}$ varies as $(\sin x'/x')^2$ and has a first minimum at $(\lambda/A)f$ from the central maximum. $I_{p'}$ is generally called the diffraction pattern corresponding to the aperture A.

Eq. (3.12) which expresses the diffraction at infinity by an aperture, can be made heuristically plausible on the basis of superposition and of the Huygens' principle as it is generally understood. Consider a pupil in the x, y plane and a point P centered on an element of area dxdy in that plane. The element of area centered on P emits an elementary wave $f(x,y)dxdy$, where $f(x,y)$ describes the scalar component of the E vector in the pupil. In the direction defined by the direction cosines α, β, γ the wave from P is dephased by $(2\pi/\lambda)(\alpha x + \beta y)$ with respect to an elementary wave emitted from the center of the pupil 0. By superposition, one has in the direction α, β, γ the sum of these waves

$$\int\int^{\text{pupil}} f(x,y)\exp\left[i\frac{2\pi}{\lambda}(\alpha x + \beta y)\right] dxdy \tag{3.13}$$

All elementary waves emitted by the pupil (or transmitted by the pupil) in the direction α, β, γ come to focus at a single point $P'(x',y')$ in the focal plane of a perfect focussing system, such that

$$x' = \frac{f\alpha}{\gamma}; \quad y' = \frac{f\beta}{\gamma} \tag{3.14}$$

Inasmuch as $\gamma \simeq 1$ and $\alpha^2 + \beta^2 + \gamma^2$ one has

$$x' = f\alpha; \quad y' = f\beta \tag{3.15}$$

If one defines the coordinates in terms of reduced coordinates

$$u = \frac{x'}{\lambda f} \quad ; \quad v = \frac{y'}{\lambda f} \tag{3.16}$$

and if one takes λ for the unit of length, one obtains

$$F(u,v) = \iint_{\text{pupil}} f(x,y) e^{2\pi i (ux+vy)} dx dy \tag{3.17}$$

which is an expression for the diffraction pattern identical to that obtained from Maxwell's equations in Eq. (3.1) by using the there stated approximations.

In summary, a knowledge of the distribution of complex amplitude (amplitude and phase) of the electromagnetic field (or more exactly of a component of the electric field) within an aperture, no matter how this field is created in this aperture, permits to compute the diffraction patterns corresponding to this aperture and light distribution. A unique relation between the field distribution in the aperture and the light distribution in the diffraction patterns exists, within the stated approximations, and takes the form of a Fourier transformation. The powerful techniques of operational calculus have been extensively applied to optical image formation problems with very fruitful results. In particular, a basic similarity has been recognized between problems in electrical engineering and problems of optical image formation and spectroscopy, whenever superposition and operational methods are appropriate

4. <u>Restrictions in applicability of the Fourier transform formulation of the diffraction problem.</u> We have noted that Eq. (3.1) is identical in form to an expression of the diffraction problem, Eq. (3.13) which can be derived heuristically on the basis of scalar-theory approximations, starting from a postulation of Huygens' principle. According to Huygens' principle an aperture can be broken up into a family of "source elements" which all emit "spherical wavelets" that conveniently sum and cancel out by "interference" at appropriate regions in the image space. However, the weakness of Huygens' principle derivation resides in the fact that the "principle" has no formal physical basis to start with, and, moreover, that the equations derived contain no warning as to the domain of applicability, thus permitting laborious extensions to situations where the equations are inapplicable.

For instance, it has been traditional in grating theory to attempt calculating the energy distribution among orders by assuming the grooves to be simple slits, or by "improving" on this assumption by taking into account the detailed groove shape with the introduction of appropriate phase retardation resulting from the path difference within the groove and so on. It should be clear that such attempts should be fruitless, as they are found to be, experimentally, except in special cases where it happens that electromagnetic solution of the energy distribution problem coincides with the qualitative results that may be inferred simply from geometrical optics.

Indeed, one of the important approximations involved in deriving Eq. (3.1) is that it is applicable only to the near vicinity of the quasi-spherical wave fronts originating from the pupil. Eq. (3.1) therefore is eminently suitable for dealing separately with each of the waves diffracted by an optical grating: it permits very simple calculation of the complex amplitude distribution $E_{p'}$ and of the intensity distribution $I_{p'} = E_{p'} \cdot E_{p'}^*$, in the diffraction pattern formed by the diffracted wave fronts of various orders and wavelengths, in the focal plane of the focussing optical system. Therein resides the interest of Eq. (3.1) in optical image formation.

5. Background and References. Great simplicity in the mathematical theory of image formation in general, and grating image formation in particular, results from the use of the Fourier transform formulation of physical optics. An important stimulus was given to this formulation by the work of M. Duffieux in 1944. Duffieux points out that the foundations for the Fourier transform formulation of image formation were apparently given in 1892 by A.A. Michelson and Lord Rayleigh. Even the operational form of the Fourier transform relation between wave fronts and images is explicitly given by Michelson in his book Studies in Optics, first published in 1927, where he refers (p. 58) to a Philosophical Magazine article of April 1905.

Very little if any use of this formulation appears to have been made until the 1940's, when it was newly developed by several authors, many of them starting quite independently, as it seems. Aside from the work of Duffieux (1946) [1], one finds the Fourier transform formulation given in particular by S. Silver [2] (1949), in relation to microwave antenna patterns, and a significant impetus is also due to the work of A.Maréchal [6,11,14,15,29], H.H.Hopkins [12], G. Toraldo di Francia [27], E.Ingelstam [7], D.Gabor [8], E.H.Linfoot [9,13], P.B.Fellgett [9], E.L.O'Neill [25], P.Jacquinot [10] and J. Strong [24] in the 1950's, as well as a paper by P.Elias, D.Grey and D.Robinson [4,5] in 1952. The concept and methods of spatial filtering and transfer functions comparable to frequency filtering in electronics, were introduced into optics by a number of workers, among which Luneburg (1944)[28], Duffieux (1946) [1], and more particularly Maréchal (1951) [29], Cheatham and Kohlenberg (1952) [30], Elias (1953) [4,5], O'Neill (1956) [31], H.H. Hopkins (1955), E. Ingelstam and E. Djurle (1956). It seems that the first formal description of a spatial frequency filter under the name "un filtre de fréquences spatiales pour l'amélioration du contraste des images optiques" was given by A.Maréchal and P.Croce [6] in C. R. Acad. Sci. Paris in September 1953. A more complete history of these developments may be found in E.L. O'Neill's book "An Introduction to Statistical Optics" (1963)[1] which may also serve as a reference for the mathematical developments used in this section, along with A. Maréchal's and M. Francon's "Diffraction Structure des Images" (1960)[2] and J. Arsac's "Transformation de Fourier et Théorie des Distributions" (1961)[3].

As shown by J. Arsac (loc.cit. 1961), great additional simplicity in the Fourier treatment of optical image formation results from application of the Théorie des distributions first evolved by Laurent Schwartz in 1950-1951. Well-known contributions to this theory were also made by M. J. Lighthill [19](1958) and by A.Erdelyi [20] (1962). The theory of distributions is sometimes known as the theory of generalized functions. Here we shall follow the development of J. Arsac which is proving to be very fruitful both in optics and in radio-astronomy (J. Arsac, R.N. Bracewell).

Application of the Fourier transform methods to grating image formation appears to have been made as early as 1951 by A. Camus, M. Francon, E. Ingelstam and A. Maréchal[4], and by E. Ingelstam and E. Djurle[5] in 1952 and 1953, in particular in relation to phase contrast testing of gratings. It was recognized that study of the diffraction patterns, rather than of the transfer functions, appeared to be of more particular interest in the spectroscopic use of gratings, even though A. Lohmann[6] also investigated the frequency transfer in spectrographs in 1959.

A very general treatment of grating theory in terms of Fourier transform formulation was given by G.W. Stroke[7] in 1959 and 1960, when he showed that the complex amplitude and intensity distributions in the spectral diffraction patterns formed by gratings could be easily calculated from the amplitude and phase distribution in the wavefronts, as they appear in readily usable form in the wavefront interferograms. It is in fact in the simplicity by which the wavefront interferograms (Stroke 1955, 1960)[8,9] may be obtained, and by

---

[1] Published by Addison-Wesley Publishing Co., Reading, Mass.

[2] Published by Revue d'Optique, Paris

[3] Published by Dunod, Paris

[4] A.M. Camus, M.Francon, E.Ingelstam et A. Maréchal: Rev.Opt.30,3 (1951).

[5] E. Ingelstam and E. Djurle: J. Opt. Soc. Amer. 43, 572 (1953).

[6] A. Lohmann: Optica Acta 6, 175 (1959).

[7] G. W. Stroke: Etude de théorique et expérimentale de deux aspects de la diffraction de la lumière par les réseaux optiques: l'évolution de défauts dans les figures de diffraction et l'origine électromagnétique de la répartition entre les ordres. Rev.Opt. 39, 291-398(1960).

[8] G.W. Stroke: J. Opt. Soc. Amer. 54, 30 (1955).

[9] G.W. Stroke, "Diffraction Gratings" in Handbuch der Physik (loc.cit.

which they display the grating and wavefront departures from perfection, which lends such interest to the use of Fourier methods in grating image formation.

References (Section 5).

[1] Duffieux, P.M." L'intégrale de Fourier et ses applications à l'optique, chez l'auteur, Faculté des Sciences, Besançon 1946.
[2] Silver, S.: Microwave Antenna Theory and Design, Vol. 12, M.I.T. Radiation Laboratory Series, New York: McGraw Hill Book Co. 1949.
[3] Schade, O. H.: Electro-optical Characteristics of Television Systems, RCA Rev. 9 (1948).
[4] Elias, P., D. Grey, and D. Z. Robinson: J. Opt. Soc. Amer. 42, 127 (1952).
[5] Elias, P. J. Opt. Soc. Amer. 43, 229 (1953)
[6] Maréchal, A., et P. Croce: C. R. Acad. Sci. Paris 237, 607 (1953).
[7] Ingelstam, E., E. Djurle, and B. Sjogren: J. Opt. Soc. Amer. 46 707 (1956).
[8] Gabor, D.: Lectures in Communication Theory, Report 238, Research Laboratory of Electronics, M.I.T. 1952.
[9] Linfoot, E. H., and P. B. Fellgett: Trans. Roy. Soc. Lond. 247 (1955).
[10] Jacquinot, P., P. Boughton, and B. Dossier: In: Contribution à la Théorie des Images Optiques, edit. de la Revue d'Optique 1949 p. 183.
[11] Maréchal, A.: Thèse, Revue d'Optique 1948.
[12] Hopkins, H. H.: Wave Theory of Aberrations. Oxford: Clarendon Press 1950.
[13] Linfoot, E. H.: Recent Advances in Optics. Oxford: Clarendon Press 1955.
[14] Maréchal, A.: Optique Géométrique Générale. In: Handbuch der Physics, Vol. XXIV, edit. by S.Flügge. Berlin-Göttingen-Heidelberg: Springer 1956.
[15] Marechal, A., et M. Francon: Diffraction, Structure des Images, Influence de la Cohérence de la lumière, edit. de la Revue d' Optique, Paris 1960.
[16] Born, M., and E. Wolf: Principles of Optics. London: Pergamon Press 1959.
[17] Stroke, G.W.: Ruling, Testing and Use of Optical Gratings for High-Resolution Spectroscopy. In: Progress in Optics, Vol. II edit. by E. Wolfl. Amsterdam: North-Holland Publishing Company 1963.
[18] Arsac, J.: Transformation de Fourier et Théorie des Distributions. Paris: Dunod 1961.
[19] Lighthill, M. J.: An Introduction to Fourier Analysis and Generalized Functions. Cambridge: Cambridge University Press 1958.
[20] Erdélyi, A.: Operational Calculus and Generalized Functions. New York: Holt, Rinehart and Winston 1962.
[21] Guillemin, E. A.: The Mathematics of Circuit Analysis. New York: John Wiley and Sons, 1951.

[22] Angot, A.: Compléments de Mathématiques, édit. de la Revue d'Optique 1957.
[23] Stroke, G. W.: Etude théorique et expérimentale de deux aspects de la diffraction par les réseaux optiques: l'évolution des défauts dans les figures de diffraction et l'origine électromagnétique de la répartition entre les ordres. Rev. Opt. 39, 291-398 (1960). Thèse, Sorbonne 1960.
[24] Strong, J.: Concepts in Classical Optics. San Francisco. W.H. Freeman & Co. 1958.
[25] O'Neill, E. L.: Introduction to Statistical Optics, Reading, Mass.: Addison-Wesley 1963.
[26] Stone, J. M.: Radiation and Optics. New York: McGraw Hill Book Co. 1963.
[27] Toraldo di Francia, G.: La diffrazione della luce. Torino: Einaudi 1958.
[28] Luneburg, R. K.: Mathematical Theory of Optics. Providence, R.I. Brown University 1944.
[29] Maréchal, A.: The Contrast of Optical Images and the Influence of Aberrations. National Bureau of Standards Circular 526, U.S. Dept. of Commerce (Proceedings of the N.B.S. Semicentennial Symposium on Optical Image Evaluation, Oct. 1951).
[30] Cheatham, T. P. jr., and A. Kohlenberg: Analysis and Synthesis of Optical Systems. Technical Note 84, Boston University Physical Research Lab., March 1952.
[31] O'Neill, E. L.: Selected Topics in Optics and Communication Theory. Boston University 1956/57.
[32] Stroke, G. W.: An Introduction to Coherent Optics and Holography New York and London: Academic Press 1966 [Second Edition: 1969]

III. OPTICAL vs. DIGITAL PROCESSING

Digital computers, including most recently also microprocessors with corresponding new architectures and various "fast Fourier transform" algorithms would seem to be ideally suited for the types of signal and image processing for which optical processing is suited naturally because of the well-known Fourier-transforming properties of lenses.

The principal advantages of digital computers, in this context, is their ability to deal with non-linear equations, and, more generally, with data already in a digital form. Great "geometrical" coordinate precisions and grey-level accuracies are attainable with ready compensation for dynamic-range limitations and for other non-linearities. Unfortunately, these and other advantages of digital computers, (when presented with a great number of data and Fourier transformations) are presently attained only in the use of very large computers, such as the CDC 7600 at Los Alamos, and then only if extremely expensive, laser image scanning micro-densitometer device for the input, and correspondingly good output (printer, CRT, etc.) devices are involved.

On the other hand, many of the required processing operations can be readily carried out by a family of optical devices, based on coherent optics and holography, and now generally known as "optical computing" systems. The principal advantages of optical computers is their direct, large-capacity and extremely high-speed, parallel-processing Fourier-transform capability which may, in addition, perform a whole series of successive on-line Fourier-transformations, as often required in image processing work. Perhaps the most important advantage, of all, is the fact that images may be processed directly, as they are, without any of the signal-degrading microdensitometer scanning and digitization (characteristic of all but the very best and most expensive devices in the present state of the art). Unfortunately, when one needs to attain in optical computers some of the specific advantages in precision and accuracy of the best digital computers, as singled out above, we now know that the problems with optical computers are not the readily surmountable problems of "dynamic range' or "noise", as often claimed in some of the literature. Rather, there may be real problems in ignoring the limitations of only a linear approximation (i.e. a Fourier-transform approximation) of the actual physics, namely of the electromagnetic nature and description of the diffraction of light by lenses.

It is expected that the usefulness of optical computing systems will be greatly enhanced, by establishing theoretically and experimentally both the limitations of the simple Fourier transform formalism, on the one hand, an, on the other, means of surmounting the residual non-linearities by taking them into account, through suitable means.

Among the means of correcting for residual non-linearities, one may mention the following:
1. Holographic lens compensators
2. Aspherically-corrected new types of lens systems
3. Digitally assisted real-time compensation in spectral plane
4. Holographically assisted on-line convolution compensation in spectral plane (real time).

THE INFORMATION CONTENT OF OPTICAL DIFFRACTION PATTERNS

Brian J. Thompson

The Institute of Optics

University of Rochester, Rochester, New York 14627

INTRODUCTION

The phenomenon of diffraction has been well known and well researched for over a century. It is interesting to note that the application of the basic principles of diffraction has long been, and continues to be, an attractive and fruitful field of study. In fact, it appears that the interest in this field has intensified during the last two decades and the resulting productivity of both research results and practical applications has been considerable. Clearly being able to write down the integral equations that describe the diffraction process and then solving these equations does not always give the necessary insight into the real understanding of the phenomenon. The field of diffraction is one excellent example where physical insight and intuitive feel can be so much more meaningful than mere formal theoretical understanding of the problem.

In this short paper, I would like to take the opportunity to extend the idea of physical insight in diffraction problems and give a mixture of historical and modern results that support this hypothesis. I hope that there will be some useful contributions in this paper that will make it more than a simple review of the development of the science and technology of using the diffraction pattern associated with a given object or signal to specify, or abstract, some particular property or portion of that object or signal. This process is, of course, the major topic of this seminar and can be given a variety of names: optical information processing, spatial filtering, optical data processing, pattern recognition, optical computing, etc.

The Forerunners - Pre 1950

We tend to think that the subject of optical processing is a recently developed area of research and development but, in fact, it has many important roots that should not be ignored. The well known Foucault knife-edge test, first described in 1859, is really a method of optical processing to remove the direct image light and keep the scattered or diffracted light. Another excellent example is the field of Schlieren photography that began with the work of Töpler in 1867.

A major contribution was made by Abbe in 1873 with his theoretical treatise on vision with a microscope. He realized the important role that diffraction plays in the image forming process. The experiments conducted by Porter in 1906, to demonstrate the process that Abbe described, clearly illustrates the physical insight that was referred to earlier.

There is an excellent example of Michelson's insight in his book published in 1927. He recognized the complex nature of the diffracted field and asked whether an image could be obtained by reconstruction from the recorded diffraction information. Once the diffraction pattern has been recorded, the phase information is lost. However, Michelson made a phase mask to use with the recorded diffraction pattern and was able to get a good reconstructed image of a slit.

The phase contrast microscope is now a well known and important device. It is a tribute to Zernike's real understanding of diffraction and coherence that he was able to develop the Nobel prize-winning concepts of phase microscopy in 1934.

Perhaps the single most important development that has become the cornerstone of modern work in much of optics was the study made by Duffieux on the use of the Fourier integral in optical problems.

Optical Processing - 1950-1965

A very productive period in optical processing started in the early 1950's and really represented a major step forward. Most of the progress that was made in this period could, in principal, have been made many years before, but the physical insight and the interdisciplinary interactions had not been developed. The theoretical work of Elias, Grey and Robinson (1952), Cheatham and Kohlenberg (1954) and Rhodes (1954) was one ingredient. The other, perhaps even more important, was the experimental studies of Maréchal and Croce (1953) and Asakura and O'Neill (1956).

It would be a major task to list all the workers and their publications during this period and it isn't appropriate to do that here. Suffice it to say that many talented people worked in the field in the fourteen year period 1950-1964, and a very firm foundation of the principles and applications of optical processing was established. The major problem that hampered the further progress was the difficulty of making the necessary filters that were used to process the Fraunhofer diffraction pattern of the input signal. It was relatively easy to make the required amplitude parts of the filters but relatively difficult to make the phase portions of the filters.

Holographic Filters, 1964-

The technological problem of filter generation was solved with the introduction of the idea of creating the filter in a hologram. This concept was introduced by Vander Lugt in 1964 for a specific filter type. However, it was clear that the idea could be extended to generate any required filter. An added advantage is that the holographic filter need not be calculated but can be generated directly from the appropriate optical signal. Other workers were quick to recognize the importance of this development and to apply the holographic filter to other problems. (Leith, Upatnieks, and Vander Lugt (1965); Stroke and Zech (1967)). Again it would be impossible to list all the active researchers who have participated in these developments. Many of the ideas that were difficult, if not impossible, to implement in the early years of the development of optical processing can now be attempted successfully.

Real-Time Processing, 1972-

It is correctly claimed that a coherent optical processing system is a real-time, two-dimensional, parallel- processing device, since it operates in the time it takes for the light to travel through the optical system. However, this is only true once the required filter has been fabricated and placed in position and the appropriate optical input is available. What is really required is an input device that can act as an incoherent to coherent converter, or a material on which an incoming signal can be written. The filter itself also, needs to be created directly in the system on an erasible material.

There are now promising solutions to these problems with the development of a number of devices. These include the so-called PROM device developed by Itek Corporation (Nisenson and Iwasa 1972, Iwasa and Feinleib 1974), the RUTICON developed by Xerox (Sheridan, 1972), and the liquid crystal panel developed by Hughes (Grinberg et al 1975).

Hybrid Systems, 1974-

The current and future trends in the field of optical processing will most probably be in the use of hybrid systems. Work has already started in this field, particularly by Casasent (1974). The fundamental idea is to create the correct marriage of optical, digital and electronic methods that will allow real problems to be solved.

DIFFRACTION MEASUREMENTS

Measurement of the Intensity

Direct measurement of the intensity of diffraction patterns is often used for determining information about the diffracting object. This method becomes increasingly important as the object gets smaller since the scale of the diffraction pattern increases as the object scale decreases. The Young's eriometer (see e.g. Ditchburn 1952) is an early example of this principle. The device was first used by Young for measuring wool fiber diameters and hence the name eriometer[+]. The method works well if the fibers or particles have a narrow size distribution and the mean size is the information required. The coherence requirements are not very important since it is really only necessary for the light to be spatially coherent over a distance equivalent to a few particle diameters. Figure 1 illustrates this point; each diffraction pattern results from the composite diffraction pattern produced by a large number of particles of lycopodium powder each particle of which is approximately spherical and of mean diameter 33 microns. In this series of photographs the spatial coherence is changed from being equivalent to several thousand particle diameters, to several hundred particle diameters, and to several tens of particle diameters. Clearly there is essentially no difference between these photographs and the mean diameter can easily be determined by measuring the diameter of the first minimum in the intensity distribution.

When the diffracting sample consists of a range of particle sizes it is not immediately obvious that the distribution function can be determined. However, for the case of spherical particles this is possible as described by several independent groups of workers in the U.S.A. and the USSR. (Beissner et al 1970) . A solution can be obtained by solving a set of simultaneous equations after sampling the total intensity in the diffraction pattern by a series of annular apertures (either used sequentially or simultaneously). A given particle size contributes approximately zero intensity at one particular annular position. Figure 2 shows a test of the system in simulation (Ward 1968.)

[+] I am indebted to Prof. Brian Kaye of Laurentien University for pointing out the meaning and origin of the word eriometer.

OPTICAL DIFFRACTION PATTERNS

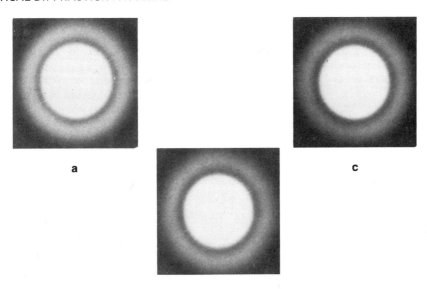

Figure 1. Diffraction by a large number of particles of mean diameter 33 microns under varying conditions of spatial coherence. (a) coherence with many 1000 particle diameters, (b) many hundred particle diameters, (c) several tens of particle diameters. (Thompson 1966)

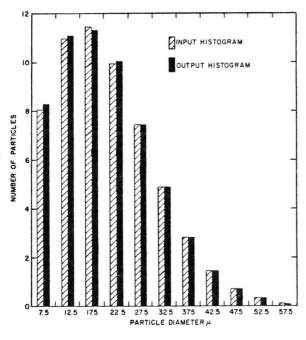

Figure 2. A comparison of an input histogram and output histogram of spherical particle size determined from the total diffraction intensity. (Ward 1968)

The Phase of the Diffraction Pattern

Naturally when the intensity is recorded the phase information is lost. A number of attempts have been made to measure (by interference) the phase associated with the pattern. Workers in optical analogues to x-ray diffraction look at diffraction patterns of representations of projections of molecules; the atoms are represented by circular holes at the appropriate locations. One method uses two diffracting masks, the first is the arrangement representing the particular projection of the molecule, the second is a similar mask but with an extra hole at the center. A comparison between the diffraction patterns of these two masks allows an estimate of the phase to be made, particulary if the molecule is centro-symmetric and the phases are either 0 or π. (see e.g. Taylor and Lipson 1964 for some excellent examples).

A second method involves the coherent superposition of a set of interference fringes upon the diffraction pattern of interest. The position of the fringes contour the phase difference in the diffraction pattern. (again Taylor and Lipson have excellent examples of this method in their book). Today, of course, we would generalize these methods and store the diffraction pattern as a Fourier transform hologram. Instead of recording the amplitude and phase, there are occasions when it is perhaps appropriate to record the real and imaginary part of the diffraction pattern - again this can be done optically. (see Taylor and Lipson 1964).

Coherence Control

In some systems it can be useful to deliberately use an incoherent primary source so that the spatial coherence of the illumination can be controlled. Some interesting examples of this control are discussed by Thompson 1966, Parrent and Thompson 1969, and Thompson 1972. Figure 3 shows how the coherence can be controlled to remove the interference term that would naturally be produced by a regular array of similar objects; however the light is still highly coherent over each unit of the array. The diffracting object is shown here as an array of many "molecules" of hexamethylbenzene; the diffraction pattern (b) was obtained by one such molecule alone while the pattern (c) was obtained from the array of molecules with coherence conditions outlined above. The two pictures are essentially identical.

An interesting example of the method of coherence control is for replication of a given input (Parrent and Thompson 1964). When a Fraunhofer diffraction pattern is formed with a lens, the diffraction plane is also the image plane of the source that produced the illumination. If the diffracting object is a large array of small circular apertures and the illumination is not fully spatially

OPTICAL DIFFRACTION PATTERNS 319

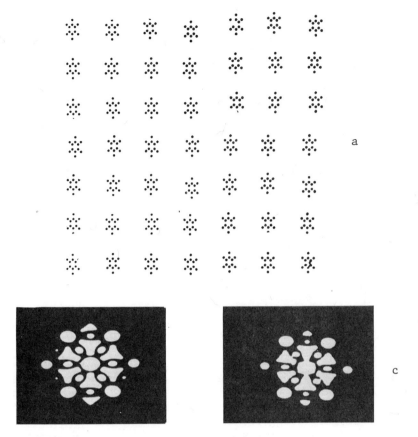

Figure 3. An illustration of coherence control (a) diffracting object (b) diffraction pattern of one group alone (c) diffraction pattern of whole array with coherence interval equal to the center-to-center spacing of the groups.

Figure 4. Source replication by diffraction method.
   (a) incoherent circular source
   (b) incoherent square source.

coherent, then the image of the source plays an important role. The diffraction pattern is still periodic but the repeated structure is the image of the source. This is illustrated in Figure 4 for a circular source in (a) and a square source (b) Thompson(1966). To replicate an input the function to be replicated is illuminated and acts as the effective source; the periodic diffraction aperture determines the dimensions of the array in diffraction space. Figure 5 shows this result. Holography has added considerably to the flexibility of this method in recent years.

Fourier Synthesis

The earliest attempts to synthesis an image from its diffraction pattern was the experiment conducted by Michelson and mentioned earlier. Bragg (1939) put forward the idea of carrying out Fourier summations by optical methods. Basically he used a diffracting mask made from a weighted reciprocal lattice of the x-ray diffraction data of a crystal. The size of the hole controlled the amount of light transmitted. If the x-ray diffraction pattern is real and positive then there is no phase difference between the varies diffraction peaks. Thus when the diffracting mask suggested above is illuminated coherently its diffraction pattern is an 'image' of the original structure. Bragg illustrated this method with the (010) projection of diopside ($CaMg(SiO_3)_2$).

In general of course, it is necessary to provide for the relative phases of the various regions of the weighted reciprocal lattice. For centro-symmetric structures discrete values of the phase of 0 and π are required. Bragg (1942) suggested a method for accomplishing this using mica. An improvement of this method was demonstrated by Buerger 1950 and his result on marcasite (FeS_2) is well known. Considerable progress has been made with this concept particularly with respect to amplitudes and phase control. The non-centro-symmetric structure of sodium nitrate was demonstrated by Harburn and Taylor 1962. (see Fig. 6(b)) An excellent summary of this topic can be found in a recent review by Harburn (1972).

OPTICAL PROCESSING

The diffraction pattern does not have to be detected directly but can be modified and this modified complex amplitude allowed to propagate to form an image. It is this process that we will be discussing in some great detail as the week progresses. I would like to merely make some comments on the general topic and refer the reader to some more detailed reviews of the subject that adequately represent my own point of view (Thompson 1972, 1975).

It is possible, of course, to operate on either the amplitude or the phase or in general on the complex amplitude of the diffraction

OPTICAL DIFFRACTION PATTERNS 321

Figure 5. Replication of a source function   B   by the diffraction method.

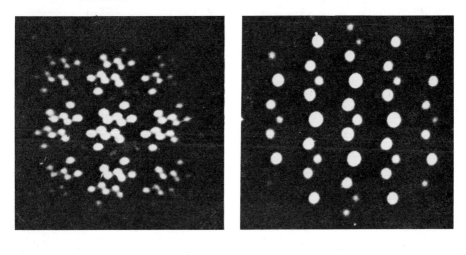

a   b

Figure 6. Optical Fourier synthesis of (a) hexamethylbenzene as illustrated by Hanson 1952 and the non centro-symmetric molecules of sodium nitrate by Harburn and Taylor 1962.

pattern. Generally speaking the simpler the filter the better the method works.

BAND-PASS FILTERS

Perhaps the simplest filters are blocking filters that allow certain regions of the diffraction pattern to pass and hence remove the remaining portions. There is an interesting illustrative example of this again in the field of optical analogues to x-ray diffraction - a demonstration of the effect of series termination errors. (Thompson 1959, 1972). For this illustration the [110] projection of the molecule of bishydroxydurylmethane, as determined by Chaudhuri and Hargeaves 1955, was chosen as a suitable example. A diffracting mask was made containing 28 molecules of bishydroxydurymethane on a scale of 1 cm = 1Å and each hole that represents an atom is 0.5 mm in diameter. The resulting diffraction pattern was about 3 mm in diameter which is large enough to make the modifications practicable. Figure 7 shows the results of truncating the diffraction pattern in a controlled way; in each illustration (a) is the image formed by using that portion of the total diffraction pattern shown in (b) ; c represents the electron density contour map produced by a calculated Fourier synthesis from the equivalent region of the weighted reciprocal lattice (d); the amplitude and phase information for the calculation was taken directly from the published data of Chaudhuri and Hargeaves and the calculation carried out by Morley (1958).

There is one distinct difference between the optical image and the calculated 'image', and that is that optically the atoms are represented by circular holes having sharp edges, whereas in the calculation the information is derived from real atoms which have no distinct edges.

The transform was limited in successive steps until $\sin\theta < 0.22$. ($\sin\theta = 1$ is equivalent to using all the information in the central maximum of the diffraction pattern due to the holes representing an atom). As the amount of information used to form the image is reduced the atomic positions become less well defined and some atoms cease to be resolved; spurious detail arises in some instances. Eventually the only information that is retained is the rough shape of the molecule.

Another interesting experiment conducted with this same molecule of bishydroxydury methane was to make a mask with two of the atoms misplaced. The diffraction pattern produced by this mask is then modified by passing it through a filter representing the weighted reciprocal lattice determined from the x-ray diffraction intensities. This filter was made by etching the weighted reciprocal lattice in copper foil. Thus the amplitude of the transform is somewhat controlled by this filter and the phases are those produced by the

OPTICAL DIFFRACTION PATTERNS

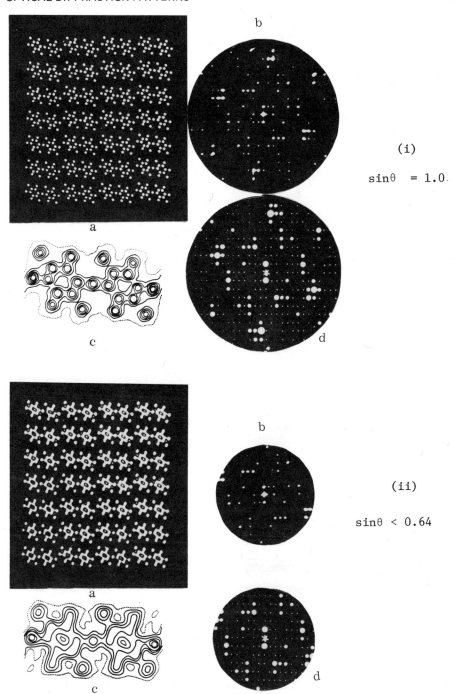

(i) $\sin\theta = 1.0$

(ii) $\sin\theta < 0.64$

Figure 7. Series Termination Effects

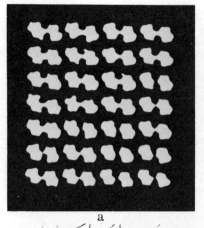

(iii)

$\sin\theta < 0.54$

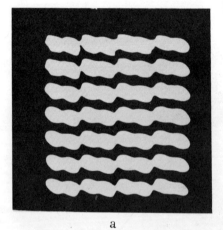

  (iv)

$\sin\theta < 0.42$

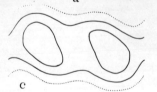

OPTICAL DIFFRACTION PATTERNS   325

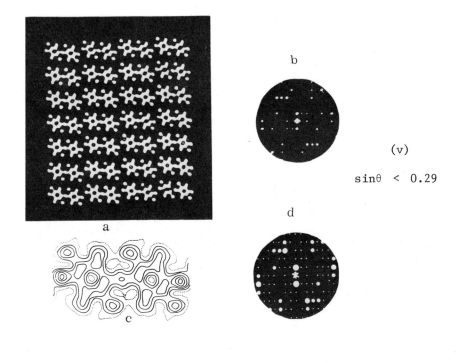

(v)

$\sin\theta < 0.29$

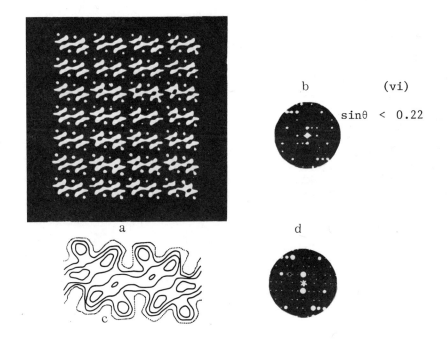

(vi)

$\sin\theta < 0.22$

diffraction pattern of the input mask. This is a crude experiment but the image produced is interesting. The image of the atoms which were incorrectly positioned show much lower intensities (A) and a peak occurs at the correct location of that particular atom (B). (Fig. 8). A new mask can be prepared which is a better approximation than the original input mask. A second example is shown in which the starting mask has all the atoms slightly misplaced. The processed image allows a better estimate of the atomic position to be obtained. It is doubtful if this process can be generalized.

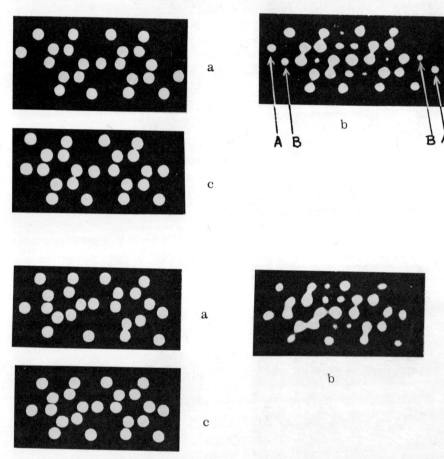

Figure 8. An illustration of filtering using a weighted reciprocal lattice.
(a) is the original mask and (b) is the image after processing by modifying the diffraction pattern of (a) with the correct weighted reciprocal lattice; (c) new atomic positions chosen from (b)
(i) two atoms misplaced.
(ii) all atoms slightly misplaced.

Raster and Half-Tone Removal

Band-pass filtering is a particularly attractive technique when periodic signals are considered. A popular and sometimes useful application is to remove a raster from a television-type picture. The input can be considered to be a periodic function whose envelope function is the appropriate band-limited continuous scene; or equivalently the periodic function samples the scene. Hence the Fourier transform is also periodic with a periodicity reciprocally related to the raster periodicity. At each of these periodic locations in the transform, the transform of the scene itself is located. If the scene is correctly sampled then the adjacent transforms of the scene do not overlap. Thus if a band-pass filter allowed one of these regions to pass, then the information about the periodicity is lost but the total scene information is retained. Then the final image is a continuous image without any periodic structure.

The two-dimensional equivalent of this example is the removal of the halftone to produce a continuous image. In this example a half-tone input produces a two-dimensional array of diffraction orders with the transform of the scene centered at the location of each other. Selection of any order will produce an image without the half-tone. There are, of course, some other methods of effectivly removing the half-tone. The observer may half close his eye and thus remove the visual effect of the half-tone or move the page further away until again the half-tone is not visible. Another method is to image the half-tone scene with a lens that is slightly out of focus. Again the half-tone is blurred sufficiently that it is not visible.

The question naturally occurs, why remove the half-tone at all, since no extra information is added by the processing? Cosmetically the image is more pleasing with the half-tone removed. There is, however, a more valid reason. If the half-tone scene is viewed under high magnification so that the field of view only contains a few half-tone dots, it becomes extremely difficult to interpret the image. We can't see the image for the half-tone! A real half-tone is actually more complicated than indicated here. A recent study of the detailed spectrum of half-tone images has been published by Kermisch and Roetling (1975).

Signal Separation by Orientation

The above example requires knowledge of the signal. It is often of equal value to know something about the unwanted part of the input. An interesting example of this, which also illustrates the importance of having a two-dimensional transform is that of selecting a signal from a set of recorder traces. This is an interesting example that Becherer and I did some years ago (see, e.g., Thompson 1970, 1972). Figure 9(a) shows a set of twenty nine pen recorder traces, one of which contains a signal of importance. The unimportant traces are essentially information in the horizontal direction, whereas when a signal is present, the pen trace contains strong vertical and near vertical components. The transform of this input is displayed in Figure 9(b). The required filtering operation can be carried out by an opaque vertical strip that removes most of the transform that is visible. The processed image is now shown in Figure 9(c); this signal remains since its transform exists in a cone of angles around the horizontal axis. The signal of interest is retained as are the timing marks. This result was achieved without any detailed knowledge of the signal or its position in the field.

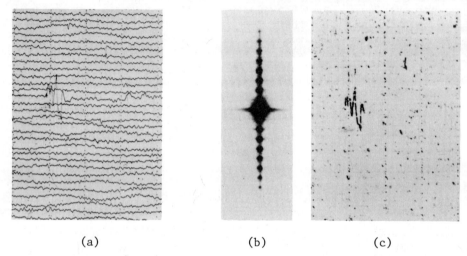

(a) (b) (c)

Figure 9. Signal separation by orientation. (a) the input; (b) its transform; (c) the processed image (Becherer and Thompson 1965, Thompson 1970, 1972).

Application of the Dark Field Method

One of the earliest techniques in optical processing was well known by Foucault and others since it is really the classical dark field method. Modern technologists often call this d.c. blocking. The reason for this terminology is rather obvious since the Fourier transform is the spatial frequency spectrum of the input and hence the region of this spectrum at the origin is the zero (or d.c.) spatial frequency component. If the input consists of an object in a uniform bright (transparent) field, then that field can be removed by removing the d.c. component of the spatial frequency spectrum and the image is then the object appearing bright against a dark background. This method has considerable value when phase objects are to be viewed.

We will consider a phase object as an input given by

$$\psi(\xi,\eta) = \exp i\, \phi(\xi,\eta), \qquad (1)$$

where $\phi(\xi,\eta)$ is the phase distribution. The transform of this input can be written in the following form

$$\Psi(x,y) = \Psi_0 + \Psi_1(x,y),$$

where

$$\Psi_0 = \text{const} \iint_{-\infty}^{\infty} \exp \frac{-ik}{f}(x\xi+y\eta)\, d\xi d\eta, \qquad (2)$$

$$\Psi_1(x,y) = \text{const} \iint_{-\infty}^{\infty} [\Psi(\xi,\eta)-1] \exp \frac{-ik}{f}(x\xi+y\eta)\, d\xi d\eta.$$

The contribution Ψ_0 can be thought of as the zero frequency component of the transform of the input. That is, it is the distribution located in a region around the optical axis - i.e., within a resolution element. $\Psi_1(x,y)$ is the contribution from the detailed structure of the input distribution.

If Ψ_0 is removed completely by placing a small circular optical stop on the optical axis, the complex amplitude distribution after the filter is simply $\Psi_1(x,y)$. The resulting image is thus

$$\psi(\xi',\eta') = \text{const}\, [\exp i\, \phi(\xi',\eta')-1] \qquad (3)$$

and the detected intensity is

$$I(\xi',\eta') = \text{const}\, [1-\cos\phi(\xi',\eta')], \qquad (4)$$

and this intensity maps the cosine of the phase of the object. Under the small phase approximation this intensity variation becomes linearly related to the $[\phi(\xi',\eta')]^2$.

Consider producing a fingerprint on a glass plate; the oil on the finger will leave a print which is a phase object. This phase object may be converted to a rather nice intensity distribution in the image by removing the d.c. component of its transform. Figure 10 shows this result. There is, of course, only qualitative information in this image and no information about the path difference variation in the phase object.

Figure 10. Dark field method used on a fingerprint (after Hecksher and Thompson 1965).

APPLICATION OF AMPLITUDE FILTERS

There are a number of operations that can be carried out successfully by changing the relative amplitude of the various spatial frequency components in either a discrete or continuous way. The phase of the transform is left untouched in these applications. Amplitude filters can be fabricated by using photographic film or a variable thickness absorbing film produced by evaporation.

Contrast Enhancement

A low contrast input is characterized in the Fourier plane by a relatively large value at the zero-order (or d.c.) whilst the higher frequencies contain relatively little energy. This balance can be changed by placing a small, partially transmitting filter over the d.c.; this allows less light to pass through to the image plane and redresses the balance between the zero spatial frequency content and the higher spatial frequencies. The resultant images has improved contrast. Naturally this technique has value only in conjunction with other filtering operations.

Differentiation

The derivative of an input may be produced by operating upon the transform of that input with a filter function whose amplitude transmittance is

$$f(x) = \alpha(x_0 - x),$$

OPTICAL DIFFRACTION PATTERNS

where α and x_o are both constants. (Clearly without the bias term x_o we would require a complex filter). This filter is clearly a linear amplitude taper, the filter being denser in the middle (minimum transmittance) and falling off to zero density (maximum transmittance) at the highest spatial frequency of interest.

For an example, we will again use a purely phase object. This time it consists of a bleached photographic film that had been exposed to a standard tri-bar target (Sprague and Thompson 1972). This phase object can be represented, as in equation (1), by

$$\psi(\xi) = \exp i\, \phi(\xi) ,\qquad (5)$$

where $\phi(\xi)$ is the phase variation in the ξ-direction. When $\psi(\xi)$ is transformed, multiplied by $f(x)$, and retransformed, the complex amplitude in the image plane is

$$\psi(\xi') = \left[\frac{x_o}{f} + \frac{d\phi(-\xi')}{d\xi'} \right] \exp i\, \phi(-\xi') \qquad (6)$$

and the recorded intensity is linearly related to the derivative of the input phase. Figure 11 shows this derivative, which is an excellent qualitative view of the phase object which actually had phase variations over several wavelengths of path difference (Sprague and Thompson 1972). The axis of the differentiating filter is oriented at 45° to the vertical. The random pattern in the broad clear areas is the actual derivative of the surface of the phase object, which is slightly irregular due to the bleaching. The fringes inside the bars are produced by multiple reflections between the front and back surfaces of the phase object.

APPLICATION OF PHASE FILTERS

There are a few examples where phase filters are used alone, although they are more usually found in conjunction with amplitude filters to form a complex filter. The examples that will be briefly discussed here require a single-valued phase-filter to be located over some region of the spatial frequency spectrum. Again a number of methods of fabricating the required filter function exist. Early methods relied on evaporated transparent films of appropriate optical thickness; bleached photographic film, and diochromated gelatin films are alternatives.

Phase-Contrast Methods

In their simplest form the phase-contrast methods can be understood by returning to the analysis of equations (1) - (3). Specifically we consider equation (2), which describes the transform of an input phase object. Instead of removing the term Ψ_o completely,

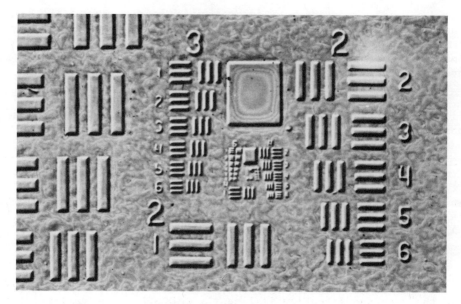

Figure 11. Derivative of a phase object that consisted of a bleached photographic record on a standard bar target (after Sprague and Thompson 1972).

we will change its phase relative to the rest of the transform $\Psi_1(x,y)$ by $\pi/2$. Hence a filter has to be inserted over a small region at the center of the transform that has an optical thickness just sufficient to produce a $\pi/2$ phase change. Hence the image field would then become

$$\psi(\xi',\eta') = \text{const} \, [\exp i\pi/2 + \exp i\phi(\xi',\eta')-1] \tag{7}$$

and the intensity is

$$I(\xi',\eta') = \text{const} \, [3 + 2\sin\phi(\xi',\eta')-2\cos(\xi',\eta')] \tag{8}$$

Under the small phase approximation that

$$\exp i\phi(\xi,\eta) \simeq 1 + i\phi(\xi,\eta), \tag{9}$$

equation (8) becomes

$$I(\xi',\eta') = \text{const} \, [1 + 2\phi(\xi',\eta')] \tag{10}$$

This is Zernike's important result that, under the small phase approximation of equation (9) the image intensity is linearly related to the object phase. The particular scheme described here is the so-called positive phase-contrast or bright contrast. If the phase filter changes the phase by $3\pi/2$, then equation (10) would become

$$I(\xi',\eta') = \text{const} \left[1 - 2\phi(\xi',\eta')\right], \tag{11}$$

which is the negative phase-contrast or dark contrast condition.

The knife-edge test in which one half of the spatial frequency distribution is removed has a variation introduced by Zernike in which one half of the spatial frequency plane has its phase reversed with respect to the other half.

APPLICATION OF COMPLEX FILTERS

In the early work in optical processing, it was possible to produce the required complex filters but only with considerable difficulty. The filter had to be constructed in two parts - an amplitude portion made as described above and a separate phase portion made by evaporation techniques.

Phase-Contrast with Contrast Enhancement

The results obtained in the previous section on phase-contrast microscopy can be combined with a method for contrast enhancement discussed earlier. The results described for phase contrast produced an image intensity that was linearly related to object phase. However, the contrast in the image was quite low since all the d.c. energy was allowed to contribute to the final image. Clearly contrast enhancement can be achieved if the amount of energy passing through the d.c. portion is reduced. Then a complex filter is used described by

$$f = a \exp i \pi/2 \text{ or } a \exp i 3\pi/2, \tag{12}$$

where a is a constant less than unity. Hence equations (10) and (11) become

$$I(\xi',\eta') = \text{const} \left[a + 2\phi(\xi',\eta')\right] \tag{13}$$

and

$$I(\xi',\eta') = \text{const} \left[a + 2\phi(\xi',\eta')\right] \tag{14}$$

Integration

Optical integration can be carried out with a complex filter that should have the following form

$$f(x) = \frac{\text{const}}{x} \tag{15}$$

Unfortunately this function has a singularity at the origin, and hence is not realizable experimentally. Instead, a truncated integration filter is used whose amplitude transmittance is given by

$$f(x) = \frac{D}{x}, x \geq D \quad D = \text{const},$$

$$= iB, x \leq D \quad B = \text{const}. \tag{16}$$

This filter requires a half-plane phase filter of $\pi/2$ and the appropriate amplitude filter which can be made separately (see Sprague and Thompson 1972). These authors used the integration technique to operate on the recorded derivative shown in Figure 11. The image amplitude is given by

$$\psi(\xi') = \text{const}\left[\frac{x_o}{f} + \phi(\xi')\right]. \tag{17}$$

Clearly the image intensity can be made proportional to the input object phase if $C x_o/f$ is chosen sufficiently large. Thus the intensity distribution of Figure 12 is linearly related to the phase of the input object.

Aberration Balancing

One of the prime motivations for the work in optical processing carried out by Maréchal and Croce (1953) was to correct aberrated images. Tsujiuchi illustrated the use of this technique in a series of papers (1960a,b 1961) which are perhaps best read in his review paper (Tsujiuchi 1963) that discusses a variety of aberrations. Clearly considerable technological advantage can be gained if photographs taken with an aberrated optical system can be corrected by subsequent coherent processing. Within some definable limits this can be accomplished. It must be stressed that the impulse response, or the transfer function, of the aberrated system must be known or determined.

Clearly if the transfer function contains any zero's the value of the required filter function has to be infinity - an impossible situation; furthermore, small values of the transfer function create equally serious problems. Hence of those spatial frequencies that were recorded, some processing can be carried out to get a flatter transfer function; both the contrast and phase of the recorded spatial frequencies can be changed. This process will become clearer if we discuss a typical example.

(a) Defocused image

An approximate form for the impulse response of a defocused imaging system is a blur circle of uniform intensity. The transfer function of this imaging system is then a first order Bessel function divided by its argument. The effect of the defocusing is very nicely seen in the image of a converging bar target shown in Figure 13. It

OPTICAL DIFFRACTION PATTERNS 335

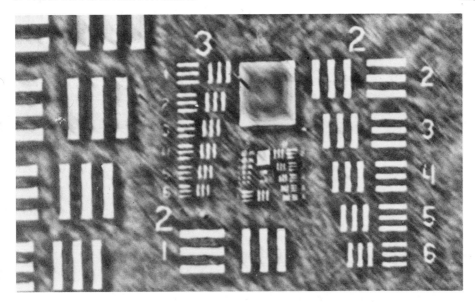

Figure 12. The integrand of the recorded derivative of a phase object. The intensity is proportional to the input phase (after Sprague and Thompson 1972).

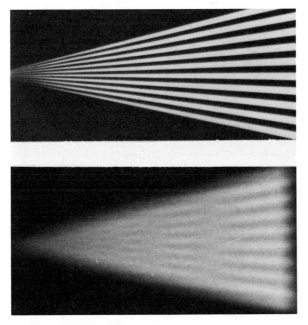

Figure 13. A converging bar target (a) and its image in a defocused optical system (b).

will be noted that the contrast of all frequencies is considerably lower than the in-focus image. Certain frequencies (approximately represented by a particular vertical plane through the converging bars) have a contrast that is zero. After the first zero position a region in which the contrast is reversed is visible, i.e., the black bars show up as white bars and vice versa. Figure 14(a) shows the transfer function that has to be corrected; this is, of course, a two-dimensional symmetric function. The negative value of the transfer function is interpreted as a contrast reversal. An appropriate contrast level is picked and the first maximum region of the transfer function flattened out. Now a phase filter is applied to invert the negative regions shown in Figure 14(b).

The techniques can be carried out, but the difficulty of manufacturing the filter makes it a somewhat impractical technique in general. However, if a large amount of data has to be processed, then the effort to make the filter is rewarded. However, this situation is changed with the use of holographic filters.

(b) Linear image motion

The problem of linear object motion during photography can be treated in a similar way to the aberration just described. It is necessary to know the impulse response of the system. The image of a point object will be essentially a line in the direction of motion. This can be treated approximately as a rectangular function and hence the transfer function to be corrected is a sinc function. The filter required to correct the motion blur in the original photograph is a complex filter comprising separate amplitude and phase portions. The procedure is almost identical to that for the defocused image except that the sinc function replaces the Bessel function and the problem is one-dimensional, not two-dimensional. Figure 15(a) shows the image of a bar target that has undergone linear motion in the vertical direction during exposure. The phase reversal determined by the sinc transfer function is very obvious. The corrected image is shown in Figure 15(b)(Considine 1963).

It is important to comment in concluding this section that the impulse response of an incoherent system is a real positive function, since it is an intensity distribution. If it is also symmetric, then its transform (i.e., the transfer function) is also real - but not necessarily positive. For an asymmetric impulse response, the transfer function is complex. Hence the correction filter may require continuous variation of phase.

The examples used above are interesting but nevertheless rather special. The improvements obtained are dramatic because of the phase reversals present in the aberrated images. It may be claimed, therefore, that in a sense resolution is improved; in general this is

OPTICAL DIFFRACTION PATTERNS 337

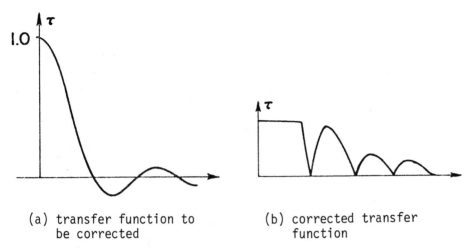

Figure 14. Aberration balancing for an out-of-focus image.

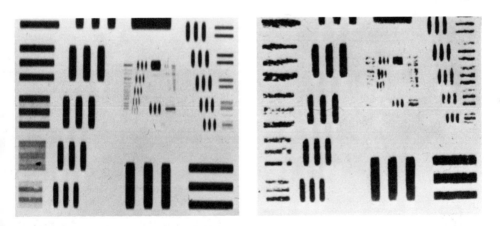

Figure 15. Aberration balancing of an image formed with linear image motion.

not true and any claims of improved resolution with aberration balancing should be viewed with suspicion.

Detection Filtering

There is a class of optical processing techniques that is often referred to as detection or inverse filtering. The problem is to design an optical processing system that will recognize the presence of a particular signal in a given scene and indicate its presence by providing a bright spot of light at its location.

Conceptually the idea can best be thought of in the following way. For the coherent processing system to produce a bright spot of light of the location of the object of interest means that the modified transform has to be essentially a constant. Thus the filter must be the inverse of the Fourier transform of the object of interest. For example, consider that we wish to select the circles A, B, and C in Figure 16(a). The Fourier transform of the circular object is, of course, a function of the form $2J_1(r)/r$ where r is the radial coordinate. The inverse of this function would be formed by first recording the intensity distribution in the Fraunhofer pattern of a circular aperture at the appropriate scale and with the appropriate γ-control of the film. This is then the filter $1/2J_1(r)/r$. The appropriate phase portion of the filter is a series of annular rings of π phase. Naturally when $2J_1(r)/r$ is zero, nothing can be done, but the filter can be effective over a considerable range of values of the function $2J_1(r)/r$. Application of this process to the input of Figure 16(a) yields the output of Figure 16(b). Clearly bright spots of light occur at the locations of the circles of a given size (Considine et al 1963).

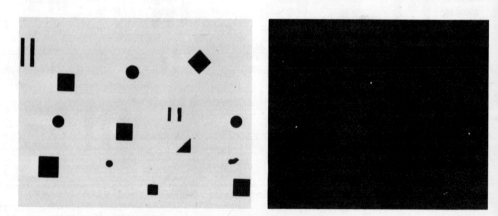

Figure 16. An example of detection filtering (after Considine et al 1963).

This is a rather important example, since it illustrates the power of the method. The size of the various geometrical shapes are very similar, so that their Fourier transforms occupy the same range of frequencies. Clearly other geometries, such as the square, could be identified by this same method. However, the triangular object has a transform whose phase is continuously variable, which makes a filter fabrication problem.

Summary Discussion of Classical (Pre 1964) Filtering Methods

While these various techniques can be implemented, the generation of complex filters is a difficult task, not only because of the difficulty of fabricating the continuous phase variations but also in determining what that phase variation should be in specific cases. For example, consider the detection filtering method described above for a non-simple object such as the recognition of a particular human face from a group photograph! These problems provided a major obstacle to the further progress of optical processing that was conveniently solved by the introduction of holographic methods of filter generation.

HOLOGRAPHIC FILTERS

Holography has had an important impact on coherent optical processing - it has provided a method for preparing a complex filter directly, often without requiring detailed knowledge of the filter function itself. Hence many of the applications described above can be accomplished much more simply than by conventional filter fabrication. The basic process, however, remains the same. The idea of preparing holographic filters for optical processing was first seriously discussed by Vander Lugt (1964) in a paper on signal detection by complex spatial filtering. This idea has since been extended by a number of workers to a variety of traditional and new applications of optical processing.

Detection Filters

Vander Lugt's idea for complex spatial filtering was to automatically produce an inverse filter or detection filter such that the output consisted of a delta function in the image plane, at the location of the object for which the filter was designed.

The principle of Vander Lugt's holographic method can best be illustrated by reverting to the simple example of detection filtering described above. It will be recalled that the necessary filter is of the form of the inverse of $2J_1(r)/r$. This filter can be generated as a Fourier transform hologram.

An appropriate circular aperture is placed in the input plane and its Fourier transform produced. A portion of the incident collimated beam is also caused to illuminate the transform plane with a collimated beam having been deviated for example by a prism. The interference pattern between the collimated beam and the transform of the input object is recorded. This record, which is a negative, is the required filter.

Aberration Balancing

As we have seen, one of the original ideas in optical processing was the correction of aberrated images. The concept of using a holographic filter for this purpose was first suggested by Leith et al (1965) and Upatnieks et al (1966). A hologram is made of the wavefronts emerging from the aberrated lens. This hologram is then used together with the lens to correct the aberration.

Once the idea was established that filters could be made holographically many of the earlier experiments on aberrated images were repeated using holographic filters. Needless to say, this process made the general method easier and more versatile. Thus the deconvolution process has been exploited for image motion and other aberrations (Stroke 1965, Stroke and Zech 1967, Stroke 1969). However, it must be remembered that the concept and limitations are the same as those described in earlier sections. Spatial frequencies not recorded in the original image cannot be generated by the processing step without additional controlled images being recorded.

Optical Subtraction

An additional advantage that the holographic filter has is that it allows the processing system to be used in a number of different modes. The filter does not have to be in the Fourier plane but could be in the Fresnel region or even in an image plane. These methods have been used for optical subtraction (Bromley et al 1971). Figure 17 illustrates the use of an image plane hologram for optical subtraction: (a) is the object for which an image plane holographic filter is made such that this object would be completely removed by the filter. Figure 17(b) shows the second input that is not identical to the original object. The filter removes the common parts of the two objects and produces an image of the differences (Monahan et al 1970).

REAL TIME PROCESSING

The holographic method of filter generation certainly solved one problem in optical data processing. However, several other problems still remained. The first of these problems relates to the input of information into the optical system. All the examples discussed so far needed a pre-recorded input to be placed in the

OPTICAL DIFFRACTION PATTERNS

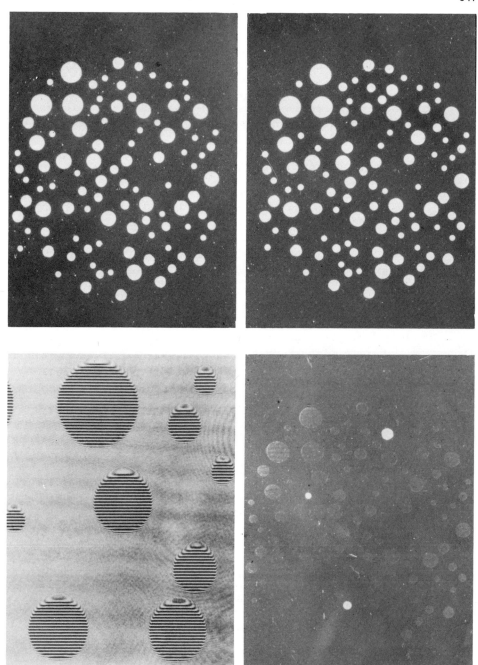

Figure 17. Example of optical subtraction using image plane holographic filtering (after Monahan et al 1970).

optical processing system. Only in very special cases can reflected coherent light be used. Hence an incoherent to coherent converter is required which will take the two-dimensional incoherent signal directly into the system and convert it into a two-dimensional coherent signal for processing. Techniques are now being developed that will allow this problem to be solved. These methods include the use of electro-optical devices such as the bismuth silicon oxide crystals (Vohl et al 1973, Iwasa and Feinleib 1974); the ruticon family of elastomeric devices developed by Sheridan 1972; the electric beam addressed potassium dihydrogen phosphate crystal methods (see, e.g., Casasent 1974). There are many other contenders for either electronically, acoustically, or optically addressed devices which are under consideration. Other electronically addressed methods include liquid crystals, thermoplastic membranes, and transparent ferroelectric ceramic materials (PLZT). Other optically addressed methods include liquid crystals, PLZT, photothermoplastics, photodichroics, electro-optic crystals like lithium niobate, etc.

Many of these materials can be used not only as an input device to coherent processes but also as a device for generating the required filter.

The details of the construction and properties of these devices is beyond the scope of this short article. However, the principles of their use can be readily understood. Take, for example, the bismuth silicon oxide device usually called a PROM (Pockels readout optical modulator). The PROM can be successfully used as an incoherent to coherent converter by writing onto the device incoherently with blue light, and reading out coherently with red light. The readout signal is then the input to a coherent optical processor.

The filter in the coherent optical processor can also be generated in one of the devices described above. We will continue with the PROM example, however; a typical system suggested and used by Iwasa and Feinleib (1974) is shown in Figure 18. The lower laser is a Krypton laser operating at 476nm that is used to write the required filter on the PROM device located in the Fourier plane of the input. The upper laser and its associated optical system comprise the coherent optical processing system.

Clearly similar configurations can be used for the other devices mentioned above. This work is only in its infancy and we should expect to see significant progress in coming years.

HYBRID SYSTEMS

If we assume that progress will continue in the directions discussed above, then there is still one very important question: what is the role of optical processing with relationship to digital

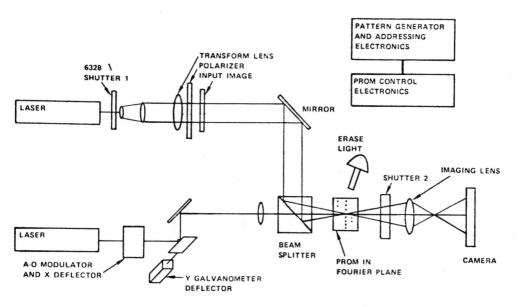

Figure 18. Basic computer controlled Fourier plane filter system using the PROM as the filter (after Iwasa and Feinleib 1974).

processing? Both methods have an independent role to play which initially depends upon the nature of the input. For recorded pictorial information an optical system should be considered first. If the information is a video signal, then digital methods are most appropriate. However, the most useful system will probably be a hybrid system that makes use of both optical and digital methods. Figure 19 shows an example of such a system used by Casasent (1974) (see also his paper in these proceedings for a more detailed discussion). The input is an electrically addressed light modulator (EALM); an optically addressed light modulator is shown in the filter plane produced by the lens L_1. A processed output is produced by lens L_2 onto the vidicon. The detected output is then further processed to produce the final system output.

CONCLUSIONS

It has only been possible in this brief review to discuss the fundamental ideas and some of the major steps forward that have been made in the science and technology of coherent optical data processing. Sufficient progress has now been made that we may reasonably

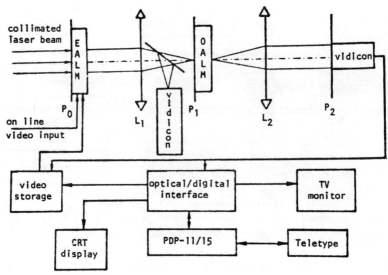

Figure 19. Hybrid optical-digital processor system block diagram (after Casasent 1974).

foresee a useful role for optical methods developing over the next few years. I hope that this paper provides a sufficient introduction to the material to follow.

Acknowledgment

I would like to thank the Society of Photo-Optical Instrumentation Engineers for permission to use portions of a recent review article. (Coherent Optical Processing - a tutorial review) that I prepared that was very recently published in the proceedings of a seminar. Coherent Optical Processing (Vol. 52 SPIE Publication).

REFERENCES

Abbe, E. (1873), Archiv Mikroskopische Anat. 9, 413.
Beissner, R.E., R.L. Bond, W.W. Bradshaw, and W.L. Anderson, J.O.S.A. 60, 741A, 1970.
Becherer, R. and Thompson, B.J. (1965), Unpublished results (see Thompson 1970, 1972).
Bragg, W.L.(1939), Nature, London 143, 678.
Bragg, W.L. (1942), Nature, London 149, 470.
Bromley, K., Monahan, M.A., Bryant, J.F., and Thompson, B.J. (1971). Applied Optics 10, 174.
Casasent, D. (1974), Optical Engineering 13, 228.
Chaudhuri, B. and Hargreaves, A. (1956), Acta Crystallogr. 9, 793.
Cheatham, T.P. and Kohlenberg, A. (1954), "IRE Convention Record IV, 6.

Considine, P. (1963), Unpublished results.
Considine, P., Parrent, G.B., and Thompson, B.J. (1963), Unpublished results.
Ditchburn, R.W. (1952), "Light", p. 171, Blackie.
Duffieux, P.M. (1946), "L'Integrale de Fourier et ses Applications a l'optique" Chez L'Auteur, Faculte de Sciences, Universite de Besancon.
Elias, P., Grey, D.S., and Robinson, D.Z. (1952), J. Opt. Soc. Amer. $\underline{42}$, 127.
Foucault, L. (1859), Ann. Obs. Imp., Paris $\underline{5}$, 197.
Grinberg, J. and Jacobson A. (1975) Optical Eng. May/June
Hanson, A.W. (1952), Nature $\underline{170}$, 58.
Harburn, G. and Taylor, C.A. (1961), Proc. Roy. Soc. A $\underline{264}$, 339.
Harburn, C. (1972) Chapter 6 in Optical Transforms.
Hecksher, H. and Thompson, B.J. (1964), Unpublished results.
Iwasa, S. and Feinleib, J. (1974), Optical Engineering $\underline{13}$, 235.
Kermisch, D. and Roetling, P.G. (1975) Jour. Opt. Soc. Amer. $\underline{65}$, 716.
Leith, E.N. and Upatnieks, J. (1962), Jour. Opt. Soc. Amer. $\underline{52}$, 1123.
Leith, E.N., Upatnieks, J., and Vander Lugt, A. (1965), Jour. Opt. Soc. Amer. $\underline{55}$, 595.
Michelson, A.A. (1927), Studies in Optics.
Maréchal, A. and Croce, P. (1953), Compt. Rendu. $\underline{237}$, 706.
Monahan, M.A., Bromley, K., Bryant, J.F., and Thompson, B.J. (1970), AGARD Conference Proceedings No. $\underline{50}$, 18.
Nisenson, P. and Iwasa, S. (1972), Applied Optics $\underline{11}$, 2760.
O'Neill, E.L. (1956), IRE Trans on Information Theory IT-2, 56.
Parrent, G.B. and Thompson, B.J. (1964). U.S. Patent No. 3,320,852
Parrent, G.B. and Thompson, B.J. (1969), Physical Optics Notebook (SPIE Publications).
Porter, A.B. (1906), Phil. Mag. $\underline{11}$, 154.
Rhodes, J.E. (1953), Amer. J. Phys. $\underline{21}$, 337.
Sheridan, N.K. (1972), IEEE Trans. ED-19, 1003.
Sprague, R. and Thompson, B.J. (1972), Applied Optics $\underline{11}$, 1469.
Stroke, G.W. (1965), Appl. Phys. Letters $\underline{6}$, 201.
Stroke, G.W. and Zech (1967), Physics Letters $\underline{25A}$, 89.
Stroke, G.W. (1969), Optica Acta $\underline{16}$, 401.
Taylor, C.A. and Lipson, H.S. (1964) Optival Transforms (Bell)
Thompson, B.J. (1959), Thesis. University of Manchester.
Thompson, B.J. (1966), Jour. Opt. Soc. Amer. $\underline{56}$, 1157.
Thompson, B.J. (1970), Proc. Electro-Optical Systems Design Conference.
Thompson, B.J. (1972), Optical Transforms, Ed. H.S. Lipson, Academic Press, Chapter $\underline{2}$, 27-69.
Thompson, B.J. (1972), "Optical Data Processing", Optical Transforms, Ed. H.S. Lipson, Academic Press, Chapter $\underline{8}$, 267-298.
Thompson, B.J. (1975), Volume 52 (SPIE Publication) 1-22.
Töpler, A. (1867), Ann. d. Physik u Chemic $\underline{131}$, 33.
Tsujiuchi, J. (1960a), Optica Acta $\underline{7}$, 243.

Tsujiuchi, J. (1960b), Optica Acta 7, 385.
Tsujiuchi, J. (1961), Optica Acta 8, 161.
Tsujiuchi, J. (1963), Progress in Optics, Vol. II (Ed. E. Wolf), North Holland Publishing Company, 130.
Upatnieks, J., Vander Lugt, A., and Leith, E.N. (1966), Applied Optics 5, 589.
Vander Lugt, A. (1964), IEEE Trans Information Theory, IT-10, 2, 139.
Vohl, P., Nisenson, P., and Oliver, D.S. (1973), IEEE Trans ED-20, 1032.
Ward, J.H. (1968), Private Communications
Zernike, F. (1935) Z. Tech, Phys. 16, 454.

QUESTIONS AND COMMENTS

1. Dr. P. Tverdokhleb (question to Dr. B. Thompson)
I am interested in the experiments with a car. They illustrate a possibility of detecting changes on pictures. But I have not understood quite clear in what way it can be realized in an optical system.

Dr. B. Thompson. At the same place of the photomedium holograms of the initial images were successively recorded. When performing the second recording the phase of the reference wave was shifted by π. Then the recorded images were reconstructed simultaneously.

HOLOGRAPHIC MEMORIES

A. Vander Lugt

Electro-Optics Operation
Harris Electronic Systems Division
P. O. Box 37
Melbourne, Florida 32901

INTRODUCTION

Since Leith and Upatnieks[1] showed how holographically stored information can be reconstructed separately from the direct and conjugate image beams, holography has provided potential solutions to many problems. Through continued development of electro-optic devices needed to implement practical systems, considerable progress has been made in realizing useful hardware for data storage and retrieval systems.

Holographic techniques offer several advantages over conventional techniques for archivally storing and retrieving both analog and digital information. The primary advantage stems from the increase in packing density that can be achieved holographically, particularly when the information is recorded as a Fourier transform hologram; the packing density is then maximized (except for object fields containing fine detail) for both analog and digital information.[2] Furthermore, if the information is recorded as a Fourier transform hologram, other advantages automatically accrue: (1) each picture element or bit contributes to every part of the hologram, providing a natural encoding process which causes the recorded information to be retrieved with greater freedom from the effects of imperfections such as dust, scratches or blemishes; (2) the shift-invarient theorem for Fourier transforms means that the readout beam does not have to track perfectly the recorded information so that alignment tolerances are substantially independent of packing density; (3) both the readin and readout can be arranged for low packing density, while retaining high packing density on the storage medium, which again relieves tolerances on alignment for input/output devices; (4) very high recording rates can be achieved for storing digital information by using the parallel, high-speed nature of optical systems.

In this paper we discuss several holographic storage and retrieval systems. Although each system is intended to satisfy a particular need, considerable variations can be envisioned to meet other requirements. These systems are (1) an analog storage and retrieval system for graphical information such as maps, charts, and engineering drawings, (2) a digital storage and retrieval system for rapid, random access to blocks of data without moving parts, (3) a mass memory system having the capability for storing both analog and digital information, and (4) a high-speed digital system in which the data is recorded on a roll of film.

HOLOGRAPHIC STORAGE AND RETRIEVAL OF ANALOG INFORMATION

Large graphics that contain fine detail (such as maps, charts and engineering drawings) are particularly difficult to store at high densities using conventional micrographic techniques. For example, an engineering drawing that is 24" x 36" may contain detail as small as 0.004 inches; if this graphic is reduced 60X using conventional techniques so that the micro-image is 10 mm x 15 mm, the resolution required at the storage medium is 1.7 microns. Not only is this resolution difficult to preserve, it also leads to severe tolerances on depth of focus both in the recording and the retrieving processes and on lateral motion in retrieving.

Holographic techniques offer an attractive and practical solution to these problems. Suppose we first reduce the graphic approximately 20X onto 35 mm film, using conventional micrographic techniques (see Figure 1a). The resolution for this intermediate image is then approximately 5µm. Other initial reduction ratios are possible; the ratio does not impact the ultimate packing density that can be achieved holographically. One advantage of this initial photoreduction is that smaller diameter

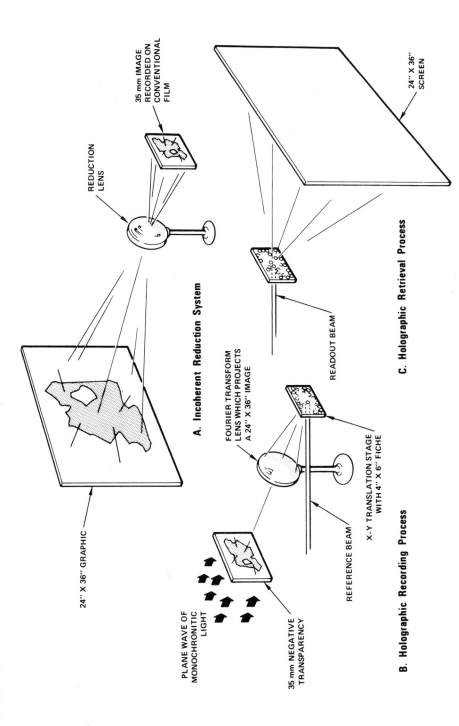

Figure 1. Holographic Storage and Retrieval of Analog Information

optics can be used in the subsequent holographic systems; a second advantage is that most graphics originate as black lines on a white background so that a negative image provides an optimum redundancy removal process, i.e., the information (black lines) now appears as transparent regions within an opaque surrounding a third advantage is that the negative transparency allows for recording of specular instead of diffuse information, thus eliminating the speckle phenomenon in the resultant display.

The negative transparency is then placed in a holographic recording system, as shown in Figure 1b, and illuminated by a plane wave of monochromatic light. Light diffracted by the transparency is collected by a Fourier transform lens which also projects the transparency onto a screen at the original size. If this Fourier transform lens is the same as the one used in the initial photo-reduction step, and if the overall magnification is 1:1, the resultant image is completely free of distortion. This results from the basic theory of optical design from which it can be shown that telecentric system is automatically free from distortion.

Suppose that we now place a photographic film at the back focal plane of the lens as shown in Figure 1b, and capture the light intensity at that plane, having added an off-axis reference beam. The reference beam interferes with the signal beam so that both the amplitude and phase of the signal beam wavefront can be recorded as a non-negative (intensity) function on photographic film. After a set of exposures are made on one 4" x 6" film chip (one hologram per graphic), the fiche can be developed in the usual way and placed in a storage and retrieval unit.

Figure 1c shows the basic retrieval process. This system is particularly simple since no lenses are required to reconstruct the image of the graphic on the screen. A hologram is selected by means of an x-y mechanism and illuminated by the readout beam (a replica of the reference beam). The signal beam wavefront is thereby released and it propagates from the hologram in exactly the same way as though it had never been recorded. This wavefront includes the distortion compensation term that ensures that the reconstructed image is free from distortion.

An alternative method for recording and retrieving information, using a two-lens holographic system, is suggested in Reference 2. It provides improved performance and simplifies the design of the lenses used.

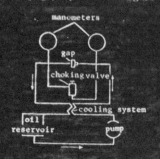
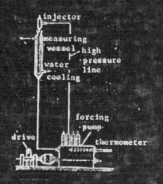

Figure 2. Reconstruction of an "A" Size Drawing

The primary factor that determines the size of the hologram is the minimum detail that must be recorded and retrieved. If we let the size of the minimum resolution element be d, the diameter of the hologram is given by

$$\bar{d} = 2\lambda F/d$$

where $\bar{d}$ is the diameter of the hologram, λ is the wavelength of light, and F is the focal length of the Fourier transform lens. If d = 5μm (corresponding to 100μm detail in the input graphic) and if λ = 514.5 nm, the diameter of the hologram is $\bar{d}$ = 0.2F. Since we want the focal length of the lens to exceed the diagonal of the input transparency, we select F = 50 mm which gives a hologram diameter of $\bar{d}$ = 10 mm. We see, therefore, that a hologram diameter of 10 mm should be more than adequate from a resolution viewpoint. We also see that the initial photoreduction ratio does not influence the hologram size; if a 10X system is used, both F and d increase by a factor of two so that $\bar{d}$ remains constant. A hologram diameter of 10 mm gives an equivalent reduction ratio of 60X for a 24" x 36" graphic.

Compared to conventional micrographic techniques, the holographic storage and retrieval system has several features:

1. Each resolution element (pixel) of the graphic contributes the light to every part of the hologram. This natural encoding phenomenon gives immunity to dust, scratches or blemishes that could obscure significant areas of micrographic images.

2. The axial position tolerances of the hologram, both recording and retrieval, are much greater. Instead of maintaining a depth of focus adequate to record a 1.7μm spot (equivalent to a f/1.7 lens system) the hologram need be positioned to an axial tolerance equivalent to an f/100 lens. Since the depth of focus is proportional to the square of the f/#, the holographic system has a tolerance that is 3,500 times less stringent (for an equivalent reduction ratio) than a conventional micrographic system.

3. An even larger reduction on the lateral positional tolerance of the hologram is possible. In conventional systems operating at a magnification of 60X, a 0.05 inch movement of the film results in a 3 inch image movement. Such systems are, therefore, sensitive to vibrations and thermal variations, and require accurate x-y selection mechanisms. A hologram recorded

in the Fourier transform plane can be moved any distance (as long as it is fully illuminated) without introducing any image motion. Although it is usually better to record in a near Fourier transform plane to reduce the dynamic range required of the recording medium, even then the sensitivity to motion is significantly less. The distance that the image can move is, at worst, the same distance that the hologram moves. But in any practical system using near Fourier transform holograms, the image motion will be approximately 1 percent of the hologram motion. Thus, if the hologram moves 0.05 inch, the image will not move more than 0.0005 inch -- a factor of 6,000 times better compared to a conventional system at the same magnification.

Figure 2 shows a reconstruction of an A-size drawing that had been stored in a hologram having a 4 mm diameter. As can be seen, resolution has been preserved throughout the storage and retrieval process. This technique can also be applied to graphics or documents having grey scale; hence, it is a holographic technique for storing analog information.

HOLOGRAPHIC STORAGE AND RETRIEVAL OF DIGITAL INFORMATION

Many of the advantages of holography cited in the previous section can also be used in systems used to store digital information. Because of its unique properties, it is not surprising that attempts have been made to apply holography to a broad range of memory and storage hierarchy. Activity has ranged from developing small-capacity, high-speed memories to large-capacity, read-only storage in the multi-terabit range. Additionally, significant activity has been directed toward solving the highly specialized problems associated with ultra-high data rate recorders and reproducers. Memory systems are now, and will continue to be, the highest single cost item in the computer hardware structure. This, at least in part, accounts for the intensive research activity in optical alternatives to computer memory and storage.

Although research continues across the broad spectrum of memory hierarchy, some strong indicators point to very specific areas where the technology has a reasonable chance of success. Before we consider the basic characteristics of holographic memories, we identify the targets at which holographic memories have been aimed. Perhaps the two most widely used performance measures for memories are capacity and access time. Clearly there are many other factors such as transfer rate, size, power consumption, interface ease, reliability and reproducibility which may play equally important roles in characterizing memory performance. Similarly, memory cost or (more commonly) cost per bit, forms one of the important criteria for memory selection. For purposes of this discussion, capacity and access time will be sufficient factors if we remember that even the ultimate in memory performance is unacceptable if the eventual costs are not consistent with what the marketplace can afford.

Figure 3 shows the present state of the art in so-called conventional memory technology in terms of capacity and access time. The technology ranges from the relatively small but fast semiconductor memory through moving head disc memory to the larger and slower bulk storage devices such as magnetic tape. Clearly most memory and storage technology is confined to magnetic phenomena. The exceptions to the magnetic dominance have been at the low-capacity slow access end with the IBM 1360 and Precision Instrument Model 190 bit-by-bit optical technology.

The trends in memory and storage technology indicate a gradual (although sometimes rapid) trend up and to the left, i.e., toward larger and faster devices. Consequently, the aim of early holographic memory researchers was toward those areas where the payoff would be largest, i.e., $10^8 - 10^{10}$ bit capacities with $1 - 10\mu$sec access time for disc replacement. Similarly, the promise of extremely high data packing density afforded by holographic encoding encouraged research activity at the upper end of the spectrum to achieve terabit capacity.

A logical distinction between various holographic memories can be made by first considering those which are truly read/write memories and are aimed at existing mainframe and peripheral technology and those which are directed at mass storage applications. While the basic technology in both applications is somewhat similar, the approaches to solutions require emphasis on different components.

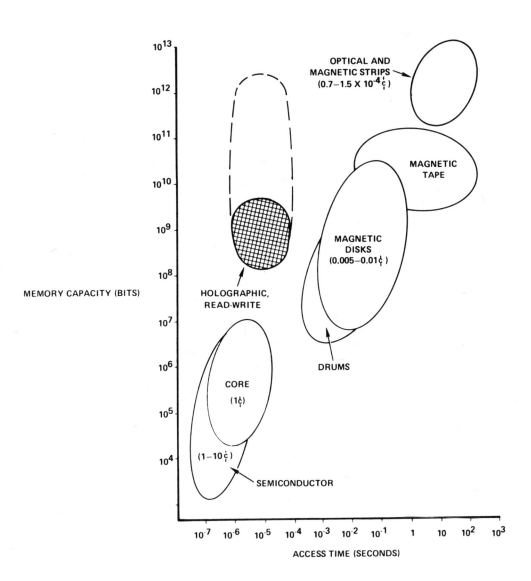

Figure 3. Current Memory Technology

Read/Write Holographic Memories

The major effort to date on read/write holographic memories has addressed capacities between 10^8 and 10^{10} bits with access times measured in microseconds. Several domestic companies, including Harris Corporation, Bell Labs, IBM, and RCA, as well as several foreign laboratories at Siemens, Nippon, Hitachi and Thomson-CSF, have either developed breadboard memories or are actively pursuing development of major components which are required in holographic memories. It is beyond the scope of this paper to discuss in detail the work of each laboratory and to dwell on the progress made in developing individual components. It is sufficient to say that, although progress on each holographic memory component has been significant over the past decade, a truly viable cost competitive memory has not emerged from the breadboard stage into the marketplace. Because of the interactive nature of all holographic memory components, a major advance in one area may produce only minor improvement in system performance.

Figure 4 shows some of the basic elements of a block oriented read/write memory. In many respects it is similar to the system shown in Fibure 1b. The differences are that a page composer which converts an electrical bit stream into a two-dimensional spatial signal is used instead of a transparency to modulate a laser beam; the hologram array is stationary and the information from successive page composer inputs is deflected to any of several recording positions by a fast, random-access acousto-optic device; the output is a photodetector array instead of a viewing screen; the recording medium has a read/write/erase instead of an archival read only capability.

The recording operation is basically the same as described before except that both the signal and reference beams are directed toward a storage position in the hologram array by an acousto-optic beam deflector which is not shown. Several trade-offs can be established to determine the optimum number of bits to be stored in each hologram and, based on fundamental optical considerations, the maximum capacity can be found.[3] For a 10^8 bit capacity system, a convenient number of bits per hologram requires a 128 x 128 element page composer and a 78 x 78 element hologram array. The memory capacity cannot be increased much beyond 10^9 bits without encountering rather

HOLOGRAPHIC MEMORIES

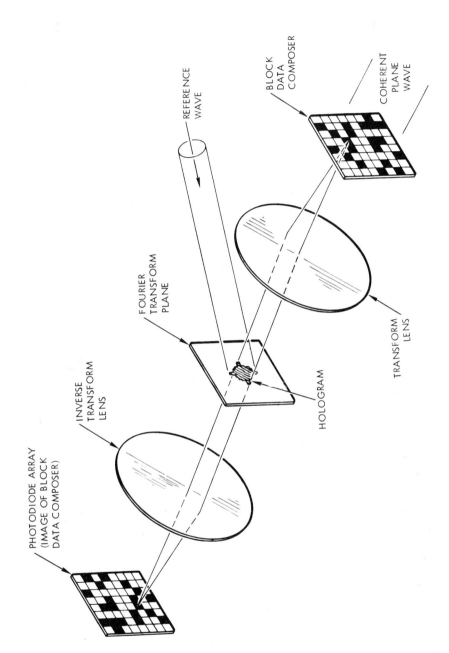

Figure 4. Basic Elements of a Block Oriented Memory

severe lens design problems. To achieve a larger capacity memory, therefore, requires either a reversible volume storage medium or a moving recording medium.

Perhaps the two components which have received the greatest attention are the input page composer and the holographic recording media. In both cases, the performance limitation has been dictated by the availability of suitable materials. The page composer should contain between 4000 and one million elements. Various materials have been or are being considered, including PLZT, liquid crystals, thin deformable membrane mirror arrays and cadmium sulfide. All currently suffer from one or more shortcomings including uniformity, speed, contrast ratio and stability. The holographic recording materials have also received considerable attention and all candidates fall short of the desired properties of high efficiency, high sensitivity and long lifetime. Significant progress has been made in developing beam deflectors and photodetector arrays, stimulated in part by the requirements imposed on these devices by holographic memory researchers. The performance of beam deflectors has nearly doubled in the past ten years and sophisticated multielement two dimensional photodetector arrays were essentially unavailable ten years ago. Research in all component areas is continuing. We can expect significant breakthroughs only by application of new materials or through better understanding and perfection of existing materials.

Until recently, heavy emphasis was placed on preserving high speed access with no moving parts while increasing capacity. Clearly the target was disc replacement. Analyses of the constraints imposed on common optical components such as lenses as well as a better appreciation of physical limitations imposed on the electro-optical components has led most investigators to revise their predictions of ultimate read/write holographic memory performance, regardless of cost considerations. The capacity of nonmechanical, practical read/write holographic memory with microsecond access time is likely to fall between 10^8 and 10^9 bits, not up to 10^{12} bits as was believed earlier. In any case, because of costly electro-optical components, this type of memory will be characterized by high cost per bit. Only in highly specialized applications where technical performance plays an overriding part will we see this memory used. Even so, it is not likely to emerge from the research laboratory before the end of this decade.

HOLOGRAPHIC MEMORIES

Read-Only Memories

Read-only holographic memories typically use film as the recording media. Once exposed, the film record is removed from the recorder, developed by normal techniques, and placed in a holding area until data retrieval is required. If any portion of the recorded data must be changed or updated, the entire record must be re-recorded and replaced within the memory. Read-only memories, therefore, are best suited for archival, non-dynamic memory applications or applications where updating is relatively infrequent.

Other than the recording media, the holographic exposure and data readout processes are similar to those used in a read/write memory. Similar devices are required to implement a read-only memory as are required to implement a read/write memory; hence, read-only memories are therefore constrained by similar device limitations. Possibly the most critical device limitation in read-only memory implementation has been the page composer. Since a one dimensional (instead of a two dimensional page composer) is typically required, device capability in this area has recently been improved enough to allow system applications.

One way to overcome the page composer limitation in holographic recording is to use a synthetic hologram approach. In this approach, a film intensity function is calculated by a special purpose digital processor and scanned onto the film by a scanning device. The resulting film exposure has nearly the same reconstruction properties as does an interferometrically generated hologram.

The recent advances in materials and components have allowed production of a few prototype holographic memories. For example, let us discuss some specific hardware systems being developed by Harris Corporation, Electronic Systems Division. Synthetic holography has been successfully applied to the recording and storage of digital data in the Human Read/Machine Read (HRMR) System developed for Rome Air Development Center. A research prototype, shown in Figure 5, was delivered in May 1973 and an engineering prototype is currently under development.

The HRMR System addresses the document storage, retrieval and dissemination problem which is impacting both government

Figure 5. Human Read/Machine Read Research Prototype

and industrial complexes having large document data bases. The HRMR concept is based upon annotating a standard microfiche with the digital equivalent of the associated images. Optical readout of the digital data directly from the microfiche facilitates storage, retrieval and dissemination of data to both local and remote locations.

A direct extension of the concept is the full utilization of the microfiche film chip for digital data recording. Thirty megabits of user data per film chip is presently being realized at a packing density exceeding one megabit per square inch. Since this packing density is significantly below theoretical limitations, considerable improvement can be anticipated as components and techniques are further refined.

Utilization of holography as the digital data recording technique in the HRMR System provides an inherent immunity to dust, scratches, and film imperfections associated with practical hardware which is capable of functioning in an operational environment. Only normal microfilm storage environmental conditions are required. The recorded data is archival and optical readout of data is nondestructructive. This results in a virtually permanent record and contrasts with magnetic media which suffers from signal loss and deterioration due to readout and long term storage. Of further benefit, the positional invariance property of holography facilitates readout and allows relatively simple and economical hardware configurations.

Microfiche generation in the HRMR System is accomplished by means of a laser recorder which scans onto a film chip both the human readable images and the synthetically generated machine readable holograms containing digital data. Digital data is recorded sequentially onto the fiche in 500 kilobit blocks and at data rates compatible with magnetic tape drives. Fiche are automatically developed and all digital data is verified by means of parity bits which are appended during the recording process.

Since the HRMR System storage media is oriented around the standard microfiche film chip, there is a maximum compatibility with commercially available microfiche handling equipment. It has been a straightforward development to configure a medium scale microfiche storage and retrieval device capable of handling approximately 7000 microfiche. Total digital store of this module is $2(10^{11})$ bits.

In the present HRMR System configuration, the mass memory is on-line to a PDP-11/45 computer. Random fetch of any 500 kilobit data block stored within the memory is provided with a maximum access time of 3 seconds and with a maximum access time of less than 15 seconds. Transfer rates are compatible with DEC Unibus cycle times, but can be tailored to any host computer's channel characteristics and data absorption rates.

Because the holographic technique used in the HRMR mass memory is read-only, utilization of the system is projected to be oriented primarily toward archival data store applications in which the data placed in the memory is non-dynamic. Such data bases are quite common in both governmental and industrial organizations and are typically characterized by large magnetic tape libraries. A magnetic tape, having typical block sizes and utilizations, can be holographically recorded on one to two fiche. The 7000 fiche storage capacity of the holographic memory provides on-line access to approximately 3500 magnetic tapes with access times a fraction of manual retrieval, mount and read times. Based upon user requirements, additional holographic memory modules can be added to increase this capacity at least an order of magnitude.

Possibly the most significant characteristic associated with the HRMR System's holographic memory is simplicity of operation. The HRMR configuration has integrated into a mini-computer system, a $2(10^{11})$ bit memory and has made this data store available at a storage cost of approximately 2.5×10^{-6} cents per bit. In contrast to many other conventional mass memory approaches, which record sequential blocks and must retrieve blocks sequentially, the HRMR approach provides random access to data blocks without a sacrifice in overall access time.

While the use of the synthetic holography for storage and retrieval of digital data on microfiche provides a solution to document-oriented mass memory requirements, different recording techniques and physical record formats are more suitable to other types of applications. For example, the storage and retrieval of digital data in very large data records at extremely fast recording and readout data rates can be best handled using roll film and interferometric holography.

HOLOGRAPHIC MEMORIES

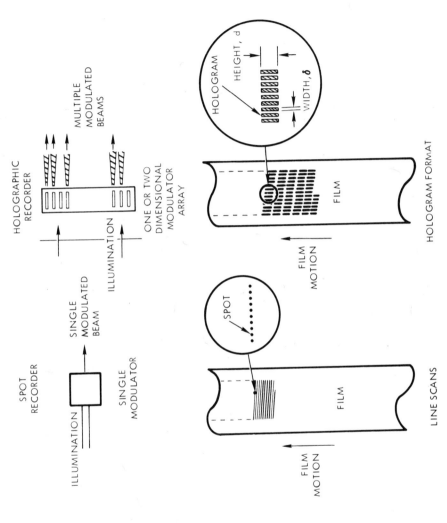

Figure 6. Data Formats for a Spot Recorder and a Holographic Recorder

High-Speed Digital Recorder

Two existing techniques for recording digital data at high speeds are multi-track longitudinal magnetic recorders and laser beam recorders which record binary spots directly onto film. In each case, as the data rate increases the packing density must increase to maintain a reasonable tape or film velocity. But as the packing density increases the tolerance on the lateral position, axial position and skew of the recording medium decreases significantly, often to the point where the data can be recovered only by the machine on which it was recorded.

Again, holographic techniques offer fundamental features for solving these problems. As we showed before, a high packing density can be achieved with relaxed tolerances on the position and angle of the recording medium. High speed recording can be obtained by using the parallel nature of the optical system. These features are noted in Figure 6 which compares the recorded data format of a spot recorder and a holographic recorder. In a spot recorder a single beam of light is temporally modulated and scanned onto film, generally by means of a rapidly rotating multi-facet mirror, to produce scan lines of binary data as the film moves. In a holographic recorder a single beam of light is both temporally and spatially modulated by a page composer to produce multiple beams. These beams are then recorded holographically on film by means of a multi-facet mirror. Because each hologram may contain up to 128 bits, the angular velocity of this mirror is lower, by a factor of 128, than that of a spot recorder for the same data rate.

Tolerances on readout are reduced because a typical hologram height is of the order of 800µm whereas a scan line is typically 5-10µm wide. Therefore, it is considerably easier to accurately address a row of holograms than a scan line. Tolerance to skew is similarly decreased because a small angular change in the film upon readout represents a much smaller change in overlap in the readout beam. Lateral and focal tolerances are also eased as described before.

Figure 7 shows the basic elements of a high speed digital recorder.[4] The recording process begins with the demultiplexing of the incoming data stream into 128 channels of time ordered data. Sufficient buffering (32 bits per channel) is

HOLOGRAPHIC MEMORIES

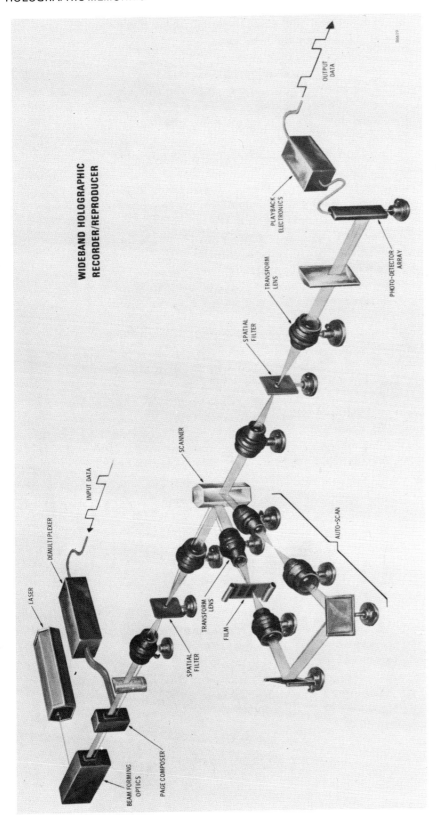

Figure 7. Basic Elements of a High Speed Digital Recorder

provided to bridge over the small deadtime of the scanner, and to absorb small fluctuations in data rate.

The parallel electronic information is converted to parallel optical signals by the acousto-optic page composer as shown in Figure 6. The presence or absence of an acoustic wave packet indicates a "logic one" or a "logic zero" respectively. After the acoustic waves have propagated into the optical aperture, the laser beam is gated on by a single acoustic modulator located immediately following the laser. This beam propagates through the beam forming optics where both reference and signal beam illuminations are produced. The signal beam is appropriately shaped to uniformly illuminate the 128 parallel acoustic beams and the reference beam. After passing through the reference and signal beam acoustic modulators, the two beams are brought to focus at a filtering slit. The slit is imaged onto the film after being reflected from a spinner facet. The rotation of the spinner generates a master scan on the moving film as shown in Figure 6.

The first hologram is recorded immediately after the data handling logic receives a data synchronization pulse from the control electronics. Data records are continually recorded in holograms across the film until a counter in the control logic reaches a fixed count. Data records are not clocked out of the data generator until the next synchronization signal is received. This process is then repeated and the next hologram row is recorded.

After the film is developed, this system can be used as a reader by illuminating only the reference beam. The same scanner is used to cause the recovered data pattern from each hologram to fall on a photodetector array. Note that, since the system is symmetrical about the axis of the scanner, the photodetector array is in the image plane of the acousto-optic page composer. Thus, 128 bits are detected in parallel and multiplexed into a replica of the input data stream.

If the readout rate is much lower than the recording rate, the recording and reading functions may be separated. The generous tolerance allowances afforded by the holographic recording technique implies significant reader simplicity relative to spot or magnetic recorders operating at high packing densities. Typical recording rates are $400(10^6)$ bits/sec and readout rates are $40(10^6)$ bits/sec.

In general, one-dimensional holograms lead to simpler system implementations because the film provides motion of the recording medium relative to the hologram recording plane. Short pulse lasers are, therefore, not required and the data can be read out with linear photodetector arrays. Roberts has described an alternative technique, using two-dimensional holograms, for recording data at high rates.[5] This technique does not require the use of a spinning mirror but does require very high-speed, two-dimensional photodetector arrays.

CONCLUSIONS

During the past several years we have witnessed a considerable effort which was and still is being undertaken by many research laboratories to apply the principles of holography to a broad spectrum of memory and storage applications. The research is directed toward both the intermediate-capacity, fast access time read/write memory market as well as the large-capacity, longer access time read-only storage devices. Effort is also continuing in the specialized area of ultra-high speed transfer of data into and out of large intermediary bulk stores.

Read/write memories have not yet emerged from the research laboratory into the commercial marketplace. Efforts have been hampered primarily by the unavailability of suitable materials which are needed to configure several key memory components. Even if we assume that material and other technologal problems are overcome, the prospects that holographic memories will seriously challenge other existing and emerging technologies before the end of this decade, indeed if ever, is unlikely. The inertia of magnetic technology coupled with remarkable yearly improvements in packing density, access time and transfer rates presents a formidable challenge to those who desire to penetrate that particular segment of the market.

The prospects for read-only holographic memories which have multi-microsecond access time and 10^8 to 10^9 bit capacity appear to be better because problems associated with high-speed page composers and reuseable storage media are obviated. Unfortunately, the read-only property will limit its usefulness to special applications where data volatility, extreme environments and data security overshadow cost considerations.

The prospects for application of holographic techniques to read-only bulk storage appear to be much better. Systems with capacity between 10^{11} and 10^{13} bits and with multi-second access time are currently being built as engineering developmental units. At these capacities, costly electro-optical components can be justified. On a more modest scale, 10^7 bit capacity, 1.5 second access time read-only storage units are already commercially available for application to the point-of-sale credit card verification problem.

REFERENCES

1. E. N. Leith and J. Upatnieks, JOSA 53, 1377 (1963); JOSA 54, 1295 (1964).

2. A. Vander Lugt, "Packing Density for Holographic Systems," Appl. Opt. 14, 1081 (1975).

3. A. Vander Lugt, Appl. Opt. 12, 1675 (1973).

4. A. M. Bardos, Appl. Opt. 13, 832 (1974).

5. H. N. Roberts, Appl. Opt. 13, 841 (1974).

QUESTIONS AND COMMENTS

Dr. P. Tverdokhleb (questions to Dr. A. Vander Lugt)
a) Have you applied randomization of elements of the data pages formed and recorded with the help of a multichannel acousto-optical modulator?

Dr. A. Vander Lugt. Yes, we have.

b) What way was the randomization performed in?

Dr. A. Vander Lugt. The elements of the page formed in each channel of the modulator were given a squared phase raid.

c) Would you tell us the type of the film used in your wide-band holographic recorder?

Dr. A. Vander Lugt. We use the Agfa-Gevaert 10E56 film.

BIOGRAPHIES OF CONTRIBUTORS *

DAVID CASASENT

David Casasent received his Ph. D. from the University of Illinois in Electrical Engineering. While there he was employed in the Digital Computer Laboratory where he started his work in optical data processing. He is presently an Associate Professor of Electrical Engineering at Carnegie-Mellon University where he is head of the Information Processing Group. His main areas of research are optical processing, real-time devices, hybrid optical/digital processors, and image and signal processing.

He is also a consultant to the U. S. Army Missile Command, Los Alamos Scientific Labs, Digital Equipment Corporation, MacDonnell Douglas, Singer Librascope, and the Office of Naval Research.

He is president of the Pittsburgh Section of the Optical Society of America, past President and on the Board of Directors of the Pittsburgh IEEE Electron Devices Group plus numerous other Societies.

He has guest edited several Journal Special Issues on Optical Computing (Society for Information Display and IEEE Transactions on Computers). He has been on the organizing committee and Technical Program Chairman of the 1974 and 1975 International Optical Computing Conferences plus OSA, EOSD and SPIE Conferences. He is the author of two textbooks on Linear Electronic Circuits and Digital Electronics and Chapters of 3 books on optics plus over 40 technical papers.

V. A. FYODOROV

In 1956 Dr. Fyodorov graduated from the Tomsk Polytechnical Institute (the department of radio engineering). Since 1960 he has been working as a researcher at the Siberian Branch of the USSR Academy of Sciences (Novosibirsk) and since 1964 at the Institute of Automatics and Electrometry. In 1968 he was given a doctor degree. He has been involved in the investigations on the development of image transformers since 1966 giving the main attention to the development of optical filters for high spatial frequencies (the formation of two-dimensional derivatives and Laplacian, the realization of isotopic Hilbert transforms). In this field he has more than 20 scientific publications and inventions.

I. S. GIBIN

I. S. Ghibin graduated from the Novosibirsk Institute of Electrical Engineering (the department of automatics and computing) in 1968. Since 1969 he has been working as a researcher at the Institute of Automatics and Electrometry of the Siberian Branch of the USSR Academy of Sciences (Novosibirsk). He is engaged in the development and investigation of holographic memories and optical systems of data processing mainly in the direction of studying the modes of holographic recording and reading, optimization of holographic memories, associative retrieval and spectral analysis of images.

He is the author of 20 scientific publications and inventions.

J. W. GOODMAN

Joseph W. Goodman received the A. B. degree in Engineering and Applied Physics from Harvard University in 1958, and the M. S. and Ph. D. degrees in Electrical Engineering from Stanford University in 1960 and 1963, respectively. He joined the full-time staff at Stanford in 1963 as a Research Associate, was promoted to Assistant Professor in 1967, Associate Professor in 1969, and since 1972 has been Professor of Electrical Engineering. During the academic year 1973-74, Dr. Goodman was a Visiting Professor at the University of Paris-South, in Orsay, France.

Dr. Goodman has served as Director-at-Large of the Optical Society of America (OSA), Chairman of the OSA Technical Group on Information Processing, Holography and Coherence, and currently serves as OSA Topical Editor for Information Processing and Holography. In addition, he served as Chairman of the ad hoc Committee for Optical and Electro-Optical Systems of the Institute of Electrical and Electronics Engineers (I. E. E. E.). He was co-editor of the journal Applied Physics, and currently serves as an associate editor of the journals Optical Engineering and Computer Graphics and Image Processing.

Dr. Goodman is a Fellow of the OSA and a Fellow of the I. E. E. E. In 1971 he received the F. E. Terman award of the American Society for Engineering Education. He is the author of more than fifty technical publications, including the widely-used textbook, Introduction to Fourier Optics.

S. B. GUREVICH

Prof. S. B. Gurevich graduated from the Leningrad State University, Leningrad, USSR, in 1945 and entered post-graduate courses at the same University in 1946. He received his first degree in 1948 and the second degree in 1964. Since 1959 he has been working at the Ioffe Physico-Technical Institute in Leningrad. In 1966 he became the Head of the Laboratory of Optoelectronics and Holography. He is also Vice President of Scientific Council on Holography of the USSR Academy of Sciences. He is the author of four monographs dealing with different problems of physics.

At present his fields of interest are the problems of optical information processing and holography.

M. HALIOUA

Dr. Maurice Halioua is currently Assistant Professor of Electrical Sciences at the State University of New York at Stony Brook where he has been working with Professor George W. Stroke at the Electro-Optical Sciences Laboratory since he immigrated to the United States in 1969. He also obtained both his Ph.D. degrees for his work with Professor Stroke, a Dr. Ing in Physics in 1971 and a Dr. ès Sc. in Physics in 1974, both presented at the University of Paris in Orsay.

Before emigrating to the United States, Dr. Halioua studied at the University of Bordeaux, France, where he obtained his B.Sc. in 1962, and also at the Institute of Optics at the University of Paris where he obtained his Ing. Dipl. (Optics) in 1968.

Since 1972, he has also held the title of Instructor in Medical Biophysics at the Health Sciences Center at Stony Brook, in addition to his position of research associate at the Electro-Optical Sciences Laboratory.

Since 1969, he has been the co-author, with Professor Stroke, in some 20 publications in the field of image deblurring most recently of the article on the "Retrieval of Good Images from Accidentally Blurred Photographs" which appeared in Science in July 1975.

WINSTON E. KOCK

Dr. Kock, E. E. (1932) and M. S. (Physics, 1933), both University of Cincinnati, Ph. D. (Physics, 1934, University of Berlin) attended the Institute for Advanced Study at Princeton and was a Fellow at the Indian Institute of Science at Bangalore in 1936. He has been Director of Electronic Research, Baldwin Piano Company (where he developed the Baldwin electronic organ, Microwave Research Engineer and Director of Acoustics Research at Bell Telephone Laboratories (where he developed several microwave and acoustic lenses, directed the research on the Navy's underwater sound Jezebel-Caesar Project, and headed the group developing the picture-phone), Director of the Bendix Research Laboratories, first Director of the National Aeronautics and Space Administration's Electronics Research Center, Cambridge, Mass. and most recently Vice President and Chief Scientist of the Bendix Corporation. In 1971 he became Consultant to the Corporation and a Visiting Professor and Director of the Herman Schneider Research Laboratory at the University of Cincinnati.

Honors include the Navy's highest civilian award, the Distinguished Public Service Medal, (1964), Honorary Fellowship in the Indian Academy of Sciences (1970), an honorary D. Sc. (U. of Cincinnati, 1952), Eta Kappa Nu's Outstanding Young Electrical Engineer (1938) and Eminent Member Award (1966), fellowship in the Acoustical Society, the Physical Society and the I. E. E. E. and membership in Tau Beta Pi, Sigma Xi and Eta Kappa Nu.

He was chairman of the Professional Group on Audio of the Institute of Radio Engineers in 1954-55 and was a member of the Governing Board of The American Institute of Physics from 1957 to 1963. He is a member of the Board of Roanwell Corporation, and Hadron, Inc., and has been Chairman of the Board of Trustees of Western College for Women, and Board member of Argonne Universities Association and the Atomic Industrial Forum. He is author of five books: SOUND WAVES AND LIGHT WAVES (1965, Doubleday), LASERS AND HOLOGRAPHY (1969, Doubleday), SEEING SOUND (1971, Wiley), RADAR, SONAR, AND HOLOGRAPHY (1973, Academic Press), and ENGINEERING APPLICATIONS OF LASERS AND HOLOGRAPHY (1975, Plenum Press).

BIOGRAPHIES

VOLDEMAR P. KORONKEVICH

In 1950 Dr. Koronkevich graduated from the Leningrad Institute of Precision Mechanics and Optics. He worked at D. I. Mendeleyev State Institute of Metrology in Leningrad. In 1956 he was given a doctor degree for his thesis "Air Dispersion in Visible Spectrum Region". In 1957 he headed the laboratory of Interference Measurements at the Siberian Institute of Metrology (Novosibirsk). Since 1968 he has been heading the laboratory of Coherent Optics of the Institute of Automatics and Electrometry of the Siberian Branch of the USSR Academy of Sciences in Novosibirsk.

He has more than 50 publications on interference measurements, high precision laser interferometers for measuring length and gravity acceleration, laser Doppler systems for determining flow velocities in fluids and gases. He is an author of the book LASER DOPPLER VELOCITY METERS.

ADRIANUS KORPEL

Adrianus Korpel was born in Rotterdam, Holland. He received the M. S. E. E. degree from the University of Delft, Holland, in 1955 and obtained his Ph. D. degree in Physics from the same University in 1969.

From 1956 to 1960 Dr. Korpel worked in the research laboratories of the Postmaster General's Department in Melbourne, Australia. His investigations included bandwidth reduction in television systems, microwave ferrites and diode parametric amplifiers. In 1960 he joined Zenith Radio Corporation where, as head of the Light Modulation group, he has been active in the field of acousto-optics and ultrasonic imaging. At present Dr. Korpel is Director of Engineering Physics in the Zenith research laboratories where, among his other activities, he is in charge of the video disc research program.

Dr. Korpel has published over fifty technical papers and has been awarded thirty patents. He is a Fellow of the Institute of Electrical and Electronic Engineers (IEEE), a member of the American Physical Society, the Institute of Engineers, Australia and the Royal Dutch Institute of Engineers.

ROLF LANDAUER

Rolf Landauer received his B. S. from Harvard University in 1945 and served briefly in the U. S. Navy as an Electronics Technician's Mate after that. After receiving his PH. D. from Harvard, in 1950, he joined the Lewis Laboratory of the NACA (now NASA). Two years later he joined IBM in Poughkeepsie to work on semiconductors.

Since then, he has held a variety of research and administrative positions. He was Director of Physical Sciences at the Yorktown Research Laboratory from 1962 to 1966. In 1966, he became Assistant Director of Research. In June 1969, he was appointed an IBM Fellow, and in that role has returned to personal research, with occasional interruptions for technological assessment chores. Dr. Landauer had a key role in initiating a number of IBM Programs, including the work on injection lasers and on large scale integration. The expression "Large Scale Integration" was originally the title of a program based on field effect transistors and initiated at IBM in early 1963. Dr. Landauer's primary personal scientific interests have been in the physics of computing devices, electron transport theory, ferroelectricity and nonlinear electromagnetic wave propagation. He has been particularly concerned with the statistical mechanics of the computational process and the ultimate limitations imposed upon information handling by the fact that information is inevitably represented by physical degrees of freedom.

SING H. LEE

Dr. Sing H. Lee received the B. S. degree in 1961 from Ohio University and the Ph. D. degree in June 1968 from the University of California, Berkeley, all in Electrical Engineering. He has been an Acting Assistant Professor at the University of California, Santa Cruz (January - June 1968), an Assistant Professor (August 1968-June 1973) and an Associate Professor (July - December 1973) at Carnegie-Mellon University, Pittsburgh, Pennsylvania. In January 1974 Dr. Lee joined the faculty of University of California, San Diego, as an Associate Professor in the Department of Applied Physics and Information Science.

While working toward his Ph. D. degree, Dr. Lee researched topics in electro-magnetic propagation. In 1968 he became interested in optical information processing and holography, and has pursued research in these areas since. Dr. Lee is broadly interested in lasers, optical computing systems, optical communications, and laser applications generally.

Dr. Lee has conducted research supported by grants and contracts from NSF and NASA. He has supervised a number of graduate Ph. D. and M. S. theses and undergraduate projects. He has taught courses in lasers and optics, high-frequency engineering, information and communications theory, and solid-state electronics. Currently he is supervising six graduate students.

YURI E. NESTERIKHIN

Professor Yuri E. Nesterikhin graduated from M. V. Lomonosov Moscow State University in 1953. Until 1961 he was engaged in thermonuclear problems at I. V. Kurchatov Institute of Atomic Energy. From 1961 to 1967 he headed the department of plasma physics of the Institute of Nuclear Physics of the Siberian Branch of the USSR Academy of Sciences. Since 1967 he has been Director of the Institute of Automatics and Electrometry of the Siberian Branch of the USSR Academy of Sciences (Novosibirsk). In 1970 he was elected a corresponding member of the USSR Academy of Sciences.

Professor Nesterikhin has more than 90 publications on the problems of plasma physics, controlled thermonuclear synthesis, diagnostics of plasma, optical data processing holographic memories.

Professor Nesterikhin is an author of the book "Methods of High-Speed Measurements in Hydrodynamics and Plasma Physics". He is the chairman of the Council on Automation of Scientific Researches, the editor-in-chief of the journal "Avtometria" and heads the department of Automation of Physico-Technical Measurements of the Novosibirsk State University.

GEORGE W. STROKE

Dr. George W. Stroke obtained his Ph.D. in Physics from the Sorbonne in Paris in 1960, and is currently Professor of Electrical Sciences and Medical Biophysics at the Stony Brook campus of the State University of New York and Director of its electro-optical sciences laboratory. Concurrently, since 1970, he has served for three consecutive years as Visiting Professor of Medical Biophysics at Harvard University Medical School.

Dr. Stroke was previously Professor of Electrical Engineering at the University of Michigan and Head of its electro-optical sciences laboratory which he founded in 1963. Before joining the University of Michigan, Professor Stroke spent 10 years at MIT where he did research work, principally devoted to the development of the method of interferometric servo-control of grating ruling, in a collaborative effort with Dean George R. Harrison. This has earned them world fame. During his tenure at MIT, he also helped in originating a method of velocity of light measurement using microwave-cavity resonance, and participated in the Office of Naval Research Fleet Ballistic Missile (Polaris) program at the Instrumentation laboratory there.

Professor Stroke's work in coherent optics and holography originated with his work in the Radar Laboratory at the University of Michigan where he helped in initiating the work on three-dimensional "lensless photography" as a consultant in 1962-1963. He wrote the first treatise on the subject under the title, <u>An Introduction to Coherent Optics and Holography</u> (Academic Press, 1966) which was immediately translated into Russian (MIR, 1967) and appeared in its second (enlarged) U.S. edition in 1969.

In recent years, Professor Stroke has been devoting his primary research interests increasingly to two new fields: optical and digital information processing and to the life sciences, including the development of new methods of image deblurring and communications and to the improvement of high-resolution electron microscopy. The method of holographic image deblurring which he originated in 1965 has recently permitted him, with his team at Stony Brook, to sharpen up electron micrographs of virus test specimens to a degree considered unattainable in the past, as well as to extract seemingly irretrievable information from accidentally blurred photographs, notably also of archeological specimens.

In addition to An Introduction to Coherent Optics and Holography , Dr. Stroke published another book (at the age of 24) as well as the 320 page "Diffraction Gratings" section of the Handbuch der Physik (Springer Verlag, Vol. 29, 1967) and approximately 100 scientific papers including about 50 on holography. He is also editor of Ultrasonic Imaging and Holography--Medical, Sonar and Optical Applications (Plenum Press, 1974). A widely traveled lecturer on the subject of holography and its scientific, industrial and biomedical applications, Dr. Stroke has served in a number of United States government and other advisory capacities including, most recently as a member of the National Science Foundation Blue Ribbon Task Force on Ultrasonic Medical Diagnostics and as a consultant to the American Cancer Society. In 1971 he served as U. S. delegate to the Popov Society meeting in Moscow under the U. S. State Department Scientific Exchange program. In 1972 he was invited by the Japan Industrial Technology Association to officially advise the Japan Ministry of International Trade and Industry on its program of large-scale development of computer pattern recognition technologies. For several years he has also been assisting the National Science Foundation in its U. S. -Japan and U. S. -Italy science cooperation programs.

Dr. Stroke was the U. S. Coordinator of the NSF-sponsored U. S. -U. S. S. R. Science Cooperation Symposium on "Optical Information Processing" held in Washington, D. C. (16 - 20 June, 1975) and of the post-conference visits, coordinated with the U. S. Department of State, to Bell Laboratories, M. I. T. and the IBM Research Laboratories in Yorktown Heights. He is the coordinator of the Seminar Proceedings. Among his most recent scientific publications are "Retrieval of Good Images from Accidentally Blurred Photographs" (SCIENCE, July 25, 1975) and "Optical Holographic Three-Dimensional Ultrasonography (SCIENCE, September 19, 1975).

BRIAN J. THOMPSON

Dr. Brian J. Thompson is currently Dean of the College of Engineering and Applied Science, Director of the Institute of Optics and Professor Optics, University of Rochester, Rochester, New York.

Dr. Thompson is President of the Society of Photo-Optical Instrumentation Engineers, a member of the Editorial Advisory

Board of Laser Focus and Optics Communications, American editor of Optica Acta, Associate Editor of the Journal of the Optical Society of America and Optical Engineering.

He is a Fellow of the Institute of Physics and Physical Society of Great Britain and a Fellow of the Optical Society of America, and a member of the U. S. Committee for the International Commission for Optics.

Dr. Thompson's education was obtained at the University of Manchester, where he obtained a B. Sc. degree in 1955 and Ph. D. in 1959.

Dr. Thompson has published over 100 papers in the general field of optics.

YURI V. TROITSKY

Dr. Yuri V. Troitsky graduated from the Gorky State University (the department of radiophysics) in 1952. Since 1955 he has been working at the Siberian Branch of the USSR Academy of Sciences (Novosibirsk). At present he heads the laboratory of quantum electronics of the Institute of Automatics and Electrometry of the Siberian Branch of the USSR Academy of Sciences. He is engaged in microwave electronics, electron optics of high-current beams, gas lasers, optical resonators, reflecting interferometers.

Dr. Troitsky is an author of about 70 scientific publications and of the monography "Single-Frequency Oscillation of Gas Lasers" (1975). In 1971 he was given a doctor degree.

P. Y. TVERDOKHLEB

Dr. Peter Y. Tverdokhleb graduated from the Polytechnical Institute in Lvov (the Ukraine) in 1958 as an electrical engineer. Since 1958 he has been working as a researcher at the Institute of Automatics and Electrometry of the Siberian Branch of the USSR Academy of Sciences (Novosibirsk). He was engaged in the problems of investigation and development of analog-digital transducers as well as in the problems of processing the measurement results. In 1965 he was given a doctor degree.

In 1968 Dr. Tverdokhleb started investigations on developing optical (in particular holographic) information storing and processing devices. He concentrated attention on the development of space-noninvariant optical systems intended for image analysis, multichannel signal processing and associative retrieval. In this field he has about 35 scientific publications and inventions. He heads the laboratory of optical information processing.

A. VANDER LUGT

Anthony Vander Lugt (M'66) was born in Dorr, Michigan, in 1937. He received the BSEE and MSEE degrees from the University of Michigan, Ann Arbor, in 1959 and 1962, respectively. In 1969 he received the Ph.D. degree in theoretical and applied optics from the University of Reading, Reading, England.

From 1959 to 1967 he was a Research Engineer at the Radar and Optics Laboratory, Institute of Science and Technology (now ERIM) at the University of Michigan. There he concentrated on the development of optical spatial filtering techniques, the construction of complex spatial filters, the use of interferometric techniques for measuring the frequency response of photosensitive materials, and the analysis and design of optical data processing systems. In 1969 he joined the Electro-Optics Center of Radiation, Inc., Ann Arbor, Michigan, as Director of Research, where he directed research activities associated with optical data processing, photosensitive materials, and holographic memories. In 1973 he became the Director, Systems and Programs, at the Electro-Optics Operation of Harris Electronic Systems Division (formerly Radiation), Melbourne, Florida, where he continues to direct the development of holographic memories, high data rate holographic recorders, optical data processors, and directs related research activities.

Dr. Vander Lugt is a member of Eta Kappa Nu, Tau Beta Pi, Sigma Xi, and is a Fellow of the Optical Society of America. He was the American Editor of Optica Acta from 1970 to 1974 and is currently a member of its Editorial Advisory Board.

---

*Biographies as available by the time of printing.

I. N. KOMPANETS

I. N. Kompanets was born in 1940. He graduated from Moscow Physical-Engineering Institute in 1968, and proved the degree of the Doctor of Physics and Mathematics in 1973. He works at Lebedev Physical Institute, USSR Academy of Sciences, since 1967, at present as the head of the research group. Dr. Kompanets is engaged in the study of opto-electronic materials, the construction, investigation and application of the controlled transparency devices.

AUTHOR INDEX

Abbe, E., 283 ff.
Adler, R., 192
Aero Service Corporation, 118
Agajanian, A., 38
Akselrod, A. A., 70
Amodei, J., 39
Ampex Corporation, 37
Anderson, A. G., 249
Anderson, L. K., 41, 70, 250
Andrews, H. C., 152, 216
Angot, A., 310 ff.
Arbuzov, V. A., 1, 11
Arecchi, F. T., 247
Arm, M., 186, ff.
Armitage, J. D., 278
Armstrong, J. A., 222 ff.
Arsac, J., 308 ff.
Artman, I. O., 73, 216
Asakura, 314 ff.
Ashar, K. G., 248
Auston, D. H., 251

Baechtold, W., 249
Baker, B. B., 298
Bardos, A. M., 40, 364 ff.
Barrekette, E. S., 250
Barrett, H. H., 102
Bartholomew, B., 274
Bartolino, R., 39
Basov, N. G., 151, ff., 222 ff.
Beard, T. D., 39
Becherer, R., 327
Beissner, R. E., 316 ff.
Bell Laboratories, 356
Belveaux, 11

Bendix Research Laboratories, 121
Bennett, C. H., 249
Berry, J., 152
Bismuth, G., 41
Blinov, L. M., 151
Bobrinev, V. I., 70
Boller, A., 40
Bordogna, J., 39
Borel, J., 40, 151
Born, M., 202, ff.
Boughton, P., 309 ff.
Bouwkamp, C. J., 298
Boyle, A. J., 249
Bracewell, R. N., 308
Bragg, W., 292 ff.
Breikss, I. P., 250
Brody, T. P., 151
Broers, A. N., 249
Bromley, K., 340 ff.
Brophy, J. J., 250
Brown, B., 40
Brown, J., 127
Bryngdahl, O., 102
Buckhardt, C. B., 39, 71, 127
Buerger, M. J., 292, 320 ff.
Burfoot, J. C., 251

Caimi, F., 40
Cannon, T. M., 293
Carnegie-Mellon University, 13 ff.
Casasayas, F., 41
Casasent, D. 13ff., 39 ff., 369
Cedarquest, J., 274

Chang, M., 249, 250
Chapman, D. W., 251
Chaudhuri, B., 322 ff.
Cheatham, T. P. Jr., 310 ff.
Chugui, Yu. V., 72, 152 ff.
Clair, J. J., 169
Collier, R. J., 39
Collins, W. E., 40, 251
Columbo, U., 250
Compaan, K., 250
Considine, P., 366 ff.
Copson, E. T., 298
Cosentino, L., 41
Croce, P., 307 ff.
Culver, W. H., 222 ff.
Cummins, S., 41
Cutrona, L. J., 127

Dallas, W. J., 169
Darmois, E.E., 298
De Cillo, J. J., 248
Dell, H. R., 249
Desmares, P., 193
De Witt, D., 248
Dickson, L. D., 247
Diemer, G., 247
di Francia, G. T., 307 ff.
Dill, F. H., 249
Ditchburn, R. W., 316 ff.
Djurle, E., 307 ff.
Donjon, J., 40, 41
Dossier, B., 309 ff.
Doyle, R., 39
Drake, M. D., 151
Duffieux, M., 283 ff.

Eastman Kodak, 39
Edgar, R. F., 152
Elias, P., 307 ff.
Ellis, B., 38
Environmental Res. Inst., 39
Erdelyi, A., 308 ff.
ERIM Company, 37
Ernstoff, M., 40
Eschenfelder, A. H., 250
Eu, J. K. T., 11

Faiman, M., 276
Fateev, V. A., 82, 153 ff.
Federov, V. A., 1, 11
Feldbaum, A. A., 71
Fellgett, P. B., 307 ff.
Feth, G., 249
Flannery, J., 39
Foss, R., 250
Foucault, L., 1 ff., 328, 345
Fourier, J. B., 283 ff.
Fowler, A. B., 222 ff.
Francon, M., 298 ff.
Fraser, D. B., 151
Fridman, G. Kh., 152
Fyodorov, V. A., 369

Gabor, D., 115 ff., 152 ff.
Gallaher, L. E., 70
Ganzhorn, K., 247
General Electric Co., 39
Ghosh, H. N., 248
Gibin, I. S., 47ff., 152, 370
Gilbert, J. C., 249
Gingerich, M. E., 251
Gitis, V. G., 71

AUTHOR INDEX

Glass, A. M., 251, 252
Glenn, W. E., 39
Goetz, G. G., 40, 122, 127
Gofman, M. A., 71, 72, 152
Golay, M. J., 152
Goldberg, L., 274
Goldina, N. D., 75, 83, 170
Goodman, J. W., 38 ff., 73 ff., 170, 370
Gordon, E. I., 192
Goto, E., 247, 249
Grandchamp, P. A., 127
Grant, D. G., 277
Grey, D., 307 ff.
Grinberg, J., 40 ff.
Groh, G., 41
Gueret, P., 249
Guillemin, E. A., 309
Gurevich, S. B., 105 ff., 371

Haertling, G. H., 151
Hahn, E. L., 249
Haken, H., 249, 251
Halioua, M., 102, 281 ff., 371
Hall, E. L., 276
Hammond, A. L., 250
Hamy, M. 75, 82
Hannay, 250
Hansen, J., 40, 88
Hanson, A. W., 322 ff.
Harburn, G., 320 ff.
Harker, J. M., 249
Hargeaves, A., 322 ff.
Harland, R., 250
Harris Corporation, 39, 347 ff.
Hatzakis, M., 249
Hecksher, H., 330 ff.

Heilmeir, G., 39
Heinz, R. A., 73, 216
Henkels, W. H., 249
Herran, J. P., 251
Hirsch, P. M., 169
Hitachi Co., 356
Hoffman, H., 127
Hoffman, W. K., 250
Hopkins, H. H., 309 ff.
Hopner, E., 250
Hughes Corp., 315 ff.
Hughes Res. Lab., 30, 37
Huignard, J. P., 251

IBM Co., 219 ff., 322, 374, 354 ff.
Inagaki, T., 41
Ingebretsen, R. B., 293 ff.
Ingelstam, E., 309 ff.
Inokuchi, S., 152
Inst. of Automation & Electrometry, USSR, 195 ff., 203 ff.
A. F. Ioffe Physical-Tech. Inst., Leningrad, 105 ff.
Ishibashi, Y., 249
Ishida, H., 249
Itek Corp., 37 ff.

Jablonowski, D. P., 216
Jacobs, J. T., 252
Jacobson, A., 40
Jacquinot, P., 297 ff.
Jakeman, E., 277
Jensen, H., 118
Johnson, E. J., 248
Jordan, J. A., Jr., 169

Kalter, H. L., 250
Kane, J., 152, 216
Karan, C., 251
Kato, H., 277
Kaye, B., 316 ff.
Kaye, D. N., 250 ff.
Kazan, B., 251
Ken, Li Si, 70
Keneman, S., 39
Kerby, H. R., 250
Kermisch, D., 345 ff.
Kestigian, M., 41
Keyes, R. W., 222 ff.
Khaikin, B. E., 152
Kharitonov, V. V., 70
Kharkevitch, A. A., 202
Kikuchi, Y., 293
King, M., 192
King, T. A., 170
Klein, H. P., 251
Knight, G. R., 70
Knoll, M., 251
Koblova, M. M., 70
Kock, W., Title ff., 117 ff., 372
Kohlenberg, A., 310 ff.
Kompanets, I. N., 130 ff., 380
Koronkevich, V. P., 82, 152, 373
Korpel, A., 40, 73, 171 ff., 373
Korshever, I. I., 71
Korsakov, V. V., 169
Kosonocky, W. F., 222 ff.
Kostsov, E. G., 195 ff.
Kottler, F., 298
Kozma, A., 250
Kramer, P., 250
Krivenkov, B. E., 203 ff.
Kruger, R. P., 276
Krutitsky, E. I., 152
Kuehler, J. D., 250
Kushtanin, K. E., 70

Labrunie, G., 40
La Macchia, J. T., 152
Land, C. E., 41, 151
Landauer, R., 45, 219 ff., 374
Larmor, J., 298
Larunie, G., 151
Lasher, G. J., 222 ff.
Lee, S. H., 252 ff., 374
Lee, T. C., 39, 73, 74, 216
Leighton, W. H., 277
Leith, E. N., 127 ff., 322 ff., 347 ff.
Lempicki, A., 248
Lenkova, G. A., 82, 153 ff.
Lescinsky, M., 40
Lighthill, M. J., 308 ff.
Lin, L. H., 39
Linfoot, E. H., 307 ff.
Lipson, S., 39 ff.
Lisem, L. B., 169
Litton Industries, 118
Lo, A. W., 202
Lohmann, A. W., 11, 102, 277
Lorentz, H. A., 298
Lotsoff, S. N., 193
Luke, T., 41
Luneburg, R. K., 310

Maiorov, S. M., 70
Maitzler, A. H., 151
Maldonado, J. R., 151
Malinovski, V. K., 195 ff.
Malitson, I. H., 170
Maloney, W., 40
Mandel, L., 102
Maréchal, A., 292 ff.
Margerum, J. D., 151
Marie, G., 40, 41

AUTHOR INDEX

Marinace, J. C., 248
Matick, R. E., 250
Mattson, R. L., 247
Matsuoka, Y., 249
Matushkin, G. G., 71
Maydan, D., 151
Mayeux, C., 251
McNichol, J., 249
Mee, D., 250
Mehran, F., 222 ff.
Mehta, R., 251, 252
Meier, R., 39
Meissner, K. W., 170
Meyer, R. A., 277
Micheron, F., 41, 251
Michel, A. E., 248
Michelson, A. A., 283 ff.
Micheron, F., 41
Middleton, D., 101
Mikaeliane, A., 40
Mikaelyan, A. L., 70
Mikhal'tsava, I. A., 82, 153 ff.
Mikhyaev, S. V., 203 ff.
Miller, M., 40
Milnes, A. G., 276
Monahan, M. A., 340 ff.
Morita, Y., 152
Moto-Oka, T., 249
Mueller, R. K., 122
Murata, K., 249

Nakagawa, K., 249
Naliwaiko, V. I., 169
Nathan, M. I., 248
NASA, 45
Naumov, S. M., 70
Negran, T. J., 251
Nemoto, T., 70, 152

Nesterikhin, Yu. E., Title ff., 375
Neureuther, A. R., 249
Nezhevenko, E. S., 71, 74, 152, 216
Nikitin, V. V., 151, 152
Nilson, N., 216
Nippon (Gakki) Co., 356
Nisenson, P., 152, 202, 315 ff.
Nishida, N., 71, 152
Norgen, P. E., 152

Oberai, A. S., 248
O'Bryan, H. M., Jr., 151
Office of Naval Res., 101
Okazaki, K., 151
Okolicsanyi, F., 187 ff.
Oliver, D. S., 346
O'Neill, E. L., 307 ff.
Orinoco River, 118
Orlov, L. A., 151

Papulis, A., 152
Parrent, G. B., 318 ff.
Pen, E. F., 71, 152
Phillips, W., 41
Plugge, S., 292
Pohm, A. V., 232 ff.
Pole, R. V., 250
Poppelbaum, W. J., 40 ff.
Porcello, L. J., 127
Porter, A. B., 314 ff.
Potapos, A. N., 195 ff.
Potaturkin, O. I., 71, 152, 216
Pratt, W. K., 152, 216
Preston, K., Jr., 38, 39, 101, 102

Rafuse, M., 40
Rajchman, J. A., 70
Ragnarsson, S. I., 98, 102
Ramberg, E., 39
Rayleigh, Lord, 283 ff.
Raynes, E. P., 277
RCA, 356
Regensberger, P., 277
Reizman, F., 39
Remesnik, V. G., 82, 153 ff.
Rhodes, J. E., 345
Riley, W. B., 249
Riseberg, L. A., 248
Roberts, H., 41
Roberts, J., 151, 367
Robieux, J., 298
Robinson, D., 307 ff.
Rodney, W. S., 170
Roetling, P. G., 345
Roizen-Dossier, B., 297
Rosenthal, A. H., 186 ff.
Rotz, F. B., 276
Rozenfeld, A., 152
Rubinowicz, A., 298
Russell, F. D., 102
Rutz, R. F., 248

Sacaguhi, M., 70
Samelson, H., 248
Sand, D. S., 276
Schade, O. H., 309
Schaefer, D. H., 278
Schawlow, A. L., 247, 251
Schneeberger, R., 40
Schneider, I., 40, 251
Schulz-DuDois, E. O., 247
Schwartz, L., 308
Scophony Television Laboratories, 187

Selezniov, V. N., 151
Shah, B., 223 ff.
Sheridan, N., 39, 345
Shuikin, N. N., 151
Siemens Company, 356
Silver, S., 307
Silverman, B. D., 252
Sjorle, E., 309 ff.
Slobodin, L., 185 ff.
Smith, W. V., 247
Smits, F. M., 70
Sobolev, A. G., 129 ff., 151
Socolov, V. K., 115
Soderstrom, R. L., 247
Sokolova, L. Z., 70
Soma, T., 249
Sorel, R., 40
Soroko, L. M., 11, 152
Spiller, E., 278
Sprague, R., 332 ff.
Srinivasan, V., 293
Staebler, D., 41
Stalker, K. T., 277
Stanford University, Joint Services Electronics Program, 101
Stark, H., 277
State University of New York, Stony Brook, 281 ff.
Sterling, W., 40
Stewart, W., 41
Stockham, T. G., 293
Stone, J. M., 310
Stroke, G. W., Title, ff., 38 ff., 45 ff., 98 ff., 313 ff., 332 ff., 376
Strong, J. P., 277 ff.
Suchki, P., 151
Svyazi, T., 216
Szentesi, O. I., 193

AUTHOR INDEX

Taylor, C. A., 345
Taylor, G. W., 151
Taylor, H., 273 ff.
Thacher, P. D., 151
Thaxter, J., 41
Thompson, B. J., 278 ff., 377
Thomson-CSF, 356
Thon, F., 289 ff.
Tichenor, D. A., 99, 102
Titchmarsh, E. C., 11
Töpler, A., 314 ff.
Triebwasser, S., 247
Troitsky, Yu. V., 75, 83, 170, 378
Trotier, J. C., 251
Tsujiuchi, I., 293 ff.
Tsukerman, V. G., 82, 153 ff.
Tufte, O. N., 39, 250
Turbovich, I. T., 71
Turner, A. F., 276
Turpin, T., 39
Tuttle, J. A., 249
Tverdokhleb, P., 47 ff., 71, 74, 103, 152, 216, 378

Vander Lugt, A., 38 ff., 71, 73, 101, 152 ff., 313 ff., 377 ff., 279
Van Raalte, J., 40
Van Santen, J. G., 247
Vareille, J. C., 11
Vasiliev, A., 129 ff.
Vivian, W. E., 127
Vohl, F., 41, 342 ff.
von der Linde, D., 251, 252
von Fraunhofer, J., 283 ff.
von Neumann, J., 247
Voskoboinik, G. A., 71

Wada, E., 249
Walton, A., 38
Ward, J. H., 316 ff.
Whitman, R. L., 184 ff.
Willasch, D., 289 ff.
Wintz, P. A., 71
Wolf, E., 298 ff.
Wollenmann, H. P., 250
Woo, J. W. F., 249

United Aircraft Research, 125
U. S. Air Force Systems Command, 126
University of California, San Diego, 255 ff.
University of Cincinnati, 117 ff.
University of Rochester, 313 ff.
Upatnieks, J., 315 ff., 332 ff., 347 ff.
Urbach, J., 39
USSR Academy of Sciences, title ff.

Xerox Corporation, 315 ff.

Yao, S. K., 276
Yuokov, E. F., 71

Zanoni, L., 40
Zappe, H., 249
Zasovin, E. A., 70
Zech, R. G., 315 ff.
Zenith Radio Corp., 171 ff.
Zernike, 314 ff.
Zhabotimska, V. A., 151
Zingg, R. J., 232 ff.

SUBJECT INDEX

Aberration balancing, 334 ff.
Access time, 354
Acoustooptic:
    delay line, 171
    interaction, 171
    page composer, 191
    signal processing, 171 ff.
    signal processing with
    second sound cell, 179
Addressing scheme, 14, 15
Alkali halide memory system, 45
    color centers in, 246
Amplifying elements, 255
Analog to binary forms, 266
Analog to digital conversion
    algorithm, 268
Analogy between electromagnetic wave propagation in space and the propatation of current and voltage waves in a transmission line, 77
Aperture shaping, 282 ff.
Apodization, 297
Arrays:
    photodetector, 358
    hologram, 356
    deformable membrane
    mirror, 358
Asymmetric deformation,
    method of, 137
Asymmetric interference
    fringes, 82
    generation of in reflected light, 75

Beams:
    automatically formed, 119
    electron, 224
Belveaux system base, 4
Bit, cost per, 354 ff.
Blind deconvolution, 286, 287
Blur function, 96 ff.
    library of typical blurrings, 287
Boolean algebra using
    light transmission, 140
Bragg angle, 173
Bragg diffraction, 172 ff.
    upshifted and downshifted:
    processor, 176
    imaging, 190
Bubble lattice array, 229 ff.

Cadmium sulfide, 358
Capacitive accumulators, 198
Cathode poisoning, 30
CdS(cadmium sulphide), 266 ff.
Ceramics:
    PLZT, electro-optic effects
    in, 133
    PLZT switching, 137
    segneto, optically transparent, 132
Charge leakage, 18
Circuit analysis, mathematics of, 309
Circular polarization, 117, 123
    backscattered, rejection of
    interference pattern of, 126
    reduction of specular backscatter, 123

Circular polarization(cont'd):
    rejection of backscattered
      waves, 123
    rotation of, 123
Coherence control, 318
Coherent:
    addition, 88
    optical processing systems, informational capacity of, 105 ff.
    optics, 295
    optics and holography, 310
    processing, 19
    radar, 117
Communication:
    digital, 281 ff.
    optical, 281 ff.
    theory, 309
Computation, optical and digital comparison, 283 ff.
Computer:
    architecture, 282
    CDC 7600, 310
    digital electronic, 255 ff.
    holographic, 282 ff.
    laser, 282 ff.
    mini, 282 ff.
    optical, 282 ff.
    optical image deblurring, 284 ff.
    software, 282
    TSE, 45
Contrast:
    enhancement, 330 ff.
    inversions, 289 ff.
    transfer functions(CTF), 289 ff.
Controlled transparencies, 131, 132
Conversion, analog to digital, 282 ff.

Convolution, 255
Correlation functions, calculation of, 59
Correlator:
    digital-optical hybrid, 277
    early optical, 185, 186
    with offset pinhole, 186
Coulomb interactions, 222
Cross-correlation, 255
Cross-shaped light source, 5
Cryptocyanine dye, 257

Dark field method, 328 ff.
Data retrieval, associative, 65, 73
Deblurred image, 97, 100
Deblurring, 283 ff.
    of accidentally-blurred photos, 286, 287
    optical image(restoration), 283 ff.
Deconvolution, 289 ff.
    for image motion and other aberrations, 340
    optical computing, 289
Defocusing phase contrast, 291
Depolarization, 123, 124
    effects, 252
Devices:
    acousto-optic, 31
    B-scan, 122
    charge coupled, 239 ff.
    deformable target, 19, 26
    ferroelectric-photoconductor, 3
    input-output, 219
    liquid crystal, 26, 38
    magneto-optic, 31

SUBJECT INDEX

Devices (cont'd)
    nonlinear, 266 ff.
    nonlinear optical, 266 ff.
    optical, 13 ff.
    optical density charge, 31
    opto-electronic arithmetic, 140
    planar logic, 266 ff.
    planar optical logic, 266
    Pockels effect, 33, 34
    real time, 13, 34
    real-time nonlinear, 255 ff.
    real-time, spectral analysis, 295
    reusable, 13
    scanning microdensitometer, 310
    semiconductor, 223
    sequentially addressed, 18
    Tepler, 7
    thermoplastic, 19
    thermoplastic, fatigue in, 21
    thermoplastic, electron beam addressed, 21
    threshold, 266
Differentiation, 330
Diffraction, 313 ff.
    efficiency, 17, 18
    gratings, 295 ff.
    intensity distribution in the spectral patterns, 308
    measurements, 316
    Prolem, Fourier-transform, formulation of, 306 ff.
Diffractograms, 289 ff.
    Thon (so-called), 289 ff.
Digital conversion hardware, 268, 282 ff.

Digital data compression, 282 ff.
Digital implementations of fundamental optical principles, 281 ff.
Digital storage, optically accessed, 219 ff.
Disc replacement, 358
Disks:
    IBM 3340, 233 ff.
Display applications, 18 ff.
    CRT, 282 ff.
    fast printers, 282 ff.
    microfilm, 282 ff.
DKDP, 120
    crystal reconstructor, 121
Documents:
    display of, 281 ff.
    storage of, 281 ff.
    transmission of, 281 ff.
Dual in-line image correlation, 15, 16
Dynamic range, 284 ff.
    photographic, 284 ff.
    limitations, 310

Earth resource observation program, 277
Elastomer, OALM, 24
Electric current density, 299
Electric Field, 134 ff.
    applied, 134
    ceramic plate birefringence, 136
    dependence of birefringence, 135
    PLZT plate thickness, change of, 134

Electromagnetic theory of
  diffraction (rigorous), 295 ff.
Electron:
  beams, 21 ff.
  beams on dielectric, 21 ff.
  contrast inversions, 289
  guns, 284 ff.
  high resolution micrographs,
    improvement in, 284 ff.
  micrographs, 283 ff.
  microscopy, 283 ff.
Electro-optic:
  ceramics, 130
  effects, 195 ff.
  effects in reading, 199
  quasistate properties of
    switching, 138
  transversal effect, 134
Equivalent transmission line, 77
Erase mechanism, 18
Erasure heating, 20
Errors:
  periodic and multiply-
    periods, 296
  substrate, 296
  wavefront, 296

Fast Fourier transfer
  algorithms, 310
Feedback, 255 ff.
  optical schemes in, 266
Ferroelectrics, 251
Filter diffraction effectivity,
  114
Filtering:
  band-pass, 327
  detection, 338
  using a weighted reciprocal
    lattice, 326

Filters:
  amplitude, 330
  band-pass, 322 ff.
  binary, 151
  bleached holographic
    deblurring, 98
  blocking, 322
  complex, 333 ff.
  controlled by electric
    signals, 145
  dynamic, 149, 151
  dynamic, controlled by
    optical signals, 145
  grating, 255 ff.
  half plane phase, 334
  holographic, 315 ff.
  holographic complex
    matched, 74
  image deblurring, 287
  imperfectly adapted, 287
  phase, 331 ff.
  sandwich type deblurring, 98
  Vander Lugt matched, 149
Fingerprint on a glass plate, 329
Focus, optimum (Scherzer), 289
Foucault-Hilbert transform,
  optical realization of, 1 ff.
Foucault knife-edge test, 314
Fourier:
  analysis and generalized
    functions, 309
  integral in optical
    problems, 314
  on-line transformations, 311
  spectral analysis, 255
  synthesis, 322
  synthesyzer, 283 ff.
  transform formalism,
    heuristic, 298 ff.
  transform holograms, 347 ff.
  transform relations, 339 ff.

SUBJECT INDEX

Fourier (cont'd)
    transforms in optics, 347 ff.
    transform spectroscopy, 347 ff.
Fourier plane:
    processing, 181
    scattering, 98
Fraunhofer diffraction pattern, 318
Frequency domain coding, 188
Fresnel lens, 153
Fringe shape, 76
Functions of Haar, Walsh, Legendre, 142

Gamma conditions, 285 ff.
Gauss' theorem, 299
Gouy theorem, 303
Granularity-noise loss, 285 ff.
Gratings, 308 ff.
    optical, 309
Grey level compression, 282 ff.

Hadamard matrix, 210
Hadamard transform, 210
Halftone process, 255
Harris semiconductor, 37, 39
Heaviside-Lorentz system, 300
Heilmier, 39, 40
    classic paper, 26
Heterodyning with reference field, 178
Higher order centers, 32
High-resolution spectroscopy, 309
Hologram array, 356

Holograms:
    charging information, 121
    coherent radar, 117
    computer-generated, 99
    Fourier, 55
    liquid propellant spray, 126
    matrix, 50, 52
    matrix plane, 60
    periodically erased, 121
    real-time, 34
    repetition rate, 122
    spatially multiplexed, 16
    volume, 253
Holographic
    image improvement, 284
    improvement of motion blurred photographs, 287
    lens compensators, 311
    matched filter, 14 ff.
    memories, 347 ff.
    optical computing, 286, 287
    read-only memories, 359 ff.
    read-write, 353 ff.
    read-write memories, 356 ff.
    recorder, 363 ff.
    storage and retrieval of analog information, 348 ff.
    storage and retrieval of digital information, 353 ff.
    storage systems, 348 ff.
Holography, 281 ff.
    synthetic, 359 ff.
HRMR (Human Readable, Machine Readable) system, 359 ff.
    digital data recording technique in, 361
    microfiche generation in, 361
Huygens Principle, 294 ff.
    for electromagnetic waves, 295 ff.
    rigorous form of, 302

Hybrid
    field effect, 30
    systems, 316 ff.

IBM 3800, 220
Image:
    analysis, generalized, 67
    analysis, spectral
      "capture", 282 ff.
    deblurring, 98 ff.
    holographically improved, 289
    improvement in high resolution electron microscopy, 289 ff.
    improvement using holographic image deconvolution, 289
    improvement using optical computer processing, 289
    plane processor, 176 ff.
Image formation problem, 302
    Fourier-transform, formulation of, 302 ff.
Impulse response of a defocused imaging system, 334
Information:
    capacity, 112
    densities, 235
    density of a holographic system, 108
    formation, 129
    objective, 108
    transformation, 129
Information processing:
    digital, 279 ff.
    display, 281 ff.

Information processing(cont'd)
    holographic memory devices, in optical systems of, 47 ff.
    page-by-page recording, 48
    storage, 281 ff.
    systems, 85
    transmission, 281 ff.
Injection laser, 221 ff.
Injection laser logic, 222
Integral equations, 313 ff.
Integrated circuitry, 221
Integrated optical circuits (IOC), 271 ff.
Intensity distributions in spectral diffraction patterns, 308
Interferograms, wavefront, 296
Interferometer,
    Fabry-Perot, 75
    Max-Zender, 147
    Michelson Twyman-Green, 296
Interferometer mirrors, reflectance of, 167
Isotropic Hilbert Image pattern, 1

Josephson junction, 224

Kanji characters (Japanese ideograms), 231
Kastler coaxial phaseplate, 10
Kinoform elementary cell, 159
    recording system for, 158
Kinoform lens, 163, 170
    angular aperture of, 166

SUBJECT INDEX

Kinoform lens (cont'd)
    phase distribution in, 164
    phase profile, cylindrical
      lens, 161
    resolution of, 165
    zones of, 170
Kinoform optical elements, 153 ff.
    complex, 160
    in a real-time operation
      mode, 154
    optical technique for, 153
    production methods, 153
    simplest, 153
Kinoform recording, 156
Knife-edge test, 333
Kroneker symbols, 207

Laser:
    Fourier transformation, 287
    heterodyne scanning, 189
    logic, 227 ff.
    ruby, 258
Lasers, 220 ff.
Legendre polynomial, 211, 214
Lenslet arrays, 16
Lens systems:
    Fourier-transforming,
      properties of, 310
    Information character-
      istics of, 105
Light modulator:
    electrically addressed
      (EALM), 343
    electron beam addressed,
      15
    electronically addressed,
      15
    membrane systems, 252

Light modulator,
    optically addressed, 14, 343
    spatial, 14, 15
    spatial, comparison of, 18
Light valve:
    AC liquid crystal, 277
    electron-beam-addressed
      dielectric, 23
    G. E., 296
$LiNbO_3$, 266
Linear image motion, 336
Linear processing, 257 ff.
Liquid crystals, 13, 342
    nematic, 30, 145
    panel, 315
    PLZT, 358
Liquid-crystalline cell, 143
Lithium mobate, 17
Logic, 221 ff.
    IOC, 279
    optical, 247
    planar, and gate, 269 ff.
    technologies, 223
Lumatron, 296

Magnetic:
    bubbles, 234 ff.
    current density, 299
    disks, 355
    surface recording, 232
    tape, 232, 355
Magnification coefficient, 4
Marcasite (FeS_2), 320
Mass storage:
    Holographic memories, 241 ff.
    IBM photo digital system, 241
    IBM 3851, 233

Mass storage (cont'd)
   precision instrument's
     unison, 241

Materials:
   birefringent electro-
     optic, 266
   ceramic, 243
   chalcenogide, 156, 170
   ferro-ceramic, 129
   ferroelectric, 35
   iodine, 264
   nonlinear optical, 255 ff.
   nonlinear optical table, 265
   sodium, 264
   thermoplastic, 242

Matrix logical equation, 65

Matrix multiplication
   by optical method, 74

Matrix transforms
   by optical coherent
     techniques, 203 ff.

Maxwell-Lorentz theory, 295

Maxwell's equations, 299 ff.

Memory devices:
   associative retrieval, 73
   holographic (HMD), 47
   integrated circuit, 235
   optically accessed, 239 ff.
   opto-electronic, 195
   random access, 235
   with data processing
     orientation, 50

Memory technology, 354 ff.
   block oriented, 357
   current, 355
   IBM 1360 model
   precision instrument
     model, 354

Micrographs
   of biological interest, 289
   of thin specimens, 289
   of viruses, 289
   techniques, conventional, 352

Microprocessors, 310

Microscopes, 289 ff.
   phase contrast, 314 ff.
   resolution capabilities of, 289
   Siemens electron
     elmiskop, 102, 289

Microwaves,
   analogous lens for, 170

Mixed liquid crystal, 266

MOSFETS, 30

MTF, 16

Multilayer dielectric coating, 162

Mylar disks, 232

Nematic crystals,
   effects of electric field on, 131

Noise:
   background, 91
   dependence on various
     system parameters, 101
   filtering multiplicating, 103
   in coherent optical infor-
     mation processing, 85 ff.
   inherent in data to be
     processed, 90
   introduced by optical
     processor, 95
   prime source of, 98
   quantum, 92
   speckle, 91
   statistics in matched
     filtering, 103

SUBJECT INDEX

Non-coherent optical system
    for processing of images
    and signals, 203 ff.
Nonlinear optical processing, 255
Nonlinear two-photon process, 253

OALMS:
    electro-optic, 33
    Pockels effect, 34
On-line convolution compen-
    sation, holographically
    assisted, 311
Operation optical memory,
    some peculiarities of
    physical realization of, 195 ff.
Optical analogues to X-ray
    diffraction, 318
Optical channel,
    passage of different spatial
    frequencies through, 110
Optical computers, 297 ff.
    space invariant, 297
    space variant, 297
Optical computing, 281 ff.
    achievements in, 290 ff.
    coherent material devices
    for, 13 ff.
    convolution, 291
    electron micrographs by,
    288, 289
    improvement of high
    resolution, 288
Optical diffraction patterns,
    information content of, 313 ff.
Optical-digital techniques, 266
Optical effects, nonlinear, 222
Optical error analysis for
    spectrum analyzers, 296 ff.

Optical foundations of digital
    communication and optical
    processing, 281 ff.
Optical Fourier analysis, 321
Optical image:
    contrast of, 310
    filtering scheme, 106
Optical logic and optically
    accessed digital storage, 219 ff.
Optical matrix multiplication,
    203 ff.
Optical matrix transforms
    by non-coherent techniques,
    204 ff.
Optical memories, 13, 195 ff.
Optical processing:
    coherent, 281 ff.
    coherently illuminated
    systems, 255
    controlled transparencies
    for
    data, 129
    informational capacity for,
    105 ff.
    methods, 281 ff.
    nonlinear, 255 ff.
    of information, 313 ff.
    passive-type arrangements,
    129
    systems, 105 ff.
Optical processor:
    coherent, 86
    incoherent addition, 86, 87
    scheme, 149
Optical radar, 125
    laser, 125
    ratio of, to lightwaves, 125
Optical spectrum analyzer, 296
Optical signals, 266
Optical switch, 273

Optical system versions, 60 ff.
   defocused, 335
   for matrix by matric column multiplication, 61
   for matrix-row by matrix multiplication, 62
   for multichannel calculation of the correlation functions, 64
   for parallel calculation, 308
   for rectangular matrices multiplication, 63
   non-coherent, for processing of images and signals, 203 ff.
Optical versus digital processing, 310 ff.
Optics, 282 ff.
   and communication, 310
   computational capabilities of, 282
Optimum (Scherzer focus), 291
Organic dyes, 257 ff.

Page composer, 242, 356 ff.
   acousto-optic, 366
Parallel plate charging, 20
Parallel processing, 119 ff.
Parameter effects, 222
Particle size, 317 ff.
Pattern coding, 141
   recognition, 313
Permalloy pattern, 235
Phase:
   cholesteric, 27
   nematic, 27
Phase contrast:
   methods, 331 ff.

Phase contrast (cont'd)
   microscopy, 333
   optical visualization schemes of, 148
   with contrast enhancement, 333
Phase light delay versus electric field, 133
Phase multipliers, 53
Phase object, visualized, 150
Phase objects, 328
Phonograph records, television equivalent of, 234
Photoconductors, 20 ff.
   FC layer, 20
   PLZT, 132
Photodicroic:
   crystals, 251
   nature, 32
Photodiode crystals, 251
Photoelectric effects, 266
Photographs, 284 ff.
   motion-blurred, 290
Photoresist, 224
Photosemiconductor structure, liquid-crystal, 148
Photosensitive medium, digital, 246
Phototitus, 35
Picosecond, 224
Pixel, 348
Pockels readout optical modulator (PROM), 35 ff.
   multilayer structures, 199
Polarization, switchable, 252 ff.
Poly-n-vinyl carbazole(PVK), 266
Power spectrum, 296
Prism operation, 160

SUBJECT INDEX

Processor:
    correlation, 185 ff.
    Fourier-plane, 182 ff.
    hybrid, 255
    hybrid optical-digital, 344 ff.
    image plane, 182 ff.
Proportionality between light power and exposure time, 156

Quantization problems, 285 ff.
Quarter-wave phaseplates, 2, 124

Radar views, 122 ff.
    all weather, 122
    in commercial aviation, 122
    real-time technique, 122
    side-looking synthetic aperture, 283 ff.
    successive real time during fog, 122
    synthetic aperture coherent, 284 ff.
Radiations, ultrasonic, 284 ff.
Raindrop echoes:
    enhancement of, 124
    suppression of, 124
Ramon-Nath:
    diffraction, 175
    regime, 175
Random walk in the complex plane, 88
Raster and half-tone removal, 327
Real time:
    coherent optical computer, 15
    digitally assisted compensation, 311

Real time (cont'd)
    holograms, 34
    processing, 315 ff.
    radar processing, 34
    reconstruction, 120 ff.
Recorders:
    holographic, 363 ff.
    high-speed digital, 364
Recording:
    digital magnetic, 234
    ferromagnetic, 244
    holographic in bistable medium, 246
    magnetic, 234
Reduced coordinates, 305
Region:
    use of finer quantization, 99
    zero frequency in, 99
Resolution, 120 ff.
    element, 352
    limit, 289
Resolution constant:
    angular versus range, 120
    metric, 120
Restorability:
    absolute, 94
    image, 94
    relative, 94
Ruticon, 24 ff.

Sandwiches, ferroelectric-photoconducting, 252 ff.
Saturable absorbers, 257
Saturable resonator, 267
Scalar methods, 74
Scanning, 221 ff.
    for reading purposes, 221
Schlieren photography, 314

Semiconductors:
    amorphous, 250
    chalcogenide amorphous, 149
    chalcogenide vitreous, 154
    memories, 355
    technology, 149 ff.
Sequential one-photon processes, 253
Series termination effects, 323
Signal separation by orientation, 328 ff.
Signal-to-noise RMS ratio, 87
Spatial filtering, characteristics of, 113 ff.
Spatial Fourier transformation, 294
Spatial-frequency spectrum, 51
Spatial frequencies, 3
Spatial information loss, 108
Specimens, 289 ff.
    carbon-foil, 289
    contrast in, 289
    of biological interest, 289
    unstained, 289
Spectral dependence of refractivity vatiation, 157
Storage, 229 ff.
    digital, optically accessed, 252
    electron beam recorder, 282
    electronic image, 251
    holographic, 282
    magnetic, 282
    mass memory, 282
    media, read, write, erase, 230 ff.
    requirements, general, 230 ff.
    solid state, 282

Strips, 355
    magnetic, 355
    optical, 355
Structure synthesis, 292
Surface relief patterns, 17
Synergetics, 249
Synthetic aperture, 117 ff.
    antenna, 119
    holographic, 117
    radar, 117
    radar information processing, extensions of, 117 ff.
    real-time operation of, 117
Synthetized acoustic fields, 191

Target crystal:
    DKDP, 33
    two-year, 34
Thermoplastic membranes, 342
Thermoplastics, 17
Theory of districtions, 308
Theta modulation technique, 255
Three-dimensions, suppressed or accentuated, 126
Time compressor, 187
Transducer wavefront imperfections, 295
Transform coefficients, Walsh diagonal, 146
Transformer, solid state energy, 201
Transforms, 2 ff.
    Foucault-Hilbert, 1 ff.
    Fourier, 84, 142
    Fourier direct, 5
    Fresnel, 49

SUBJECT INDEX

Transforms (cont'd)
    Girare, 142
    Hilbert one-dimensional, 9
    inverse, 2 ff.
    isotropic, 6
    kernel, 70, 142
    linear integral, 67
    logical, 73
    two-D, 18
    Walsh-Hilbert integral, 151
Transparency applications,
    field of controlled, 132
Transistor, 223
    ultra-high speed switching, 224
Twisted nematic effect, 28
Two-mirror interferometer, 78

Ultrasonic medical diagnosis, 122

Video disks, 234 ff.
Video tape, 232
Voltage controlled,
    birefringence effect, 28

Walsh function, 147
Walsh transform, 210
Wavefront:
    aberration, 300
    imperfections, 297
    interferograms, 308
Wave fronts, curved, 119
Waveguides, 271 ff.
    of electro-optic materials, 271
Wavelength:
    100 mcm ceramic sample, 130
    transmittance vs light, 130
Wave theory of aberrations, 309

X-rays, 276 ff.
    automatic processing of, 276
    chest, 276
    crystallography, 292

Young's eriometer, 316

Zone plates, one-dimensional, 119